Elementary Algebra: Structure and Skills

Elementary Algebra: Structure and Skills

THIRD EDITION

Irving Drooyan

Walter Hadel

Frank Fleming

Los Angeles Pierce College

John Wiley & Sons

New York   London   Sydney   Toronto

Library of Congress Cataloging in Publication Data
Drooyan, Irving
  Elementary algebra.

  1. Algebra.   I. Hadel, Walter, joint author.
II. Fleming, Frank J., joint author.   III. Title.

QA152.D73 1973   512.9'042   72-4498
ISBN 0-471-22251-8

Printed in the United States of America

10 9 8 7 6 5 4

Preface

This third edition of *Elementary Algebra: Structure and Skills* reflects seven years of classroom experience with the previous editions of the text. Textual material has been rewritten and in some cases the sequence of topics has been changed to improve the presentation. In particular, additional emphasis has been given to techniques of graphing. Also, new sections concerning order of operations, numerical evaluations, and the distance between two points have been added.

Proofs of theorems have been deemphasized in the text. However, proofs of most theorems are included in the B exercises in a semiprogrammed form to accommodate those instructors who wish to consider them. Exercise sets have been carefully reviewed and revised as necessary to assure that they are appropriately graded.

Many of the useful features of the format of the second edition have been retained. Functional use of color highlights important processes; chapter summaries and chapter reviews are cross-referenced to appropriate sections in the text; and the axioms and theorems which are introduced in the text are listed in an appendix for convenient reference.

This textbook has been written for students who are beginning the study of algebra. It is designed to cover in one semester the equivalent of the first two semesters of high school work.

The point of view and the topics covered reflect the recommendations of nationally recognized groups. Unifying ideas are stressed. In particular, the structure of the real number system is introduced at an appropriate level of sophistication. This approach enables those students who have not had the opportunity to study modern texts to gain this background early in the semester as a foundation for the remainder of the course.

No sharp distinction is made between arithmetic and algebra. Algebra is treated simply as a generalized arithmetic.

In the first five chapters the axioms of a field and definitions applicable to the set of real numbers are introduced and explicitly stated. A number of theorems about the set of real numbers are shown to be logical consequences of these axioms and definitions. When theorems appear in the exercise sets, the student is carefully led through the proof one step at a time so that he can attain an understanding of structure and, in particular, the reason why one expression can be written in another form and still represent the same real number. These presentations are designed to give the student a sense of what might constitute a formal argument and to facilitate his progress in future mathematics courses. The emphasis on structure appears throughout the text as skills are developed. The notion of a set, certain operations on sets, and set-building notation introduced in Chapter 1, provide a basis for the discussion of solution sets of equations and inequalities and also for the introduction of relations and functions.

Chapters 6 and 7 on polynomials and rational expressions continue the emphasis on the "why" as well as the "how" in developing necessary manipulative skills. The notation $P(x)$ for polynomials is introduced in Chapter 6 and is used thereafter. Students will become familiar with this notation by the time they encounter function notation in Chapter 9.

Chapter 8 includes additional first-degree equations and inequalities which are models for a variety of word problems. Emphasis is given to setting up the models (a separate section) before the student obtains solutions to the word problems by solving the equation or inequality.

The important concept of a function is introduced in Chapter 9 through the more general notation of a relation as a set of ordered pairs. Emphasis is given to the nature of the graphs of both relations and functions. In Chapter 10 the solutions of systems of equations and inequalities are considered.

In Chapter 11 square roots that are elements of the set of real numbers are studied. With this background, the quadratic equation in one variable and the quadratic function are discussed in Chapter 12. Different methods of solving quadratic equations in one variable and the applications of these equations as mathematical models for physical problems are treated.

The problems in the exercise sets are designated as **A** or **B**. The **B** problems are more challenging and should provide the instructor with flexibility in making assignments, depending on the time available and the aptitude of the students.

A large number of examples are included for explanatory purposes both in the text and in the exercise sets. The subject matter of the text is continuously reviewed through the use of chapter summaries, chapter reviews, and periodic cumulative reviews. The reviews also serve to provide the student with a means of self-evaluation.

Answers are provided for the odd-numbered problems in each section and for all of the problems in the chapter reviews and in the cumulative reviews. All answers in the form of graphs and proofs have also been included for the odd-numbered problems. The answers to the even-numbered problems are available to instructors.

We wish to express our appreciation to Professors O. Robert Brown, Jr. of Federal City College and Bill D. New of Cerritos College for their many good suggestions for improving this edition.

Woodland Hills, California

Irving Drooyan
Walter Hadel
Frank Fleming

Contents

chapter 3

The Set of Integers   63

chapter 4

The Set of Rational Numbers   89

chapter 5

The Set of Real Numbers   123

Cumulative Review, Chapters 1–5   138

chapter 6

Polynomials   143

chapter 7
Rational Expressions   191

chapter 8
First-Degree Equations and Inequalities—   217
One Variable

chapter 9

Relations, Functions, and Their Graphs   260

chapter 10

Systems of Linear Equations and Inequalities   301

Appendices   385

Answers   395

Index   459

1
chapter

Sets

The branch of mathematics with which we are concerned in this book is **algebra.** Because this subject is an *extension* or *generalization* of arithmetic, we shall start our study of algebra by investigating the properties of a number system that is most familiar to you from your previous work in arithmetic. We shall do this through a consideration of "sets."

1.1 Set Construction and Subsets

Although we consider a set to be undefined, we can describe the concept informally. A **set** is merely a collection of things. The objects in the collection may have some obvious property in common or they may not.

The "things" or "objects" belonging to a set are called **members** or **elements** of the set. Braces, such as { }, are commonly used to enclose the members of a set. For example, {John, Sue, Mike} is read: "The set whose elements are John, Sue, and Mike" and {letters in the English alphabet} is read: "The set whose elements are the letters in the English alphabet." These examples illustrate two ways in which we can represent sets. The first is a *listing* of the names of the elements and the second is a *statement*, or *rule*, describing the elements in a set.

At this time we shall not be concerned with the order of the members in a set. For example, we shall consider {0, 1, 2}, {1, 2, 0}, {1, 0, 2}, {0, 2, 1}, {2, 0, 1}, and {2, 1, 0} to be the same set because they contain the same members. When listing the names of (or symbols for) members of a set, each name is listed only once because if the name of an element were repeated, each of the repetitions would simply refer to the same member. Thus the set of letters in the word "Mississippi" is {M, i, s, p}.

Sometimes we may state a rule for a set only to discover that there are no elements fitting the rule. For example, {women presidents of the United States} is such a set. Although the concept of a set without any members may seem strange at first, it is important as you will see. Such a set is called the **empty set** or **null set** and is denoted by the symbol $\varnothing$.

The symbol $\in$ is used to represent the phrase "is a member of," "is an element of," or "belongs to," and the symbol $\notin$ represents the phrase "is not a member of," "is not an element of," or "does not belong to."

Examples.   a. Mary $\in$ {Mary, John, Bill}.
　　　　　　b. Sue $\notin$ {Mary, John, Bill}.
　　　　　　c. $a \in$ {vowels in the English alphabet}.
　　　　　　d. $r \notin$ {vowels in the English alphabet}.

We now assume a property of sets called "countability." In counting the elements of a set, if there is a last element, the set is said to be **finite**; if there is no last element, the set is **infinite**.

Examples.   a. {6, 7, 8, 9} is finite.
　　　　　　b. {Points on a line} is infinite.
　　　　　　c. {0, 1, 2, 3, · · ·} is infinite.

In Example c above, the three dots indicate that the sequence of numbers continues in the same pattern without end.

Two sets are said to be **equal** if every element of the first set is an element of the second set and every element of the second set is an element of the first. The symbol $=$ is used to indicate the equality relation of sets and is read "equals," or "is equal to." The symbol $\neq$ is used to represent the phrase "is not equal to."

Examples.   a. {5, 10, 15} = {10, 15, 5}　　　b. {5, 10, 15} $\neq$ {5, 10}

Capital letters, A, B, C, etc., are often used to name sets. For example, we can arbitrarily let $A = \{6, 7, 8\}$, $B = \{6, 7, 8, 9\}$, $C = \{8, 7, 6\}$, and $D = \{0, 1, 2, 3, · · ·\}$. Then we can write $A = C$, $A \neq B$, $B \neq D$, etc.

If every element of a set A is also an element of a set B, then A is said to be a **subset** of B. From this definition, it follows that every set is a subset of itself. Furthermore, $\varnothing$, the null set, is considered to be a subset of every set, because if $\varnothing$ were not a subset, this would imply that $\varnothing$ contains at least one element not in the set. This is impossible since $\varnothing$ has no elements.

Examples.   a. $\{2, 4\}$ is a subset of $\{2, 4, 6, 8\}$.
b. $\{2, 4\}$ is a subset of $\{2, 4\}$.
c. $\varnothing$ is a subset of $\{2, 4\}$.

The symbol $\subset$ is used to denote that one set is a subset of another. Thus, $A \subset B$ is read: "A is a subset of B" or "A is contained in B." $A \not\subset B$, means "A is not a subset of B," or "A is not contained in B."

Examples.   a. If $A = \{1, 3, 5, 7\}$, $B = \{2, 4, 6\}$, and $C = \{1, 2, 3, 4, 5, 6, 7\}$, then $A \subset C$ and $B \subset C$, but $A \not\subset B$ and $B \not\subset A$.
b. For every set S, $S \subset S$ and $\varnothing \subset S$.

Care should be taken to distinguish between the symbols $\in$ and $\subset$. The symbol $\in$ refers to a relation between an *element* and a *set*, while $\subset$ refers to a relation between *sets*. For example, if $A = \{1, 3, 5, 7\}$, then $5 \in A$ and $\{5\} \subset A$ are true, but $5 \subset A$ is *not true* since 5 is an element, not a set.

Examples.   a. $\{4\} \subset \{4, 5, 6\}$ b. $4 \in \{4, 5, 6\}$ c. $\varnothing \subset \{4, 5, 6\}$

Exercise 1.1

A

Write each set by listing the elements.

Example.   {The first five letters in the English alphabet}.

Solution. $\{a, b, c, d, e\}$.

1. {The first four months of the year}

2. {The days of the week}

3. {The letters in the word *represents*}

4. {The letters in the word *possibilities*}

5. {The odd numbers between 3 and 15}

6. {The even numbers between 14 and 30}

7. {The first five digits in the decimal number system}

8. {The second five digits in the decimal number system}

Let U = {letters of the English alphabet} and A = {vowels in the English alphabet}. Replace the comma in each pair with either $\in$ or $\notin$.

Examples.   a. *h, U*   b. *h, A*   c. *p, {k, m}*

Solutions.   a. $h \in U$ b. $h \notin A$ c. $p \notin \{k, m\}$

9. *t, U*   10. *t, A*   11. *a, U*   12. *a, A*

13. *c, {b, c}*   14. *s, {b, c}*   15. *f, {a, b, c}*   16. *b, {a, b, c}*

Let K = {a, b, c}, L = {c, b, a}, M = {b, c}, N = {a, b}, and P = {a}. Replace the comma in each pair with either $\in$ or $\subset$.

Examples.   a. *b, N*   b. *P, K*

Solutions.   a. $b \in N$ b. $P \subset K$

17. *a, L*   18. *{b}, M*   19. *c, K*   20. *{a}, N*

21. *P, K*   22. *N, N*   23. $\varnothing$, *L*   24. $\varnothing$, *M*

Let C = {a, b, c, d}. List the subsets of C that contain:

25. Four members   26. Three members

27. Two members   28. One member

29. No members

30. What is the total number of subsets of {1, 2, 3}?

Let K = {a, b, c}, L = {c, a, b}, M = {b, c}, N = {a, b}, P = {a}, and Q = {b, a}. Replace the comma in each pair with either = or $\neq$.

Examples.   a. *N, Q*   b. *L, N*

Solutions.   a. $N = Q$ b. $L \neq N$

31. K, L 32. M, N 33. N, P

34. $\{a\}, P$ 35. $K, 3$ 36. $M, 2$

Let $U = \{letters\ of\ the\ English\ alphabet\}$, $A = \{a, b, c, d\}$, $B = \{c, d\}$, $C = \{c, d, e\}$, $D = \{e, f, g, h\}$, and $E = \{b, c, d, e\}$. *Which statements are true and which are false?*

Examples.   a. $A = D$ b. $B \subset A$ c. $\{b\} \in C$

Solutions.   a. False   b. True   c. False

37. $D = E$ 38. $A = E$ 39. $C \neq D$ 40. $B \subset C$

41. $A \subset B$ 42. $\varnothing \subset E$ 43. $B \subset D$ 44. $C \subset A$

45. $c \in A$ 46. $j \notin U$ 47. $\{d, c\} = B$ 48. $e \in A$

B

49. If $A \subset B$ and $3 \in B$, must 3 be an element of A?

50. If $A \subset B$ and $3 \in A$, must 3 be an element of B?

51. If $A \subset B$ and $B \subset A$, what other relationship exists between A and B?

52. If $A \subset B$ and $B \subset C$, what relationship always exists between A and C?

53. If $A = B$ and $B = C$, what relationship always exists between A and C?

54. If $A = B$ and $B \subset C$, what relationship always exists between A and C?

1.2 Set Operations

Sometimes we may be interested in forming sets from two given sets. For example, if $R = \{$John, Mary, Jane, Bill$\}$, the set of students who received an "A" on a first test, and $S = \{$Bill, Sue, Karen$\}$, the set of students who received an "A" on a second test, then the set consisting of all the students who received a grade of "A" on either the first test *or* the second test is $\{$John, Mary, Jane, Bill, Sue, Karen$\}$. The set of students who received an "A" on the first test *and* the second test is $\{$Bill$\}$, assuming, of course, that "Bill" is the name for the same person in each set. These examples illustrate two operations which are concerned with forming a third set from two specified sets. We define these operations for any sets A and B.

Definition 1.1. The **union** of two sets, A and B, is the set of all elements that belong either to A or B, or to both.

The symbol $\cup$ is used to designate the operation of union. Thus, $A \cup B$ is read: "The union of set A and set B."

Example. If $A = \{0, 1, 2, 3, 4\}$ and $B = \{3, 4, 5, 6, 7\}$, then

$$A \cup B = \{0, 1, 2, 3, 4, 5, 6, 7\}.$$

Note that each element is written only once in $A \cup B$ even though the elements 3 and 4 appear in both A and B.

Definition 1.2. The **intersection** of two sets, A and B, is the set of all elements that belong to both A and B.

Note that the intersection of two sets consists of those elements *common* to both sets.

The symbol $\cap$ is used to indicate the operation of intersection. Thus, $A \cap B$ is read: "The intersection of set A and set B."

Example. If $A = \{a, b, c, d, e\}$ and $B = \{c, d, e, f, g\}$, then

$$A \cap B = \{c, d, e\}.$$

The elements c, d, and e belong to, or are common to, both sets.

The symbols ( ), called "parentheses," are used in a collection of symbols such as $(A \cap B) \cup C$ to indicate which of two or more operations is to be done first. In this case the symbols mean the union of the set $A \cap B$ (the intersection of A and B) and the set C.

Examples. Consider $A = \{1, 2, 3\}$, $B = \{2, 3, 4\}$, and $C = \{3, 4, 5\}$. List the elements in each set.

$$\text{a. } (A \cap B) \cup C \qquad \text{b. } (A \cup B) \cup C$$

Solutions. a. $(A \cap B) = \{2, 3\}$ and $C = \{3, 4, 5\}$;

$(A \cap B) \cup C = \{2, 3, 4, 5\}$

b. $(A \cup B) = \{1, 2, 3, 4\}$ and $C = \{3, 4, 5\}$;

$(A \cup B) \cup C = \{1, 2, 3, 4, 5\}$

If two sets do not contain any members in common, for example, $\{1, 2\}$ and $\{3, 4\}$, the sets are said to be **disjoint.** The intersection of any two disjoint sets is the empty, or null, set.

Examples. a. $\{1, 2\} \cap \{3, 4\} = \varnothing$ b. $\{2, 4, 6\} \cap \{9, 11, 13\} = \varnothing$

The set which contains all of the elements involved in any particular discussion is called the **universal set** and is designated by the capital letter U. Another name for such a set is the **universe of discourse** or, more simply, the **universe.** Suppose we wished to discuss all the students enrolled in algebra in the state of Ohio. The universal set would be

$U = \{$students studying algebra in Ohio$\}$.

Then, any group of algebra students in this state would be a subset of U.

Definition 1.3. The **complement** of a set A in a universal set is the set of all elements of the universal set that do not belong to A.

The complement of the set A is designated by the symbol A' which is read "A prime."

Examples. a. If $U = \{$students studying algebra in Ohio$\}$ and $M = \{$male students studying algebra in Ohio$\}$, then

$M' = \{$female students studying algebra in Ohio$\}$.

b. If $U = \{0, 1, 2, 3, 4, 5, 6, 7, 8, 9\}$ and $A = \{0, 1, 2, 3, 4, 5\}$,

then

$A' = \{6, 7, 8, 9\}$.

Exercise 1.2

A

Let $A = \{a, b, c, d, e, f, g, h\}$, $B = \{a, b, c, d\}$, $C = \{e, f, g, h\}$, $D = \{a, c, e, g\}$, *and* $E = \{b, d, f, h\}$. *List the elements in each union or intersection.*

Examples.   a. $A \cap B$ b. $C \cup E$

Solutions.   a. $\{a, b, c, d\}$ or simply, B b. $\{b, d, e, f, g, h\}$

1. $A \cup C$ 2. $B \cup D$ 3. $A \cup B$ 4. $B \cup E$

5. $A \cap D$ 6. $B \cap E$ 7. $D \cap B$ 8. $C \cap B$

9. $B \cup C$ 10. $C \cup D$ 11. $E \cap D$ 12. $A \cap E$

Let $U = \{j, k, m, n, p\}$, $A = \{j, k\}$, $B = \{k, m\}$, $C = \{n, p\}$, and $D = \{p\}$. List the elements in each set.

Examples.   a. B' b. $A' \cap C$

Solutions.   a. $\{j, n, p\}$ b. $\{m, n, p\} \cap \{n, p\} = \{n, p\}$

13. A' 14. C' 15. $A \cup B'$ 16. $B \cup C'$

17. $B \cap A'$ 18. $C' \cap D$ 19. $B' \cap C'$ 20. $A' \cap D'$

Let $U = \{a, b, c, d, e, f, g, h, i, j\}$, $A = \{b, d, f, h, j\}$, $B = \{a, b, c, d, e\}$,
$C = \{a, c, e, g, i\}$, and $D = \{f, g, h, i, j\}$. List the elements in each set.

21. B' 22. C' 23. $A' \cap B$ 24. $B' \cap D$

25. $A' \cup D'$ 26. $B' \cup C'$ 27. $(A \cap B')'$ 28. $(C' \cup D)'$

Which pairs of sets in each problem are disjoint?

29. $A = \{a\}$, $B = \{a, b\}$, and $C = \{b, c\}$
30. $A = \{m, n, o\}$, $B = \{o, p, q\}$, and $C = \{m, n\}$
31. $R = \{2, 4, 6\}$, $S = \{1, 3, 5\}$, and $T = \{6\}$
32. $R = \{5\}$, $S = \{10\}$, and $T = \{15\}$

B

33. If $5 \in A$, must 5 be an element of $A \cap B$ for every B?
34. If $5 \in A$, must 5 be an element of $A \cup B$ for every B?
35. Must $A \cup B = B \cup A$ for all sets A and B?
36. Must $A \cap B = B \cap A$ for all sets A and B?

State the conditions on sets S and T under which each statement would be true.

37. $S \cap T = \varnothing$ 38. $S \cup T = S$ 39. $S \cap T = S$

40. $S \cup T = \varnothing$ 41. $S \cap T = T$ 42. $S \cup T = T$

1.3   Venn Diagrams

So far in our discussion of sets, we have either listed members or described the membership of a set by a rule. Sometimes relationships between sets and the result of operations on sets are clarified by the use of **Venn diagrams,** named after their originator. These are geometric representations of sets in which the universal set, U, is represented by *any* closed plane geometric figure. Rectangles are commonly used. Subsets are represented in a Venn diagram by closed plane regions which are contained in the universal rectangle. Sometimes the figures representing subsets are shaded; at other times, they simply appear without shading. For example a Venn diagram of $A \subset U$ usually appears as in Figure 1.1.

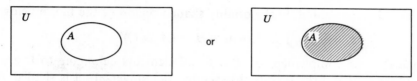

Figure 1.1

The union and intersection of two sets and the complement of a set can be represented in Venn diagrams by various types of shading, including double shading.

Examples.   a. $A \cup B$

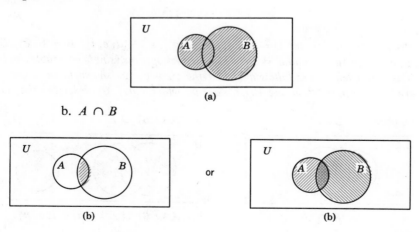

(a)

b. $A \cap B$

(b)   (b)

In the first diagram of Example b above, only that portion where sets *A* and *B* overlap is shaded. In the second diagram, *A* is shaded in one direction while *B* is shaded in another, so the intersection appears as a doubly shaded region.

Examples.   a. *A′*, the complement of *A*.

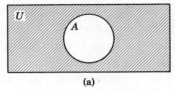

(a)

A′ is represented by the shaded region of the figure.

b. *A′* ∩ *C*, where *A* and *C* are disjoint.

A′ ∩ *C* is represented by the doubly shaded region of the figure below.

c. (*A* ∪ *C*) ∪ *B*, where *B* = (*A* ∪ *C*)′.

Since *B* is the complement of *A* ∪ *C*, all elements belonging to *U* are in (*A* ∪ *C*) ∪ *B* and the region representing the universal set is shaded.

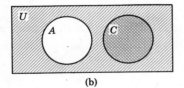

(b)

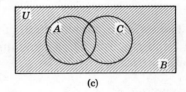

(c)

Exercise 1.3

In Problems 1 to 16, let U = {a, e, i, o, u, y}, A = {a, e, i}, B = {i, o}, C = {o, u, y}, D = {u, y}, and E = {y}. Use Venn diagrams (single or double shading) to illustrate each of the following. List the elements as shown in the examples. There may be more than one Venn diagram which correctly describes the set(s).

Examples.   a. *B* ∩ *C*   b. (*A* ∪ *B*) ∪ *E*

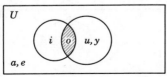

a. *B* ∩ *C* = {*o*}

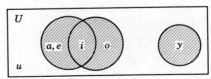

b. (*A* ∪ *B*) ∪ *E* = {*a, e, i, o, y*}

1. $A \cap B$ 2. $B \cup C$ 3. A' 4. C'

5. $A \cup E$ 6. $B \cup E$ 7. $(A \cup B)'$ 8. $(B \cap C)'$

9. $(C \cup D)'$ 10. $(A \cup E)'$ 11. $(A \cap E)'$ 12. $(A \cap B)'$

13. $(A \cup B) \cup C$ 14. $(B \cup C) \cap D$ 15. $(B \cup C) \cap A$ 16. $(C \cap D) \cup B$

Use Venn diagrams (single or double shading as required) to illustrate each of the following.

Examples.   a. $A \subset B \subset C$ b. $(A \cup B)'$, A, B disjoint

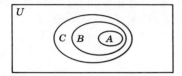
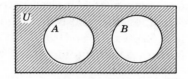

17. A and B are disjoint   18. $C \cap D = \emptyset$ 19. $A \subset B$

20. $A \subset B$, $C \subset B$, and $A \cap C = \emptyset$ 21. $A \cup B$ where $A \cap B = \emptyset$

22. $(A \cup B)'$ where $A \cap B = \emptyset$ 23. $A \cap B$ where $A \cap B \neq \emptyset$

24. $(A \cup B) \cup C$ where no two sets are disjoint

List the elements in each indicated set if the sets are specified as shown in the Venn diagram below.

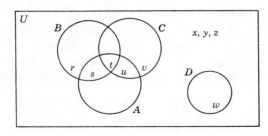

25. B 26. C 27. A 28. U

29. $B \cap C$ 30. $B \cup C$ 31. $D \cup A$ 32. $D \cap A$

33. C' 34. $(B \cup A)'$ 35. $(A \cup B) \cup C$ 36. $(A \cap B) \cap C$

1.4 Cartesian Products

Another operation on sets will be useful in our discussion.

First, consider an idea that you will encounter many times in your study of mathematics, namely, a pair of numbers which are considered in a specified order. For example, in a tennis match, a score of 40-30 does not have the same meaning as a score of 30-40, since the points earned by the player who serves are always given first. A pair of numbers, or any elements, where the order of the elements is of importance is called an **ordered pair.** Each of the elements is called a **component.** The first element is called the **first component** and the second element is called the **second component.** Ordered pairs are written with a comma between the two components, and both components are enclosed in parentheses, such as $(40, 30)$, $(x, 3)$, and (a, b).

We now define another set operation.

Definition 1.4.   The **Cartesian product** of two sets, A and B, denoted by $A \times B$, is the set of all possible ordered pairs such that the first component is an element of A and the second component is an element of B.

For example, consider the sets $A = \{a, b\}$ and $B = \{x, y\}$. Then

$$A \times B = \{(a, x), (a, y), (b, x), (b, y)\}.$$

Notice that the first component of each ordered pair in the Cartesian product $A \times B$ is an element of set A and that the second component of each ordered pair is an element of set B. This order was established by the order in which the Cartesian product, $A \times B$, was indicated. If we interchange the position of the sets in the indicated operation, then the order of the pairs in the Cartesian product is reversed. Thus,

$$B \times A = \{(x, a), (x, b), (y, a), (y, b)\}.$$

As we remarked earlier, the elements of a set may be listed in any order. Thus,

$$\begin{aligned}
B \times A &= \{(x, a), (x, b), (y, a), (y, b)\} \\
&= \{(y, b), (x, a), (y, a), (x, b)\} \\
&= \{(y, b), (x, b), (y, a), (x, a)\}, \text{ etc.}
\end{aligned}$$

Example. If $A = \{a, b\}$ and $B = \{1, 2, 3\}$, then

$$A \times B = \{(a, 1), (a, 2), (a, 3), (b, 1), (b, 2), (b, 3)\}.$$

Observe that the number of elements in set A is 2 and that in set B is 3 and the number of elements in the Cartesian product is 6.

Example. If $A = \{a, 1\}$, then $A \times A = \{a, 1\} \times \{a, 1\}$, or

$$A \times A = \{(a, a), (a, 1), (1, a), (1, 1)\}.$$

Observe that the number of elements in set A is 2 and that the number of elements in the Cartesian product $A \times A$ is 4.

Exercise 1.4

A

Let $A = \{a, e, i, o\}$, $B = \{r, s, t\}$, $C = \{a, e, t\}$, and $D = \{i, o, s\}$. Write the indicated Cartesian products.

Example. $C \times D = \{(a, i), (a, o), (a, s), (e, i), (e, o), (e, s), (t, i), (t, o), (t, s)\}.$

1. $A \times B$ 2. $B \times C$ 3. $A \times C$ 4. $A \times D$

5. $D \times B$ 6. $B \times A$ 7. $C \times B$ 8. $C \times A$

9. $A \times A$ 10. $B \times B$ 11. $C \times C$ 12. $D \times D$

13. $A \times \varnothing$ 14. $\varnothing \times B$

If K contains 4 elements, L contains 7 elements, M contains 3 elements, and P contains 2 elements, how many elements (ordered pairs) are in each Cartesian product?

15. $K \times L$ 16. $K \times M$ 17. $K \times P$

18. $L \times M$ 19. $L \times P$ 20. $M \times P$

21. If $A \times B = \{(r, x), (r, y), (r, z)\}$, list the member(s) of A.

22. If $A \times B = \{(r, x), (r, y), (r, z), (s, x), (s, y), (s, z)\}$, list the member(s) of B.

B

23. Under what conditions will $A \times B$ equal $\varnothing$?

24. Under what conditions will $A \times B = B \times A$?

If $A = \{(a, r), (a, s), (b, r), (b, s), (c, r), (c, s)\}$ and
$B = \{(a, r), (a, t), (b, r), (b, t)\}$, list the members in each set.

25. $A \cap B$ 26. $A \cup B$ 27. $A \cup \varnothing$ 28. $B \cap \varnothing$

CHAPTER SUMMARY

1.1 A set may be described by *listing the names of the members* or *elements* or by *a rule* which determines precisely the elements in the set.

1.2 The **union** of two sets, A and B, is the set of all elements that belong either to A or B, or to both. The **intersection** of two sets, A and B, is the set of all elements that belong to both A and B; that is, elements common to both sets. The **complement** of a set A in a universal set is the set of all elements in the universal set that do not belong to A.

1.3 **Venn diagrams** are geometric representations showing relationships between sets and the results of operations on sets.

1.4 The **Cartesian product** of two sets, A and B, is a set which contains all ordered pairs (a, b) that can be formed by using all $a \in A$ and $b \in B$.

CHAPTER REVIEW

A

In Problems 1 to 22, let $U = \{a, b, c, d, e, f\}$, $A = \{a, c, e\}$, $B = \{c, d, e\}$, $C = \{e, f\}$, $D = \{a\}$, and $E = \{d\}$.

1.1 *Replace each comma with either $\in$ or $\subset$.*

1. A, U 2. f, C 3. A, A

4. D, A 5. b, U 6. E, U

7. List all subsets of C. How many subsets are there?

8. List all subsets of A. How many subsets are there?

1.2–1.4 *Replace the comma in each pair on the left with $\cup$, $\cap$, or $\times$.*

9. $C, D = \{a, e, f\}$ 10. $C, D = \varnothing$

11. $C, D = \{(e, a), (f, a)\}$ 12. $E, D = \{(d, a)\}$

13. $B, C = \{e\}$ 14. $A, B = \{a, c, d, e\}$

List the elements in each set.

15. $A \cup C$ 16. $A \cap C'$ 17. $(A \cup B)'$ 18. $A \times C$

19. $C \times B$ 20. $B' \times D$ 21. $A \cup \varnothing$ 22. $A \cap \varnothing$

B

State the conditions on sets S and T for which each statement is true.

23. $S \cap T = S$ 24. $S \cup T = S$

25. $S \cap T = \varnothing$ 26. $S \times T = T \times S$

In problems 27 to 30, $A \subset B$, $B \subset C$, $3 \in A$, and $4 \in B$.

27. Must $3 \in C$?   28. Can $4 \in A$?

29. Must $(3, 4) \in A \times B$?   30. Must $(3, 4) \in B \times A$?

2
chapter

The Set of
Whole Numbers

Have you ever tried to define the word "number"? If you have, you probably said that it has something to do with the idea of "how many." Such a statement is only partly correct because many different kinds of numbers are assigned different meanings. Numbers are abstractions and we *give* them meanings that we wish them to have. In this chapter, we discuss the particular numbers that make up the **set of whole numbers.**

2.1   Cardinality and Ordinality

Numbers associated with "how many" elements there are in a set are said to be used in a **cardinal sense** and the number itself is called the **cardinality** of the set. To talk about a number used in a cardinal sense, the number is given a name, such as zero, one, two, three, etc., and is represented by a symbol such as 0, 1, 2, 3, etc. For example, the cardinality of {John, Mary} is 2, the cardinality of {John, Mary, Jim} is 3, and the cardinality of $\varnothing$ is 0.

You should note here that the words used are merely the *names* that we have given certain numbers and that the symbols, called **numerals,** are not numbers but only *representations* of them. Although the numerals only represent the *concept* of number, they do permit us to discuss the properties we conceive to be associated with numbers.

Many different symbols for numbers have been used through the ages. For example, the number *five* has been represented by the Arabic symbol 5, the Roman symbol V, and the tally notation 卌. The distinction between a number and its name, or symbol, should be kept clearly in mind.

Numbers such as 0, 1, 2, 735, and 2,683 which are associated with the cardinalities of sets are called **whole numbers,** and the set of all such numbers is represented symbolically by

$$W = \{0, 1, 2, 3, \cdots\}.$$

This set is infinite, since there is no *last* member in the set.

The set of numbers associated with the cardinality of *non-empty* sets is called the set of **natural numbers** and is represented symbolically by

$$N = \{1, 2, 3, \cdots\}.$$

The members of this set are often referred to as **counting numbers.**

Since all members of N are also members of W, the set of natural numbers is a subset of the set of whole numbers; that is, $N \subset W$.

There are other kinds of numbers besides whole numbers. Numbers such as 0.3, 3/4, $\sqrt{7}$, and -2 are quite important in mathematics but they do not belong to the set of whole numbers. They cannot be associated with the cardinality of sets. The properties of these numbers are discussed in the following chapters.

Of two sets that do not have the same cardinality, such as

$$A = \{\neq, \&, *\} \quad \text{and} \quad B = \{\bigcirc, !, \triangle, \square\},$$

the cardinality of the set with fewer elements is said to be **less than** the cardinality of the set with more elements. The cardinality of the set with more elements is said to be **greater than** the cardinality of the set with fewer elements. Thus, in the example above, the cardinality of A is less than the cardinality of B and the cardinality of B is greater than the cardinality of A.

Ideas concerning the comparison of two numbers can be expressed conveniently by using the following symbols:

$<$ means "is less than,"
$>$ means "is greater than,"
$=$ means "is equal to,"
$\neq$ means "is not equal to."

Examples.   Replace the word phrase with the proper symbol.

> a. 4 is less than 7   b. 9 is not equal to 6

Solutions.   a. $4 < 7$ b. $9 \neq 6$

Note that the symbols $=$ and $\neq$ have been used now in two ways: first, as showing a relationship between two sets, and second, as showing a relationship between two numbers.

If the elements of a set are listed so that each element is greater than those that follow it, or so that each element is less than those that follow it, the result is called an **ordered set.** If the elements of the set of natural numbers are ordered as $\{1, 2, 3, 4, 5, \cdots\}$, we can use this set in a special way. Let the members of two sets be paired as shown in Figure 2.1 in

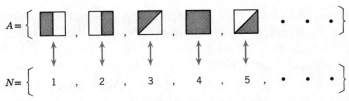

Figure 2.1

which the position of an element in A is noted by the associated number in N. The completely shaded square in A is in the *fourth* position. When a number is used to indicate the position of an element in an ordered set, the number is used in an **ordinal sense.** Thus, if Albert has a rank of 5 in a class of 30 students, 30 is a *cardinal use* of the number since it tells us *how many* students are in the class, and 5 is used in an *ordinal sense* because it refers to Albert's *position* in the class.

Exercise 2.1

A

State the cardinality of each set.

Examples.   a. $\{x, y, z\}$ b. $\{5, 10, 15, 20\}$

Solutions.   a. 3   b. 4

1. {Days of the week}   2. {Months of the year}

3. $\{a, e, i, o, u, y\}$ 4. $\{k, l, m, n, p, q\}$

5. $\{*, \triangle\}$ 6. $\{\square, \triangle, *\}$ 7. $\{2, 4, 6, 8, 10, 12\}$

8. $\{1, 3, 5, 7, 9\}$ 9. $\varnothing$ 10. $\{0\}$

11. $\{3\}$ 12. $\{5\}$

Let $A = \{1, 3, 5\}$, $B = \{2, 4, 6\}$, $C = \{1, 2, 3, 4\}$, and $D = \{1, 6\}$. State the cardinality of each set.

Example. $A \cap C$

Solution. $A \cap C = \{1, 3\}$; hence, the cardinality is 2.

13. $A \cap B$ 14. $B \cap C$ 15. $C \cap D$

16. $A \cap D$ 17. $A \times C$ 18. $B \times B$

Replace each word phrase with the correct symbol: $>, <, =,$ or $\neq$.

Examples.   a. 5 is less than 8   b. 3 is greater than 1

Solutions.   a. $5 < 8$ b. $3 > 1$

19. 6 is less than 16   20. 14 is greater than 5

21. 4 is not equal to 9   22. 8 is not equal to 2

23. 7 is greater than 3   24. 10 is less than 13

25. $2 + 3$ is equal to 5   26. 8 is equal to $4 + 4$

Replace the comma in each pair with the correct order symbol: $<, >,$ or $=$.

27. 5, 2   28. 9, 1   29. 7, 10

30. 3, 8   31. 12, 12   32. 6, 6

Indicate whether statements 33 to 40 refer to numbers in a cardinal or ordinal sense.

33. There are *four* books on the table.

34. He is number *one* in my opinion.

35. Your rank is *two* in a class of *twenty-five*.

36. Of the *five* puppies, I like the *second* from this end.

37. Seabiscuit finished *third* in a *nine*-horse race.

38. *Two* astronauts flipped a coin to see which would be *first*.

39. Of the *nine* planets in the solar system, the earth is the *fifth* largest.

40. Of the fifty states in the United States, California ranks third in area.

B

Consider $A = \{1, 2, 3\}$, $B = \{3, 5, 7, 9\}$, $C = \{2, 4, 6, 8, 10\}$, $D = \{0\}$, *and* $E = \{1, 11\}$. *State the cardinality of each of the following.*

41. $(A \cap B) \cap C$ 42. $B \cap (C \cap D)$ 43. $(C \cup D) \cap E$

44. $D \cup (E \cap A)$ 45. $(E \cap A) \cup B$ 46. $A \cap (B \cup D)$

47. $(B \cup E) \cup \varnothing$ 48. $A \cap (D \cap \varnothing)$

49. Describe the difference in meaning of the symbols $\varnothing$ and $\{0\}$.

50. Explain the difference in meaning of the words *number* and *numeral*.

2.2 Variables; Axioms of Equality and Order

So far, in the discussion of sets, we have considered whether or not a *particular* element is a member of a set. We also wish to discuss the characteristics of the members of sets in general without specifying a *particular* member. For example, the word "student" can be used to apply to any member of a class, rather than to a specific member. When it is desired to discuss some unspecified member of a set, a symbol, called a **variable,** is used and the set is called the **replacement set** of the variable. Many kinds of symbols may be used as variables, but it is usually more convenient to use letters such as a, b, c, x, y, and z. Thus, if a represents any unspecified member of the set $\{1, 2, 3, 4\}$, a is a variable and $\{1, 2, 3, 4\}$ is the replacement set of a. A specified member of a set, whose elements are numbers such as 2, is called a **constant.**

When the replacement set of a variable contains a large number of elements, it is not practical to list all the elements. Instead, a collection of symbols called "**set-builder notation**" is used. For example, we may wish to consider the set of whole numbers between 1 and 1,000, without listing all of them. If we use a as a variable for this set, we can specify how the values of a are to be determined and limited.

Since the set contains numbers between 1 and 1,000, the numbers 1 and 1,000 are not in the set. Thus, a is greater than 1 and less than 1,000. These conditions on a can be symbolized as

$$a \mid 1 < a < 1,000,$$

which is read: "all a such that 1 is less than a and a is less than 1,000," or "all a such that a is between 1 and 1,000," where the vertical bar between a and 1 is read: "such that." Since we shall be studying numbers other than whole numbers, we must include a reference to the replacement set in the symbolic statement above. Thus we include $a \in W$, read: "a represents a member of the set of whole numbers," or more simply, "a is a whole number."

Finally, we are considering a *set* of numbers. Hence, we enclose the entire statement in braces, and write

$$\{a \mid 1 < a < 1,000, a \in W\},$$

which is then read: "the set of all a such that a is between 1 and 1,000, and a is a whole number."

When variables are involved in expressions, the following symbols are also quite useful:

$\geq$ which means "is greater than or equal to," and

$\leq$ which means "is less than or equal to."

Thus the set in the example above can also be written as

$$\{a \mid 2 \leq a \leq 999, a \in W\},$$

which is read: "the set of all a such that 2 is less than or equal to a, a is less than or equal to 999, and a is a whole number."

Thus we can now specify a set by listing names of the elements, by the use of a rule, or by the use of set-builder notation.

Examples. Use set-builder notation to describe each set.

  a. $\{6, 7, 8, 9, 10, 11, 12\}$

  b. {natural numbers less than 10}

Solutions. a. $\{x \mid 6 \leq x \leq 12, x \in W\}$ or $\{x \mid 5 < x < 13, x \in W\}$

  b. $\{x \mid x < 10, x \in N\}$ or $\{x \mid x \leq 9, x \in N\}$

In Section 1.1 we discussed the equality of sets and established the use of the symbol $=$ in $A = B$ to indicate that A and B have precisely the same elements. We now use variables to discuss the "equals" relation as it pertains to elements in a set and to make some *assumptions* about this relation.

When we use a variable such as a or x as an unspecified member in a set of numbers, we wish to be assured that every time a is used in a discussion, it represents exactly the same number. In other words, $a = a$. This assumption is called the **reflexive law of equality.**

Sometimes different symbols, a and b, for example, may be used for the same number. If so, we can write $a = b$. Then we assume that if $a = b$, it is also true that $b = a$. This assumption is called the **symmetric law of equality.**

If $a = b$, which means that a and b are the same number, and if $b = c$, which means that b and c are the same number, then we can assume that it is also true that $a = c$. This assumption is called the **transitive law of equality.**

If $a = b$, then we assume that a can be substituted for b and b substituted for a in any expression without affecting the truth or falsity of the statement. This assumption is called the **substitution law.**

When an assumption is made in mathematics, it is customary to call the assumption an **axiom** or a **postulate.** In general, the terms "assumption," "axiom," and "postulate" can be used interchangeably. At this time, then, we have introduced four axioms concerning the *equals relation:*

For all a, b, c $\in$ W,

$a = a$.	Reflexive law
If $a = b$, then $b = a$.	Symmetric law
If $a = b$ and $b = c$, then $a = c$.	Transitive law
If $a = b$, then a may be replaced by b or b may be replaced by a in any collection of symbols without affecting the truth or falsity of the statement.	Substitution law

These axioms are certainly in accord with our concepts of the whole numbers. You will note that the transitive law and the substitution law are quite similar. In fact, you may use the substitution law in any situation where the transitive law is applicable.

We now make two assumptions concerning the relations $<$, $=$, $>$.

For all a, b ∈ W, exactly one of the following is true,

$$a < b, \quad a = b, \quad or \quad a > b. \qquad \text{Trichotomy law}$$

For all a, b, c ∈ W,

$$if \ a < b \ and \ b < c, then \ a < c. \qquad \text{Transitive law}$$

The first of these order axioms, commonly called the **axioms of inequality,** simply guarantees that one whole number is less than, equal to, or greater than a second whole number. The second axiom is similar to the transitive law of equality.

Exercise 2.2

A

In each of the following:

a. Translate the symbolic statements into words (rule form).
b. List the elements in the set.

Example. $\{x \mid 3 < x < 9, x \in N\}$

Solution.   a. The set of all x such that 3 is less than x, x is less than 9 (or x is between 3 and 9), and x is a natural number.

b. $\{4, 5, 6, 7, 8\}$

1. $\{x \mid 2 < x < 6, x \in N\}$
2. $\{y \mid 4 < y < 9, y \in N\}$
3. $\{x \mid x < 7, x \in W\}$
4. $\{y \mid y < 5, y \in W\}$
5. $\{z \mid 0 < z < 8, z \in W\}$
6. $\{x \mid 0 < x < 6, x \in W\}$
7. $\{x \mid x > 3, x \in N\}$
8. $\{x \mid x > 5, x \in N\}$
9. $\{y \mid y > 0, y \in W\}$
10. $\{x \mid x > 0, x \in W\}$

Write each set in set-builder notation. Use x as the variable. (There may be more than one representation.)

Example. $\{1, 2, 3\}$

Solution. $\{x \mid 0 < x < 4, x \in W\}$ or $\{x \mid 1 \leq x \leq 3, x \in N\}$

11. $\{1, 2\}$
12. $\{3, 4\}$
13. $\{5, 6, 7\}$
14. $\{9, 10, 11\}$
15. $\{0, 1, 2, 3\}$
16. $\{1, 2, 3, 4\}$

17. {4, 5, 6, 7, 8}   18. {6, 7, 8, 9, 10}   19. {0, 1, 2, 3, · · ·}

20. {1, 2, 3, · · ·}   21. {5, 6, 7, · · ·}   22. {9, 10, 11, · · ·}

Name the axiom that justifies each statement.

23. If $x = 7$, then $7 = x$.   24. If $x < 3$ and $3 < y$, then $x < y$.

25. If $x = 3$ and $3 = y$, then $x = y$.   26. $x < 3$, $x = 3$ or $x > 3$.

27. $y = y$.   28. If $y = 6$, then $6 = y$.

29. If $x < y$ and $y < 10$, then $x < 10$.   30. If $6 = y$ and $y = x$, then $6 = x$.

31. $y < 3$, $y = 3$, or $y > 3$.   32. $x = x$.

33. If $x = y$ and $4 < x$, then $4 < y$.   34. If $y < x$ and $x < 8$, then $y < 8$.

35. $x < 9$, $x = 9$, or $x > 9$.   36. If $x = 2$ and $y < 2$, then $y < x$.

37. If $x = 10$, then $10 = x$.   38. If $x < 5$ and $5 < y$, then $x < y$.

39. If $x = y$ and $x \not< 15$, then $y \not< 15$.   40. If $y = 3$ and $x \not< y$, then $x \not< 3$.

B

List the elements in each set. If the set is infinite, use three dots to indicate this fact.

41. $\{x \mid 1 < x < 5, x \in N\} \cup \{x \mid x < 3, x \in W\}$

42. $\{x \mid 0 < x < 9, x \in W\} \cap \{2, 4, 6, 8, 10\}$

43. $\{1, 3, 5, 7\} \cup \{y \mid 0 < y < 2, y \in N\}$

44. $\{x \mid x > 6, x \in N\} \cap \{x \mid x < 9, x \in W\}$

45. $\{z \mid z \in N\} \cup \{0\}$

46. $\{x \mid x \in W\} \cap \{x \mid x \in N\}$

2.3 The Number Line

Numbers have been associated with the concept of "how many." In this sense, they can be used as measures of things if the unit of measure is defined. Of particular interest at this time is the association of the natural numbers with the lengths of line segments through the use of a **number line.**

A number line, Figure 2.2, is constructed by drawing a straight line and

Figure 2.2

placing a small mark above it at any arbitrary point which is called the **origin** and which we associate with the number zero.

Next, a second mark is placed at some convenient distance to the right of the first mark, forming a line segment from the origin to the second mark. The second mark is associated with the number 1. The line segment from 0 to 1 is said to be *one unit* in length.

The right end-points of line segments, which are formed by repeating the above process and duplicating the unit of length from 0 to 1, are in turn associated with the numbers 2, 3, etc., as in Figure 2.3.

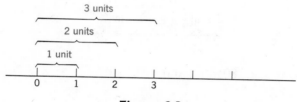

Figure 2.3

Any unit of length may be used to construct a number line and, in general, it is only necessary to show the numbers associated with a few selected unit markings to establish the scale of a number line. Furthermore, it is not necessary to start each line with a mark to indicate the number 0.

Examples

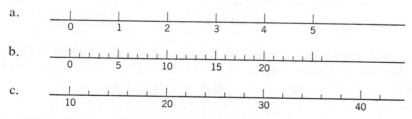

The numbers written below a number line to establish the scale are sometimes called scale numbers. In Example c above, the scale numbers are 10, 20, 30, and 40, and each division of the number line indicated by the marks represents two units.

If a number line is extended sufficiently far to the right, or if an appropriate scale is chosen, any given element of the set of whole numbers can be associated with a *unique* (one and only one) position on the number line. The point on the number line associated with a number is called the **graph** of the number.

Examples.   a.

includes the graphs of 0, 1, 6, 11, and 14.

b.

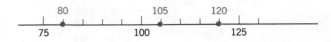

includes the graphs of 80, 105, and 120.

The number associated with a point on a number line is called the **coordinate** of the point. In Example a above, 0, 1, 6, 11, and 14 are the respective coordinates of the points marked on the line.

The order of two numbers can be shown very clearly on a number line. If the numbers are arranged in increasing order to the right, then the point associated with the smaller of two numbers is always to the left of the point associated with the larger number. Figure 2.4 shows two points

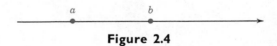

Figure 2.4

associated with the numbers a and b, and an arrowhead on the right to indicate the direction of increasing order. If a and b are whole numbers, that is, if $a, b \in W$, then we can say that $a < b$ or $b > a$.

Exercise 2.3

A

Graph each set of numbers. Use a different number line for each set.

Examples.   a. $\{2, 3, 5\}$ b. $\{x \mid x > 7, x \in W\}$

Solutions.

a.   b.

1. {1, 2, 3}   2. {4, 5, 6}   3. {3, 7, 10}

4. {2, 6, 9}   5. {13, 15, 17}   6. {16, 27, 31}

7. {1, 2, 10, 11}   8. {3, 5, 12, 15}

9. {Natural numbers less than 9}   10. {Natural numbers less than 5}

11. {Whole numbers less than 7}   12. {Whole numbers greater than 2}

13. $\{x \mid 1 < x < 5, x \in N\}$ 14. $\{x \mid 3 < x \le 9, x \in N\}$

15. $\{x \mid x \le 7, x \in N\}$ 16. $\{x \mid x \le 4, x \in W\}$

17. $\{x \mid x \ge 5, x \in N\}$ 18. $\{x \mid x > 8, x \in W\}$

19. $\{x \mid x \in N\}$ 20. $\{x \mid x \in W\}$

Let a, b, c, d represent whole numbers. Their graphs are shown on the line.

Replace the comma in each pair with the proper order symbol: $<$, $>$, *or* $=$.

21. *b, c*   22. *c, a*   23. *a, d*   24. *b, b*   25. *0, b*

26. *c, 0*   27. *a, b*   28. *c, d*   29. *c, c*   30. *c, b*

For each graph below:

   a. *List the members of the set if possible.*
   b. *Write the set in set-builder notation. Use x as the variable.*

31.

32.

33.

34.

B

Graph each of the following. If a graph cannot be drawn, so state.

35. $\{3, 7, 9, 11\} \cap \{1, 2, 3, 4\}$

36. $\{x \mid 0 < x \leq 8, x \in N\} \cap \{9, 10, 11\}$

37. $\{1, 2, 4, 5\} \cup \{y \mid y > 3, y \in N\}$

38. $\{\text{Whole numbers less than 9}\} \cup \{x \mid 1 \leq x \leq 4, x \in N\}$

2.4   Addition

The familiar operation of addition, which pairs two numbers to yield a *unique* third number, can be defined through the use of sets. Consider the two disjoint sets

$$A = \{m, n, o\} \quad \text{and} \quad B = \{p, q, r, s\}$$

and their union

$$A \cup B = \{m, n, o, p, q, r, s\}.$$

Note that the cardinality of A is 3, the cardinality of B is 4, and the cardinality of $A \cup B$ is 7. This example suggests the following definition.

Definition 2.1.   The **sum** of two whole numbers a and b is that number, $a + b$, which is the cardinality of the set formed by the union of two disjoint sets whose cardinalities are respectively a and b.

Because the operation of addition pairs *two* numbers, it is called a **binary operation.** Although the sum of two numbers, say, 3 and 4, can be written as 7, we shall also wish to consider the collection of symbols, $3 + 4$, as a representation for the *sum*. These forms are simply different names (or symbols) for the concept of the cardinality of the same set. Since a number such as 7 can be named in many ways $(7, 4 + 3, 5 + 2, 6 + 1$, etc.), it is convenient to designate one of these as basic and we shall call it the **basic numeral** of the number. We designate the basic numeral of a whole number to be one of the symbols in $\{0, 1, 2, 3, \cdots\}$. Thus, while $3 + 4$ is the sum of 3 and 4, the basic numeral for this sum is 7. Where the sum of two numbers, say, x and 9, cannot be written as a basic numeral, the sum can be indicated as $x + 9$.

There are four basic properties concerning the operation of addition which are in accord with our experiences for the sum of *any* two whole numbers. Although these properties are logical consequences of certain properties of sets, we shall simply take the following statements about the properties as assumptions.

1. *The sum of any two whole numbers is a whole number.*

$$\text{For all } a, b \in W, \qquad a + b \in W.$$

This assumption is called the **closure law for addition,** and the set of whole numbers is said to be **closed** with respect to addition.

    a. Since $34{,}187 \in W$ and $7{,}319 \in W$, then $34{,}187 + 7{,}319 \in W$.

    b. If $x \in W$, then $x + 14 \in W$.

To say that a set is closed with respect to addition is to imply that the sum of any two members of the set (including the sum of a number and itself) is also a member of the set. Not all sets of whole numbers are closed for addition. For example, the set $\{1, 2, 3\}$ is *not closed* for addition because $1 + 3 = 4$ and 4 is not a member of the set. Also, $2 + 2, 2 + 3$, and $3 + 3$ are not members of the set.

2. *The sum of any two whole numbers is the same number regardless of the order in which the two numbers are paired.*

$$\text{For all } a, b \in W, \qquad a + b = b + a.$$

This assumption is called the **commutative law of addition.**

Examples. a. $905 + 416 = 416 + 905$.

           b. If $x \in W$, then $x + 34 = 34 + x$.

3. *The sum of the whole numbers $(a + b)$ and c is the same as the sum of a and $(b + c)$. The way in which the numbers are grouped for the binary operation of addition does not affect the sum.*

$$\text{For all } a, b, c \in W, \qquad (a + b) + c = a + (b + c).$$

This assumption is called the **associative law of addition.** Observe that this assumption is concerned only with the *grouping* of the terms in a sum and not with the *order* of the terms. The order remains the same. We consider a collection of symbols such as $a + b + c$ to mean either $(a + b) + c$ or $a + (b + c)$.

Examples.   a. $2 + 3 + 4 = (2 + 3) + 4 = 2 + (3 + 4)$.

b. If $x \in W$, then $(x + 3) + 5 = x + (3 + 5)$.

c. If $x, y \in W$, then $[(x + 2) + 3] + y = (x + 2) + (3 + y)$.

Parentheses are employed here to show that two numbers are to be grouped together for the purpose of addition and the collection of symbols within the parentheses is to be considered first. In general, we do not go through the process of using parentheses in such a collection of symbols even though we do pair the numbers mentally. We use the parentheses here to emphasize that addition is a binary operation. In Example c above, the symbols [ ], called "brackets," are used in the same way.

4. *A unique whole number exists, called "zero," such that the sum of this number and any whole number a is the number a.*

$$\text{For all } a \in W, \qquad a + 0 = a \quad and \quad 0 + a = a.$$

This assumption is called the **identity law of addition;** the number 0 is called the **identity element of addition.**

Examples.   a. $0 + 5 = 5 + 0 = 5$.

b. If $x \in W$, then $x + 0 = x$ and $0 + x = x$.

The fact that 0 is said to be unique in Statement 4 means that we are assuming that there is *only one identity element of addition* in the set. Thus, if $x + a = x$, then a must equal 0.

You have seen now that the operation of addition pairs two whole numbers with a unique third whole number and is governed by four basic assumptions:

For all a, b, c $\in W$,

$a + b \in W$	Closure law
$a + b = b + a$	Commutative law
$(a + b) + c = a + (b + c)$	Associative law
$a + 0 = a \quad and \quad 0 + a = a$	Identity law

These axioms, together with the axioms of equality, justify our making other assertions about whole numbers which we will look at in Section 2.7.

Exercise 2.4

A

Name the axiom that best justifies each statement. $x, y, z \in W.$

Examples. a. $(x + y) + z = x + (y + z)$ b. $(x + y) + z = z + (x + y)$

Solutions. a. Associative law of addition   b. Commutative law of addition

1. $2 + 3 \in W$

2. $5 + 4 = 4 + 5$

3. $(6 + 1) + 7 = 7 + (6 + 1)$

4. $8 + 0 = 8$

5. $0 + y = y$

6. $9 + x \in W$

7. $x + 1 = 1 + x$

8. $(y + 2) + 5 = y + (2 + 5)$

9. $4 + (x + y) = (4 + x) + y$

10. $(x + y) + 0 = x + y$

11. $(x + y) + 6 \in W$

12. $3 + (7 + x) = (3 + 7) + x$

13. $[4 + (y + z)] + 0 = 4 + [(y + z) + 0]$

14. $[5 + (x + y)] + 2 = [(5 + x) + y] + 2$

15. If $x + y = 8$ and $y = 3$, then $x + 3 = 8$.

16. If $(x + y) + z = 6$ and $x = 2$ and $z = 3$, then $(2 + y) + 3 = 6$.

17. $(3 + y) + (2 + z) = (2 + z) + (3 + y)$

18. $(3 + y) + (2 + z) = [(3 + y) + 2] + z$

19. $[2 + (3 + y)] + z = [(2 + 3) + y] + z$

20. $[(3 + y) + 2] + z = [2 + (3 + y)] + z$

21. $(4 + x) + (5 + y) \in W$

22. $[(z + 3) + (x + 1)] + (y + 2) \in W$

23. $[(5 + x) + (3 + y)] + (z + 4) = [(3 + y) + (5 + x)] + (z + 4)$

24. $(x + 6) + [(z + 1) + (y + 7)] = (x + 6) + [z + (1 + y + 7)]$

25. If $(x + 1) + (y + 2) = z + 3$ and $y = z + 3$, then $(x + 1) + (y + 2) = y$.

26. $[(z + 7) + (x + 3)] + (y + 1) = [(x + 3) + (z + 7)] + (y + 1)$

Regroup the collection of symbols in each sum in a way that will facilitate finding a basic numeral for the sum. Find the basic numeral.

27. $(7 + 5) + 95$

28. $7 + (3 + 28)$

29. $997 + (3 + 36)$

30. $(11 + 17) + 83$

31. $87 + (13 + 3) + 7$

32. $52 + (48 + 27) + 73$

33. $(22 + 14) + (78 + 86)$

34. $(31 + 47) + (53 + 69)$

35. $(4 + 65) + (20 + 96) + 35$

36. $(19 + 5) + (6 + 44) + (81 + 56)$

B

State which of the following sets are closed under addition. If any set is not closed, give an example to justify your conclusion.

37. $\{0, 2, 4, 6, 8\}$

38. $\{1, 2, 3, 4, 5\}$

39. $\{2, 4, 6, 8, \cdots\}$

40. $\{3, 5, 7, 9, \cdots\}$

41. $\{x \mid x > 8, x \in N\}$

42. $\{y \mid y \geq 4, y \in W\}$

43. $\{0\}$

44. $\{0, 1\}$

45. Write each statement in symbolic form. Use the variables a, b, and c.

    a. Closure law for addition

    b. Commutative law of addition

    c. Associative law of addition

    d. Identity law of addition

2.5 Multiplication

The operation of multiplication is a *binary operation* that assigns to each pair of whole numbers, a and b, called **factors,** a third unique whole number $a \cdot b$, called the **product.** The operation is sometimes designated by the symbol $\times$. The $\times$ should be used with caution since it can easily be mistaken for the letter x if variables are being multiplied. The product of two variables or a number and a variable can also be indicated simply by writing the variable and the numeral representing the number adjacently.

Examples.   a. a times $b = a \times b = a \cdot b = ab = a(b) = (a)b = (a)(b)$.

            b. 3 times $c = 3 \times c = 3 \cdot c = 3c$.

The product of two whole numbers can be defined in several ways. It can be considered as the number obtained as the sum of successive additions.

In this way, $3 \cdot 2$ would be defined as the sum $2 + 2 + 2$. A product can also be defined through the use of sets. This definition of product is analogous to the definition of sum in terms of the union of two sets. It is consistent with the definition of product in terms of successive addition.

Recall that if $R = \{a, b, c\}$ and $S = \{x, y\}$, then the Cartesian product

$$R \times S = \{(a, x), (a, y), (b, x), (b, y), (c, x), (c, y)\}.$$

Notice that the cardinality of R is 3, the cardinality of S is 2, and the cardinality of $R \times S$ is 6. This example suggests the following definition.

Definition 2.2. The **product** of two whole numbers a and b is that number $a \cdot b$ which is the cardinality of $A \times B$, the Cartesian product of A and B, where A has cardinality a and B has cardinality b.

This interpretation of multiplication can be illustrated graphically by using a geometric device known as a **lattice** or an **array**, which is simply an orderly arrangement of objects (called **elements**) in rows and columns. A lattice of the above Cartesian product, $R \times S$, appears in Figure 2.5a. Instead of actually writing the ordered pairs, we could substitute dots in the position of each ordered pair, as in Figure 2.5b.

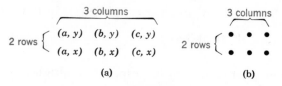

Figure 2.5

Observe that there are 6 elements in an array of 2 rows and 3 columns. In general, we can think of a product as the number of elements in an array in which the number of rows and the number of columns are the factors of the product. The array also illustrates that the product can be considered as successive addition since we can add the number of elements in each row as many times as we have rows.

The empty set has no elements. Therefore, no pairs can be formed for a Cartesian product in which one set is $\varnothing$; its array contains no elements. Hence, the product of any whole number a and zero is zero. For example,

$$3 \cdot 0 = 0, \qquad 0 \cdot 7 = 0, \qquad \text{and} \qquad 0 \cdot 0 = 0.$$

Although the product of two numbers, say, 3 and 4, can be written as 12, we shall also consider the collection of symbols, $3 \cdot 4$, as a representation for the product. Since a number such as 12 can be named in many ways [12, $3 \cdot 4$, $6 \cdot 2$, $2 \cdot (3 \cdot 2)$, etc.], we designate one of these as basic in the same way that we designated a basic numeral for a sum. We again call one of the elements in $\{0, 1, 2, 3, \cdots\}$ a basic numeral.

Where the product of two numbers, say, 3 and x, cannot be written as a basic numeral, the product can only be written as $3 \cdot x$, or simply $3x$. In this example, and in similar cases where numbers and variables occur, the numeral for the number is usually written as the first factor.

We have assumed certain properties about the operation of addition. We make similar assumptions about closure, commutativity, and associativity for the operation of multiplication on elements of the set of whole numbers. There is also a special element (analogous to zero for addition) called the **identity element of multiplication.**

1. *The product of any two whole numbers is a whole number.*

$$\text{For all } a, b \in W, \qquad a \cdot b \in W.$$

This assumption is called the **closure law for multiplication.**

Examples.   a. Since $345 \in W$ and $106 \in W$, then $345 \cdot 106 \in W$.

   b. If $x \in W$, then $3 \cdot x \in W$.

As with addition, a set is closed with respect to multiplication if and only if the products of all possible pairs of numbers are members of the set. Not all sets of whole numbers are closed for multiplication. For example, the set $\{1, 2\}$ is *not closed* for multiplication because $2 \cdot 2 = 4$ is not a member of the set.

2. *The product of any two whole numbers is the same number regardless of the order in which the two numbers are paired.*

$$\text{For all } a, b \in W, \qquad a \cdot b = b \cdot a.$$

This assumption is called the **commutative law of multiplication.**

Examples.   a. $621 \cdot 294 = 294 \cdot 621$.

   b. If $x \in W$, then $3 \cdot x = x \cdot 3$.

3. *The product of two whole numbers* $(a \cdot b)$ *and* c *is the same as the product of* a *and* $(b \cdot c)$. *The way in which the numbers are grouped for the binary operation of multiplication does not affect the product.*

$$\text{For all } a, b, c \in W, \qquad (a \cdot b) \cdot c = a \cdot (b \cdot c).$$

This assumption is called the **associative law of multiplication.** As with the similar law for addition, this assumption is concerned only with the *grouping* of the terms in a product and not with the *order* of the terms. We consider a collection of symbols such as $a \cdot b \cdot c$ to mean either $(a \cdot b) \cdot c$ or $a \cdot (b \cdot c)$.

Examples. a. $2 \cdot 3 \cdot 4 = (2 \cdot 3) \cdot 4 = 2 \cdot (3 \cdot 4)$.

b. If $x \in W$, then $(3 \cdot x) \cdot 5 = 3 \cdot (x \cdot 5)$.

4. *A unique whole number exists, called "one," such that the product of this number and any whole number* a *is the number* a.

$$\text{For all } a \in W, \qquad 1 \cdot a = a \quad and \quad a \cdot 1 = a.$$

This assumption is called the **identity law of multiplication;** the number 1 is called the **identity element of multiplication.**

Examples. a. $1 \cdot 5 = 5 \cdot 1 = 5$.

b. If $x \in W$, then $x \cdot 1 = x \quad$ and $\quad 1 \cdot x = x$.

The fact that 1 is said to be unique in Statement 4 means that we are assuming that there is *only one identity element of multiplication* in the set. Thus, if $x \cdot a = x$, then a must equal 1.

We have now defined the operation of multiplication which pairs two whole numbers with a unique third whole number and we have made four basic assumptions concerning this operation:

For all $a, b, c \in W$,

$a \cdot b \in W$	Closure law
$a \cdot b = b \cdot a$	Commutative law
$(a \cdot b) \cdot c = a \cdot (b \cdot c)$	Associative law
$1 \cdot a = a \quad$ and $\quad a \cdot 1 = a$	Identity law

These axioms, together with the axioms of equality, justify making other assertions which we shall discuss in Section 2.7.

Exercise 2.5

A

Name the axiom that best justifies each statement. $x, y, z \in W$.

Examples.   a. $3 \cdot 4 = 4 \cdot 3$.   b. $3 \cdot y \in W$.

Solutions.   a. Commutative law of multiplication.

b. Closure law for multiplication.

1. $3 \cdot 4 \in W$
2. $6 \cdot 2 = 2 \cdot 6$
3. $(5 \cdot 1) \cdot 4 = 5 \cdot (1 \cdot 4)$
4. $7 \cdot 1 = 7$
5. $1 \cdot x = x$
6. $8 \cdot y \in W$
7. $y \cdot 2 = 2 \cdot y$
8. $(3 \cdot y) \cdot z = 3 \cdot (y \cdot z)$
9. $5 \cdot (x \cdot y) = (5 \cdot x) \cdot y$
10. $(x \cdot z) \cdot 1 = x \cdot z$
11. $(x \cdot y) \cdot 4 \in W$
12. $2 \cdot (8 \cdot y) = 2 \cdot (y \cdot 8)$
13. If $x \cdot (y + z) = 6$ and $y = 2$, then $x \cdot (2 + z) = 6$.
14. If $x \cdot y = 9$ and $x = 3$, then $3 \cdot y = 9$.
15. $(x + y) \cdot (x + z) = (x + z) \cdot (x + y)$
16. If $x \cdot (y + z) = 6$ and $x = 2$, then $2 \cdot (y + z) = 6$.
17. $(3 \cdot x) \cdot (4 \cdot y) \in W$
18. $(5 \cdot z) \cdot (1 \cdot x) \cdot (2 \cdot y) \in W$
19. $[(5 \cdot x) \cdot (3 \cdot y)] \cdot (4 \cdot z) = [(3 \cdot y) \cdot (5 \cdot x)] \cdot (4 \cdot z)$
20. $(6 \cdot x) + [(8 \cdot z) \cdot (7 \cdot y)] = (6 \cdot x) + [8 \cdot (z \cdot 7 \cdot y)]$
21. If $(2 \cdot x) \cdot (3 \cdot y) = 3 \cdot z$ and $y = 3 \cdot z$, then $(2 \cdot x) \cdot (3 \cdot y) = y$.
22. $[(3 \cdot y) \cdot 2] \cdot z = [2 \cdot (3 \cdot y)] \cdot z$

Regroup the collection of symbols in each product in a way that will facilitate finding a basic numeral for the product. Find the basic numeral.

23. $5 \times (2 \times 16)$ 24. $20 \times (5 \times 29)$ 25. $(2 \times 8) \times 50$

26. $(25 \times 9) \times 4$ 27. $(13 \times 5) \times 20$ 28. $(8 \times 27) \times 125$

29. $4 \times (5 \times 25) \times 3$ 30. $2 \times (7 \times 50) \times 7$ 31. $(5 \times 8) \times (3 \times 2)$

32. $(8 \times 6) \times (125 \times 3)$

B

State which of the following sets are closed under multiplication. If any set is not closed, give an example to justify your conclusion.

33. $\{0, 1, 2, 3, 4, 5\}$ 34. $\{9, 10, 11, 12, \cdots\}$ 35. $\{x \mid x > 13, x \in N\}$

36. $\{2, 5, 8, 11, 14, 17\}$ 37. $\{0, 1\}$ 38. $\{0\}$

Graph the array for each Cartesian product.

39. $\{1, 2, 3\} \times \{5, 10\}$ 40. $\{2, 4, 6, 8\} \times \{9, 10\}$

41. $\{1, 3, 5\} \times \{2, 4, 6, 8\}$ 42. $\{0, 1, 2\} \times \{1, 2, 3, 4, 5\}$

43. Write each statement in symbolic form. Use the variables a, b, and c.

   a. Closure law for multiplication
   b. Commutative law of multiplication
   c. Associative law of multiplication
   d. Identity law of multiplication

2.6 The Distributive Law

In Sections 2.4 and 2.5 the laws of closure, commutativity, and associativity for both addition and multiplication in the set of whole numbers were introduced. We now discuss a law which establishes a relationship between these two operations, addition and multiplication. First, however, we make an agreement about a collection of symbols which involves both the sign for addition and the sign for multiplication. Recall that

$$a + b + c = (a + b) + c = a + (b + c)$$

and

$$a \cdot b \cdot c = (a \cdot b) \cdot c = a \cdot (b \cdot c).$$

How shall we interpret an expression such as $5 + 3 \cdot 6$? Perhaps it can be considered as $(5 + 3) \cdot 6$, which is equal to $8 \cdot 6$ or 48; or as $5 + (3 \cdot 6)$, which is equal to $5 + 18$ or 23. To prevent such confusion, we make the agreement that

$$a + b \cdot c = a + (b \cdot c)$$

and

$$a \cdot b + c = (a \cdot b) + c.$$

Thus, operations are to be performed in the following order:

1. Operations within parentheses.

2. Multiplication operations.

3. Addition operations.

Examples.   a. $3 \cdot (5 + 2) = 3 \cdot (7)$ b. $3 \cdot 5 + 4 \cdot 7 = 15 + 28$
$$= 21 \qquad\qquad\qquad\qquad = 43$$

An array illustrating the product $3 \cdot (5 + 2)$ in Example a is shown in Figure 2.6.

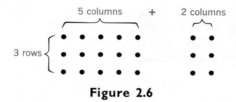

Figure 2.6

There are 21 elements in the complete array. If we separate the array into two parts as indicated, the left-hand portion is clearly an array representing the product $3 \cdot 5$ and the right-hand portion represents $3 \cdot 2$. This suggests the statement $3 \cdot (5 + 2) = 3 \cdot 5 + 3 \cdot 2$, and we can think of the factor 3 as being *distributed* over the sum $5 + 2$. Similarly,

$$5 \cdot (1 + 4) = 5 \cdot 1 + 5 \cdot 4$$

and

$$4 \cdot (8 + 2) = 4 \cdot 8 + 4 \cdot 2.$$

This relationship between sums and products is very important and quite useful in mathematics. In general, we make the assumption:

For all a, b, c $\in W$, $a \cdot (b + c) = a \cdot b + a \cdot c.$

This statement is called the **distributive law.** The way we have written the law might be termed the "**left-hand distributive law**" because the factor a, which is distributed over the sum, $b + c$, appears on the left. The statement

$$(b + c) \cdot a = b \cdot a + c \cdot a$$

is referred to as the "**right-hand distributive law.**"

Examples. a. $7 \cdot (9 + 2) = 7 \cdot 9 + 7 \cdot 2$

b. If $a, c \in W$, then $(a + 4) \cdot c = a \cdot c + 4 \cdot c$.

As a consequence of the symmetric law of equality, the two distributive laws (left- and right-hand) can be written

$$a \cdot b + a \cdot c = a \cdot (b + c)$$

and

$$b \cdot a + c \cdot a = (b + c) \cdot a.$$

Application of the associative law of addition permits us to extend the distributive law. For example,

$$a \cdot (b + c + d) = a \cdot [(b + c) + d]$$
$$= a \cdot (b + c) + a \cdot d = a \cdot b + a \cdot c + a \cdot d,$$

and, in general,

$$a \cdot (b + c + d + \cdots + m) = a \cdot b + a \cdot c + a \cdot d + \cdots + a \cdot m.$$

The distributive law is the foundation upon which many arithmetic computations are based. In the example below both the horizontal and the vertical form employ the distributive law to express a product as a basic numeral.

Examples. $3 \cdot 76 = 3 \cdot (70 + 6)$

$$= 3 \cdot 70 + 3 \cdot 6$$

$$= 210 + 18 = 228$$

$$
\begin{array}{ll}
76 & \\
3 & \\
\hline
18 & (3 \cdot 6) \\
210 & (3 \cdot 70) \\
\hline
228 & (3 \cdot 70) + (3 \cdot 6)
\end{array}
$$

Exercise 2.6

A

Apply the distributive law and rewrite each expression without parentheses. Leave all terms in factored form. All variables denote elements of W.

Examples. a. $2 \cdot (3 + 4)$ b. $y \cdot (x + 3)$ c. $(1 + y) \cdot 2$

Solutions. a. $2 \cdot 3 + 2 \cdot 4$ b. $y \cdot x + y \cdot 3$ c. $1 \cdot 2 + y \cdot 2$

1. $4 \cdot (2 + 8)$ 2. $3 \cdot (4 + 9)$ 3. $6 \cdot (7 + 9)$

4. $1 \cdot (5 + 3)$ 5. $(1 + 5) \cdot 2$ 6. $(3 + 6) \cdot 3$

7. $(6 + 5) \cdot 9$ 8. $(2 + 10) \cdot 6$ 9. $5 \cdot (6 + x)$

10. $3 \cdot (y + 3)$ 11. $y \cdot (8 + x)$ 12. $x \cdot (y + 4)$

13. $(a + b) \cdot c$ 14. $(m + n) \cdot p$

Examples. a. $(10 + 3) \cdot (10 + 7)$ b. $(x + y) \cdot (z + w)$

Solutions. a. $(10 + 3) \cdot 10 + (10 + 3) \cdot 7 = 10 \cdot 10 + 3 \cdot 10 + 10 \cdot 7 + 3 \cdot 7$

b. $(x + y) \cdot z + (x + y) \cdot w = x \cdot z + y \cdot z + x \cdot w + y \cdot w$

15. $(5 + 1) \cdot (7 + 9)$ 16. $(2 + 4) \cdot (5 + 7)$

17. $(x + 3) \cdot (y + 5)$ 18. $(x + 1) \cdot (y + 6)$

19. $(6 + y) \cdot (8 + x)$ 20. $(2 + y) \cdot (10 + x)$

21. $(x + y) \cdot (a + b)$ 22. $(p + q) \cdot (c + d)$

Illustrate the process of multiplication (a) in horizontal form by using an expanded form of the basic numeral and (b) in vertical form, as illustrated below.

Example. $15 \cdot 25$

Solution. a. $15 \cdot 25 = (10 + 5) \cdot (20 + 5)$

$$= (10 + 5) \cdot 20 + (10 + 5) \cdot 5$$
$$= 10 \cdot 20 + 5 \cdot 20 + 10 \cdot 5 + 5 \cdot 5$$
$$= 200 + 100 + 50 + 25$$
$$= 375$$

b. 15
   25
   $\overline{}$
   75   $(5 \cdot 15) = (5 \cdot 10 + 5 \cdot 5)$
   300 $(20 \cdot 15) = (20 \cdot 10 + 20 \cdot 5)$
   $\overline{}$
   375 $(5 \cdot 10 + 5 \cdot 5) + (20 \cdot 10 + 20 \cdot 5)$

23. $3 \cdot 17$ 24. $16 \cdot 9$ 25. $13 \cdot 17$ 26. $19 \cdot 19$

27. $12 \cdot 18$ 28. $11 \cdot 16$ 29. $15 \cdot 19$ 30. $13 \cdot 15$

B

Apply the distributive law and rewrite each expression without parentheses. Leave all terms in factored form. All variables denote elements of W.

Example. $(x + 3) \cdot (y + z + 2) = (x + 3) \cdot [(y + z) + 2]$
$$= (x + 3) \cdot (y + z) + (x + 3) \cdot 2$$
$$= (x + 3) \cdot y + (x + 3) \cdot z + (x + 3) \cdot 2$$
$$= x \cdot y + 3 \cdot y + x \cdot z + 3 \cdot z + x \cdot 2 + 3 \cdot 2$$

31. $(2 + 3) \cdot (4 + 5 + 6)$ 32. $(10 + 1) \cdot (2 + 3 + 4)$

33. $(x + 2) \cdot (5 + y + z)$ 34. $(y + 7) \cdot (x + 4 + z)$

35. $(x + y + 8) \cdot (w + z + 6)$ 36. $(x + y + z) \cdot (a + b + c)$

37. Write the distributive law in symbolic form.

38. Use arrays to illustrate the distributive law for the product
$$3 \cdot (5 + 7) = 3 \cdot 5 + 3 \cdot 7.$$

2.7 Consequences of the Axioms

The sets of numbers with which we have been working, that is, the set of whole numbers and the set of natural numbers, are infinite sets. By making some assumptions concerning the operations of addition and multiplication we have laid a foundation for certain logical consequences. For example, if we add 2 to each member of the true statement $3 = 3$, the resulting statement $3 + 2 = 3 + 2$ is also true. Can you assume that similar statements are true for all elements in the set of whole numbers?

Is it true that if a, b, and c are any elements of the set of whole numbers and if $a = b$, then $a + c = b + c$? Perhaps you are willing to accept this statement as an assumption or an axiom. However, we can *prove* that the statement is true for all whole numbers; that is, the statement follows logically from the axioms and definitions we have previously made. Such provable statements about numbers are called **theorems.** Consider the following theorem.

Theorem 2.1.   If a, b, $c \in W$, and if $a = b$, then $a + c = b + c$.

Notice that the theorem consists of two parts. First, a part that gives statements that are assumed to be true, namely, that a, b, and c are whole numbers and that $a = b$. This portion of a theorem is called the **hypothesis.** Second, an assertion that the conditions assumed in the hypothesis *imply* the truth of an additional statement, namely, that $a + c = b + c$. This portion of a theorem that follows from the given hypothesis is called the **conclusion.**

While it is not our primary objective to teach you to make proofs, it is very important that you be able to follow a logical argument. Observe the following argument carefully.

If a, b, and c are whole numbers, the closure law for addition asserts that $a + c$ and $b + c$ are whole numbers. The reflexive law of equality asserts that

$$a + c = a + c. \tag{1}$$

Now, if $a = b$, the substitution law permits writing

$$a + c = b + c,$$

where a has been replaced with b on the right-hand side of Statement (1). The conclusion of the stated theorem has thus been demonstrated.

The foregoing argument was developed in an informal style. A more formal style consists of a concise *statement-reason* arrangement in which each statement made is supported by citing an axiom or definition or a theorem that has previously been proved. The following is a demonstration of the proof of the above theorem in a statement-reason format.

Theorem 2.1.   If a, b, $c \in W$, and if $a = b$, then $a + c = b + c$.

Proof:

Statements	**Reasons**
1. $a, b, c \in W$, and $a = b$	1. Hypothesis
2. $a + c \in W$	2. Closure law for addition
3. $a + c = a + c$	3. Reflexive law of equality
4. $a + c = b + c$	4. Substitution law (b is substituted for a in the right-hand member of the equality in Step 3)

Since the validity of this theorem has now been demonstrated for all whole numbers, it can be used to justify writing statements as $r + 6 = s + 6$, if r and s are elements of W and $r = s$.

To prove that a statement is *not* true for all cases is often a simpler task. All you need to do is to produce one example which contradicts a suggested theorem. Such an example is called a **counterexample**. For example, suppose we theorize that "$2 \cdot a = 2 + a$ for all whole number replacements for a." This statement can be shown to be false simply by finding a replacement for a that contradicts the statement. Thus, if a is replaced by 1, we have the statement $2 \cdot 1 = 2 + 1$, or $2 = 3$, which is not true.

One consequence of the axioms concerning multiplication is a theorem similar to that proved above for addition.

Theorem 2.2. If $a, b, c \in W$ and if $a = b$, then $a \cdot c = b \cdot c$.

The proof follows from the fact that if a, b, and c are whole numbers. We know by the Closure law for multiplication that $a \cdot c$ is a whole number, and by the Reflexive law of equality that

$$a \cdot c = a \cdot c.$$

The hypothesis states that $a = b$. Hence, by the Substitution law, we can replace a with b to obtain

$$a \cdot c = b \cdot c.$$

A statement-reason proof of Theorem 2.2 is left as an exercise, as are proofs of many other theorems to which you will be introduced.

Theorems are not always named. However, the two theorems above are used so frequently that it is convenient to refer to them as the **addition law of equality** and the **multiplication law of equality,** respectively. These theorems follow directly from the substitution axiom. In fact, statements such as "If $a = b$, then $a + c = b + c$ and $a \cdot c = b \cdot c$" are sometimes viewed as an application of the substitution axiom.

In Section 2.5, page 33 we referred to the product of 0 and any whole number a. A formal statement of this property of 0 is given by the following theorem.

Theorem 2.3.   If $a \in W$, then $a \cdot 0 = 0$ and $0 \cdot a = 0$.

A proof for this theorem, which we call the **zero factor law,** is presented as an exercise.

We have introduced you to proofs that require a number of steps. It is desirable that you be able to follow the logical arguments involved. As you continue your work in mathematics, you will undoubtedly develop your ability to make logical arguments. At this time, however, we simply want you to recognize that the routine manipulations that we perform in mathematics are based on *definitions we have made, axioms that we have adopted,* and *theorems proved from these definitions and axioms.*

If a theorem has been given a name, we will in general use this name. If not, the theorem number will be given. You should develop the habit of checking the original statement of the theorem each time it is used.

Exercise 2.7

A

The following important theorems about whole numbers have been discussed:

If $a, b, c \in W$, and $a = b$, then $a + c = b + c$.   Addition law of equality

If $a, b, c \in W$, and $a = b$, then $a \cdot c = b \cdot c$.   Multiplication law of equality

If $a \in W$, then $a \cdot 0 = 0$, and $0 \cdot a = 0$.   Zero factor law

State which of these theorems is demonstrated by each statement. All variables are elements of W.

Examples. a. $5 \cdot 0 = 0$ b. If $x = y$, then $3 \cdot x = 3 \cdot y$

Solutions. a. Zero factor law   b. Multiplication law of equality

1. If $a = b$, then $a + 2 = b + 2$ 2. If $x = y$, then $5 \cdot x = 5 \cdot y$

3. $0 \cdot 75 = 0$ 4. $15 \cdot 0 = 0$

5. If $x = 1$, then $4 \cdot x = 4 \cdot 1$ 6. If $y = 2$, then $y + 5 = 2 + 5$

7. $x \cdot 0 = 0$ 8. $0 \cdot y = 0$

9. If $q = 3$, then $8 + q = 8 + 3$ 10. If $p = 6$, then $p \cdot 4 = 6 \cdot 4$

11. If $x = 2$, then $x + y = 2 + y$ 12. If $y = 8$, then $x \cdot y = x \cdot 8$

The truth of each of the following statements depends on either an axiom of equality, an axiom concerning the operation of addition or multiplication, or one of the theorems listed above. Give the name of the axiom or theorem which best justifies each statement. $x, y \in W$.

Examples. a. $3 + 5 = 5 + 3$ b. If $x = y$, then $y = x$.

Solutions. a. Commutative law of addition   b. Symmetric law of equality

13. $x \cdot 1 = x$ 14. $y + 4 \in W$

15. $5 \cdot (3 + x) = 5 \cdot 3 + 5 \cdot x$ 16. $x + 7 = x + 7$

17. $4 \cdot x \in W$ 18. If $y + 5 = x$, then $x = y + 5$

19. $(1 + 3) + 5 = 1 + (3 + 5)$ 20. $9 + 0 = 9$

21. $0 \cdot x = 0$ 22. $(2 \cdot 6) \cdot y = y \cdot (2 \cdot 6)$

23. $(x + y) + 4 = (y + x) + 4$ 24. $(x \cdot y) \cdot z = x \cdot (y \cdot z)$

25. $(y + 5) \cdot 4 = (5 + y) \cdot 4$ 26. $(x + y + 3) \cdot 0 = 0$

27. $(y + 5) \cdot x = y \cdot x + 5 \cdot x$ 28. If $x = 6$, then $2 \cdot x = 2 \cdot 6$.

29. $1 \cdot (x + y) = x + y$ 30. $(x + 1) + (y + 8) \in W$

31. Supply the missing reasons for the proof of the theorem: If $a, b, c \in W$ and $a = b$, then $a \cdot c = b \cdot c$. Multiplication law of equality.

Statements	**Reasons**
a. $a, b, c \in W$	a. Hypothesis
b. $a \cdot c \in W$	b. _____

Continued on page 46.

c. $a \cdot c = a \cdot c$ c. Reflexive law of equality

d. $a = b$ d. Hypothesis

e. $a \cdot c = b \cdot c$ e. _____

32. Supply the missing reasons for the proof of the theorem: If $a \in W$, then $a \cdot 0 = 0$ and $0 \cdot a = 0$. Zero factor law.

Statements	*Reasons*
a. $a \in W$	a. Hypothesis
b. $0 = 0$	b. _____
c. $0 + 0 = 0$	c. _____
d. $a \cdot (0 + 0) = a \cdot 0$	d. Multiplication law of equality
e. $a \cdot (0 + 0) = a \cdot 0 + a \cdot 0$	e. _____
f. $a \cdot 0 + a \cdot 0 = a \cdot 0$	f. Substitution law, using Statements d and e
g. $a \cdot 0 = 0$	g. Identity element is unique

2.8 Inverse Operations

So far, our discussion of the set of whole numbers has included the operations of addition and multiplication. There are other useful operations which can be defined in terms of these basic operations. Such operations, which we now look at, are called **inverse operations.** First, we consider an operation defined in terms of addition.

In the statement $5 + 4 = 9$, 4 is the number such that the sum of the number and 5 equals 9. The number 4 is also called the **difference** of 9 and 5 and can be written as $9 - 5$. Similarly, if $a \in W$, then $a + 0 = a$ and 0 is the difference of a and a and can be written $a - a$. These examples illustrate the following definition.

> **Definition 2.3.** For all $a, b \in W$, $b - a$ is the unique whole number d, if such a whole number exists, such that $a + d = b$. The number d, or $b - a$, is called the **difference** of b and a.

Examples.   a. $15 - 7 = 8$ because $7 + 8 = 15$.

b. $12 - 2 = 10$ because $2 + 10 = 12$.

The operation designated by the symbol $-$, read "minus," is called **subtraction.** It is the inverse operation of addition.

Definition 2.3 implies that the difference of two whole numbers does not always exist. For example, consider the difference $8 - 12$. From our definition, this difference must be a whole number (if it exists) such that the sum of the number and 12 equals 8. However, such a number does not exist in the set of whole numbers. Thus the set of whole numbers is *not closed* with respect to the operation of subtraction.

Another useful operation can be defined in terms of multiplication. Consider the statement $2 \cdot 3 = 6$. The numbers 2 and 3 are factors of the product $2 \cdot 3$ or 6. The number 3 is called the **quotient** of 6 divided by 2. This quotient can be written as $\dfrac{6}{2}$, 6/2, $6 \div 2$, or as the basic numeral 3.

This example illustrates the following definition.

Definition 2.4.   For all a, $b \in W$, a/b is the unique whole number q, if such a whole number exists, such that $b \cdot q = a$. The number q, or a/b, is called the **quotient** of a divided by b.

In the quotient a/b, a is called the **dividend** and b is called the **divisor.** The operation is called **division** and is designated by the fraction bar, the solidus or slant bar, or the symbol $\div$. Division is the inverse operation of multiplication.

Examples.   a. $\dfrac{15}{3} = 5$ because $3 \cdot 5 = 15$.

b. $\dfrac{18}{6} = 3$ because $6 \cdot 3 = 18$.

Definition 2.4 suggests that the quotient of each pair of whole numbers is not always a whole number. For example, consider the quotients 20/3 and 5/0. Is there a whole number q such that $3 \cdot q = 20$? Is there a whole number q such that $0 \cdot q = 5$? The answer to both questions is "No."

The symbols 20/3 and 5/0 do not represent numbers in the set of whole numbers. Therefore, the set of whole numbers is *not closed* with respect to the operation of division. Now consider the quotient 0/0. By Definition 2.4, q is a unique whole number such that $0 \cdot q = 0$. However, by the zero factor law (page 44), the product of 0 and any whole number is 0. Thus, q is not unique. Therefore, since $a/0$ is not a unique whole number for any $a \in W$, we say that **division by 0 is undefined.**

The symbol for a quotient written in the form $\frac{a}{b}$ or a/b is called a **fraction.**

The dividend, a, is called the **numerator** and the divisor, b, is called the **denominator.**

Exercise 2.8

A

Write each difference as a basic numeral. If the difference does not exist in the set of whole numbers, indicate this as your answer.

1. $2 - 1$ 2. $6 - 3$ 3. $4 - 2$ 4. $7 - 5$

5. $4 - 8$ 6. $5 - 9$ 7. $1 - 0$ 8. $0 - 6$

Express each statement as a difference, using numerals, variables, and the minus sign. State the restrictions on the variables for the collection of symbols to represent a whole number. $x, y, z \in W$.

Examples.   a. x minus 3   b. y minus z

Solutions.   a. $x - 3$, $x \geq 3$ b. $y - z$, $y \geq z$

9. x minus 2   10. y minus 5   11. x minus y

12. y minus x 13. x minus 0   14. 0 minus y

15. $(5 + x)$ minus x 16. $(5 + x)$ minus 5

Write each quotient as a basic numeral. If the quotient does not exist in the set of whole numbers, so state.

17. $\dfrac{8}{4}$ 18. $\dfrac{15}{3}$ 19. $\dfrac{10}{2}$ 20. $\dfrac{12}{6}$ 21. $\dfrac{5}{2}$

22. $\dfrac{7}{4}$ 23. $\dfrac{0}{3}$ 24. $\dfrac{0}{10}$ 25. $\dfrac{9}{0}$ 26. $\dfrac{10}{0}$

In Problems 27 to 34, express each statement as a quotient using the fraction symbol, numerals, and variables. State the values of the variables for which the symbol represents a whole number. $x, y \in W$.

Examples.   a. 8 divided by y b. $(2 - y)$ divided by 2

Solutions.   a. $\dfrac{8}{y}$.   By inspection we observe that only if $y \in \{1, 2, 4, 8\}$ will $\dfrac{8}{y}$ represent a whole number.

b. $\dfrac{2 - y}{2}$.   By inspection we observe that only if $y \in \{0, 2\}$ will $\dfrac{2 - y}{2}$ represent a whole number. Note that the expression $(2 - y)$ is written without parentheses. In this case the fraction bar serves the same purpose and implies that if the variable is replaced with a whole number, the difference $2 - y$ should be evaluated first.

27. 4 divided by x

28. 3 divided by y

29. 12 divided by y

30. 30 divided by y

31. 2 divided by $(y - 1)$

32. 6 divided by $(y + 1)$

33. $(8 - y)$ divided by 2

34. $(6 - y)$ divided by 3

35. For what whole number replacement of x is the expression $\dfrac{x - 3}{x + 2}$ equal to 0?

36. For what whole number replacement of x is the expression $\dfrac{7 - x}{x + 1}$ equal to 0?

2.9 Factoring

In Section 2.5 we wrote a product such as $2 \cdot 3$ as a basic numeral where 2 and 3 are factors of the product 6. It is often quite useful to write basic numerals in a form in which the whole number factors of a product are distinctly exhibited, as in the products $2 \cdot 3$, $5 \cdot 7$, $2 \cdot 3 \cdot 5$, etc. First, we shall define several terms.

If one whole number divides a second whole number, yielding a whole number as a quotient, the first number is said to be an **exact divisor** of the second number.

Examples.   a. The exact divisors of 15 are 1, 3, 5, and 15.

b. The exact divisors of 8 are 1, 2, 4, and 8.

A whole number greater than 1 which is exactly divisible only by itself and 1 is called a **prime number.**

Example.   2, 3, 5, 7, 11, and 13 are prime numbers.

All whole numbers greater than 1 that are not prime numbers are called **composite numbers.**   Thus, 6 is a composite number since it is exactly divisible by 2 and 3, as well as by itself and 1.

Example.   4, 6, 8, 9, 10, and 12 are composite numbers.

Each composite number can be expressed as a product where each factor is a prime number.   This form of a product is called the **prime factor form.**

Examples.   Write each number in prime factor form.

a. 35   b. 20   c. 32

Solutions.   a. $5 \cdot 7$ b. $2 \cdot 2 \cdot 5$ c. $2 \cdot 2 \cdot 2 \cdot 2 \cdot 2$

The number 1 is not considered a prime number, because if it were, it could appear in the factored form of a product any number of times.   For example, we could factor

$$6 = 2 \cdot 3 = 1 \cdot 2 \cdot 3 = 1 \cdot 1 \cdot 2 \cdot 3 = 1 \cdot 1 \cdot 1 \cdot 2 \cdot 3, \text{ etc.,}$$

and nothing would be gained by exhibiting the factor 1 so many times.

Your ability to factor a product depends on your knowledge of multiplication facts and your being able to determine whether one whole number is or is not exactly divisible by another.

Exercise 2.9

A

Write each set by listing the members.

Example.   {Prime numbers less than 10}

Solution.   {2, 3, 5, 7}

1. {Prime numbers between 5 and 15}
2. {Composite numbers between 1 and 10}
3. {Composite numbers between 6 and 14}
4. {Prime numbers between 15 and 35}
5. {Prime numbers less than 25}
6. {Composite numbers less than 20}
7. {Composite numbers greater than 13 and less than 23}
8. {Prime numbers greater than 5 and less than 32}
9. {Prime numbers greater than 31 and less than 50}
10. {Composite numbers greater than 50 and less than 65}
11. {Prime numbers between 100 and 125}
12. {Composite numbers between 125 and 135}

Write each of the following in prime factor form.

Examples.   a. 12   b. 7   c. 16

Solutions.   a. $2 \cdot 2 \cdot 3$ b. 7 (prime)   c. $2 \cdot 2 \cdot 2 \cdot 2$

13. 21	14. 63	15. 42	16. 96
17. 128	18. 186	19. 225	20. 384
21. 496	22. 728	23. 1236	24. 1456

In each of the following pairs, find two whole number factors of the first number whose sum is the second number.

Examples.   a. 15, 8   b. 125, 30

Solutions.   a. $3 \cdot 5$; $(3 + 5 = 8)$ b. $5 \cdot 25$; $(5 + 25 = 30)$

25. 25, 10	26. 35, 12	27. 48, 26	28. 48, 19
29. 48, 16	30. 12, 13	31. 4, 5	32. 34, 19
33. 144, 24	34. 144, 25	35. 144, 26	36. 144, 30

B

Write each expression in factored form

Example. $2x + 6 = 2 \cdot (x + 3)$

37. $3x + 6 = 3 \cdot (?)$ 38. $4x + 12 = 4 \cdot (?)$

39. $2y + 10 = 2 \cdot (?)$ 40. $5y + 15 = 5 \cdot (?)$

41. $6x - 12 = 6 \cdot (?)$ 42. $3y - 9 = 3 \cdot (?)$

43. One of the following numbers is prime. Which one?

$$63, 87, 89, 91, 93$$

44. One of the following numbers is prime? Which one?

$$6410, 7114, 9718, 4115, 1997$$

45. Explain why 14 is a factor of $6 \cdot 21$.

46. Explain why 55 is a factor of $44 \cdot 15$.

2.10   Order of Operations; Numerical Evaluation

In Section 2.6 we made an agreement about the order of operations in a collection of symbols that includes both the addition sign and the multiplication sign. Now that we have defined subtraction in terms of addition (Definition 2.3), and division in terms of multiplication (Definition 2.4), we extend our agreement to include these operations. Let us agree that operations are to be performed in the following order:

1. Operations inside parentheses, or above or below a fraction bar.

2. Other multiplication and division operations in the order in which they occur from left to right.

3. Addition and subtraction operations in the order in which they occur from left to right.

For example,

$$2 \cdot (3 + 1) + 2 \cdot (11 - 2) - \frac{7 + 5}{3} = 2 \cdot 4 + 2 \cdot 9 - \frac{12}{3}$$

(Operations inside parentheses, etc.),

$$2 \cdot 4 + 2 \cdot 9 - \frac{12}{3} = 8 + 18 - 4$$

(Multiplications and divisions in order),

$$8 + 18 - 4 = 22$$

(Additions and subtractions in order).

By using the substitution law of equality and the order of operations that we have adopted, we can now find the numerical value of mathematical expressions that contain variables. For example, if we know that $x = 4$, then

$$3 \cdot (x + 2) - 2 \cdot x = 3 \cdot [(4) + 2] - 2 \cdot (4)$$
$$= 3 \cdot (6) - 2 \cdot (4)$$
$$= 18 - 8$$
$$= 10.$$

When evaluating mathematical expressions, it is helpful to use parentheses (as we did above) each time a substitution is made.

Exercise 2.10

A

Simplify each collection of symbols by performing the indicated operations in the proper order.

Example. $2 + 2 \cdot (3 + 1) + 3 \cdot 3$

Solution. $2 + 2 \cdot (3 + 1) + 3 \cdot 3 = 2 + 2 \cdot 4 + 3 \cdot 3$
$$= 2 + 8 + 9$$
$$= 19.$$

1. $4 \cdot 3 - 3$ 2. $2 \cdot 5 + 3$ 3. $4 + 3 \cdot 2$

4. $9 - 3 \cdot 2$ 5. $2 \cdot 0 + 6$ 6. $8 - 3 \cdot 0$

7. $6 + 3 \cdot 3$ 8. $4 \cdot 4 - 5$ 9. $2 \cdot (3 + 4)$

10. $7 \cdot (4 - 2)$ 11. $5 \cdot (5 - 2) + 3$ 12. $3 \cdot (4 + 1) - 5$

Example. $\dfrac{11 + 7}{5 - 2} - \dfrac{7 + 5}{6}$

Solution. $\dfrac{11 + 7}{5 - 2} - \dfrac{7 + 5}{6} = \dfrac{18}{3} - \dfrac{12}{6}$
$$= 6 - 2$$
$$= 4.$$

13. $\dfrac{6 \cdot 4}{8} - 1$

14. $8 - \dfrac{4 \cdot 3}{6}$

15. $\dfrac{2 \cdot (7 + 3)}{4} + 3$

16. $5 + \dfrac{3 \cdot (8 - 2)}{9}$

17. $\dfrac{6 \cdot 4}{8} + \dfrac{3 + 5}{2}$

18. $\dfrac{8 \cdot 4}{2} - \dfrac{8 - 6}{2}$

19. $\dfrac{3 \cdot (7 - 3)}{2} - \dfrac{8 + 4}{4}$

20. $\dfrac{3 \cdot (5 + 3)}{4} + \dfrac{5 + 4}{3}$

21. $\dfrac{3 \cdot (14 - 6)}{6} - \dfrac{5 \cdot 0}{2}$

22. $\dfrac{0 \cdot 5}{3} + \dfrac{5 + 3 \cdot 3}{7}$

23. $\dfrac{(2 \cdot 3) \cdot (2 \cdot 3)}{6} - \dfrac{2 \cdot 9}{6}$

24. $\dfrac{(4 \cdot 2) \cdot (4 \cdot 2)}{16} + \dfrac{4 \cdot 8}{16}$

25. $\dfrac{3 \cdot 3 + 4 \cdot 4}{5} - 4$

26. $\dfrac{3 \cdot 3 + 5 \cdot 5}{17} + 6$

27. $\dfrac{7 \cdot 7 - 5 \cdot 5}{6} + 2$

28. $\dfrac{6 \cdot 6 - 4 \cdot 4}{5} - 2$

29. $\dfrac{6 \cdot (3 + 4)}{14} + \dfrac{5 \cdot (5 + 3)}{10}$

30. $\dfrac{4 \cdot (3 + 7)}{5} - \dfrac{3 \cdot (9 + 1)}{6}$

Evaluate each mathematical expression for the given value of the variable.

Example. $3 \cdot x + 5 \cdot (x + 4); \ x = 2$

Solution. $3 \cdot x + 5 \cdot (x + 4) = 3 \cdot (2) + 5 \cdot [(2) + 4]$
$$= 3 \cdot (2) + 5 \cdot (6)$$
$$= 6 + 30$$
$$= 36.$$

31. $(2 + x) + 6 - 2 \cdot x; \ x = 3$

32. $(2 \cdot x + 1) - 3 \cdot (x - 2); \ x = 4$

33. $\dfrac{4 \cdot x + 1}{3} - \dfrac{3 \cdot x + 2}{4}; \ x = 2$

34. $\dfrac{4 \cdot x + 3}{5} - \dfrac{7 \cdot x - 1}{10}; \ x = 3$

35. $5 \cdot (x + 1) + 2 \cdot (x - 1); \ x = 2$

36. $3 \cdot (x + 2) - 2 \cdot (x - 1); \ x = 4$

37. $x \cdot (x + 1) - x \cdot (x - 2); \ x = 3$

38. $x \cdot (x + 3) - x \cdot (x - 2); \ x = 2$

39. $\dfrac{2 \cdot (3 + 2)}{x} + \dfrac{6 \cdot (4 + 1)}{2 \cdot x}$; $x = 5$ 40. $\dfrac{5 \cdot (10 - 4)}{x} - \dfrac{2 \cdot (7 + 2)}{2 \cdot x}$; $x = 3$

41. $\dfrac{x \cdot (x + 3)}{5} + 2$; $x = 5$ 42. $\dfrac{2 \cdot x \cdot (x + 2)}{5} - 4$; $x = 3$

2.11   Equations and Inequalities

A mathematical sentence that involves the equals relation is called an **equation;** a sentence involving an order relation, such as "less than" or "greater than," is called an **inequality.** In either an equation or inequality, the collection of symbols on the right is called the **right-hand member** and the collection of symbols on the left is called the **left-hand member.**

Many mathematical sentences contain variables. The truth or falsity of such a sentence depends on the number used as a replacement for the variable.

Examples.   a. $x + 3 = 5$ and $2 \cdot x = 6$ are equations.

b. $x < 5$ and $x + 2 > 7$ are inequalities.

In the above examples, observe that $x + 3 = 5$ is a true statement if x is replaced with 2 and is false if x is replaced with any other whole number. The number 2 is called a *solution* of the equation.

Definition 2.5.   A **solution** of an equation or inequality is a replacement for the variable that forms a true statement. The set of all solutions is called the **solution set.**

In this section we limit the replacement set of variables in open sentences to the set of whole numbers. Hence, the solution set of $x + 3 = 5$ is {2}, the solution set of $2 \cdot x = 6$ is {3}, and the solution set of $2 \cdot x = 1$ is ∅ because there is no *whole number* that would make the equation true. In these cases there is at most only one solution. On the other hand, the solution set of $x < 8$ is {0, 1, 2, 3, 4, 5, 6, 7}, because the statement is true if x is replaced by any of the members in the set.

Equations and inequalities that are true for some members of the replacement set for the variable but not for all members are called **conditional equations** or **conditional inequalities**. An equation that is true for all members of the replacement set for which both members are defined is called an **identity**.

Examples.   a. For $x \in W$, $x + 3 = 5$ is a conditional equation.

b. For $x \in W$, $x + 1 > 3$ is a conditional inequality.

c. For $x \in W$, $x + 1 = x + 1$ is an identity.

Note that all the axioms of equality and the axioms concerning sums and products of whole numbers except the laws of closure are identities.

Examples.   a. For all $a \in W$, $a + 0 = a$.

b. For all $a, b \in W$, $a + b = b + a$.

We will treat identities in more detail in Chapter 8.

Exercise 2.11

A

Find the solution set of each equation or inequality by inspection. $x \in W$.

Examples.   a. $x + 4 = 7$ b. $3 \cdot x = 4$

Solutions.   a. $\{3\}$, because $3 + 4 = 7$.

b. $\varnothing$, because there is no whole number such that 3 times that number equals 4.

1. $x + 1 = 6$ 2. $x + 4 = 7$ 3. $x + 8 = 9$ 4. $x + 12 = 15$

5. $3 + x = 7$ 6. $5 + x = 8$ 7. $6 + x = 1$ 8. $9 + x = 2$

9. $4 - x = 3$ 10. $7 - x = 5$ 11. $x - 5 = 4$ 12. $x - 3 = 7$

13. $x = 6 - 3$ 14. $x = 9 - 5$ 15. $7 - 4 = x$ 16. $8 - 3 = x$

17. $3 \cdot x = 9$ 18. $4 \cdot x = 20$ 19. $12 = 2 \cdot x$ 20. $28 = 7 \cdot x$

21. $4 \cdot x = 11$ 22. $5 \cdot x = 17$ 23. $13 = 6 \cdot x$ 24. $15 = 8 \cdot x$

25. $\dfrac{x}{4} = 1$ 26. $\dfrac{x}{3} = 4$ 27. $\dfrac{x}{9} = 0$ 28. $\dfrac{12}{x} = 4$

29. $\dfrac{8}{x} = 2$ 30. $\dfrac{7}{x} = 1$ 31. $\dfrac{5}{x} = 2$ 32. $\dfrac{4}{x} = 0$

Example. $x < 4$

Solution. $\{0, 1, 2, 3\}$, because each of these numbers is less than 4.

33. $x < 5$ 34. $x < 7$ 35. $x \le 4$

36. $x \le 6$ 37. $2 > x$ 38. $5 \ge x$

B

39. $2 \cdot x + 3 = 5$ 40. $3 \cdot x + 4 = 10$ 41. $4 \cdot x - 2 = 6$

42. $6 \cdot x - 1 = 11$ 43. $5 < x \le 8$ 44. $10 > x \ge 7$

CHAPTER SUMMARY

2.1 The elements of the set of **whole numbers** $W = \{0, 1, 2, 3, \cdots\}$, when associated with "how many" elements there are in a set, are said to be used in a **cardinal** sense; when they are used to indicate the "position" of an element in an **ordered set** they are used in an **ordinal** sense. The set of **natural numbers,** whose members are associated with the cardinalities of non-empty sets, is represented symbolically as $N = \{1, 2, 3, \cdots\}$. The elements in the set are used for counting and often are referred to as **counting numbers.** Symbols such as a, b, c, x, y, and z, which refer to an unspecified member of a specified set, are called **variables,** and the set is called the **replacement set** of the variable.

2.2 When the replacement set of a variable contains a large number of elements, the set can be represented by using **set-builder notation.**

Four axioms concerning symbolism pertaining to equality are:

For all a, b, c ∈ W,

$a = a.$	Reflexive law
If $a = b$, then $b = a$.	Symmetric law
If $a = b$ and $b = c$, then $a = c$.	Transitive law
If $a = b$, then a may be replaced by b or b may be replaced by a in any collection of symbols without affecting the truth or falsity of the statement.	Substitution law

Two assumptions concerning the order of whole numbers are:

For all a, b ∈ W, exactly one of the following is true,

$a < b, \quad a = b \quad or \quad a > b.$	Trichotomy law

For all a, b, c ∈ W,

if $a < b$ and $b < c$, then $a < c$.	Transitive law

2.3 The set of whole numbers can be associated with selected points on a **number line.**

2.4 Because the operation of addition pairs two numbers, it is called a **binary operation.** The **sum** of two whole numbers, $a + b$, is defined as the cardinality of the set formed by the *union* of two *disjoint* sets whose cardinalities are a and b. A **basic numeral,** one of the symbols in $\{0, 1, 2, 3, \cdots\}$, is one way to name a whole number.

Four axioms governing the operation of addition are:

For all a, b, c ∈ W,

$a + b \in W$	Closure law
$a + b = b + a$	Commutative law
$(a + b) + c = a + (b + c)$	Associative law
$a + 0 = a \quad and \quad 0 + a = a$	Identity law

2.5 **Multiplication** is a binary operation. The product of two numbers, $a \cdot b$, is defined as the cardinality of the **Cartesian product** of two sets whose cardinalities are a and b. Cartesian products can be illustrated graphically by means of **lattices** or **arrays.**

Four axioms governing the operation of multiplication are:

For all a, b, c ∈ W,

$$a \cdot b \in W \qquad \qquad \text{Closure law}$$
$$a \cdot b = b \cdot a \qquad \qquad \text{Commutative law}$$
$$(a \cdot b) \cdot c = a \cdot (b \cdot c) \qquad \qquad \text{Associative law}$$
$$1 \cdot a = a \text{ and } a \cdot 1 = a \qquad \qquad \text{Identity law}$$

2.6 The **distributive law** is an axiom which relates the operations of addition and multiplication.

For all a, b, c ∈ W, $a \cdot (b + c) = a \cdot b + a \cdot c.$

2.7 Statements that follow logically from axioms and definitions are called **theorems.** The part of the theorem that gives information assumed to be true is called the **hypothesis;** the part that follows logically from the hypothesis is called the **conclusion.**

2.8 **Subtraction** and **division** are the *inverse* operations of addition and multiplication, respectively. The set of whole numbers is *not closed* with respect to the operations of subtraction and division.

2.9 A **prime number** is any whole number greater than 1 which is exactly divisible only by itself or 1. A **composite number** is any whole number greater than 1 which is not a prime number.

2.10 An agreement is made about the order in which the addition, subtraction, multiplication, and division operations shall be performed. The **evaluation** of a mathematical expression consists of replacing the variable(s) with the given value(s) and then performing the indicated operations in the proper order.

2.11 A **solution** of an equation or an inequality is any replacement for the variable that makes the equation or inequality a true statement. The set of all solutions is called its **solution set.**

CHAPTER REVIEW

A

2.1 *State the cardinality of each set.*

1. {2, 3}

2. {2, 3} ∩ {4, 5}

3. {2, 3} × {4, 5}

4. {2, 3} ∪ {4, 5}

Replace the comma in each pair with correct order symbol, $<$ or $>$.

5. 6, 3

6. 2, 9

7. A number used in a _____ sense is associated with the concept of "how many" elements in a set.

8. A number used in an _____ sense is associated with the concept of the position of an element in an ordered set.

2.2   *List the numbers in each set.*

9. $\{x \mid x < 5, x \in N\}$

10. $\{x \mid 1 < x \leq 7, x \in W\}$

Write each set in set-builder notation, using x as the variable.

11. $\{0, 1, 2\}$

12. $\{1, 2, 3, 4, \cdots\}$

Name the axiom of equality or order that best justifies each statement. $x, y, z \in W$.

13. If $x < y$ and $y < 3$, then $x < 3$.

14. If $x + y = z$, then $z = x + y$.

2.3   *Graph each set of numbers.*

15. $\{x \mid x < 5, x \in W\}$

16. $\{x \mid x \geq 7, x \in N\}$

2.4, 2.5   *Name the axiom that best justifies each statement. $x, y, z \in W$.*

17. $3 + (x + 4) = 3 + (4 + x)$

18. $5 + (2 + y) = (5 + 2) + y$

19. If $(2 + x) \cdot y = 6$ and $y = 3$, then $(2 + x) \cdot 3 = 6$.

20. If $(y + 0) + x = 4$, then $y + x = 4$.

21. $5 \cdot y \in W$

22. $[(x + 2) \cdot z] + y = [z \cdot (x + 2)] + y$

2.6   *Apply the distributive law and show each product as an equal expression without parentheses. All variable are elements of W.*

23. $x \cdot (3 + y)$

24. $(c + d) \cdot 4$

25. $(x + a) \cdot (y + b)$

26. $(y + 2) \cdot (x + 3)$

2.7 *Name a theorem that best justifies each statement. All variables are elements of W.*

27. If $x + y = z$, then $(x + y) + 2 = z + 2$.

28. If $x + y = z$, then $2 \cdot (x + y) = 2 \cdot z$.

29. $0 \cdot (x + y) = 0$

30. If $x = 3$, then $x + 5 = 3 + 5$.

2.8 *Write each indicated difference or quotient as a basic numeral in the set of whole numbers. If difference or quotient does not exist, so state.*

31. $16 - 12$ 32. $4 - 9$ 33. $\dfrac{8}{5}$

34. $\dfrac{0}{2}$ 35. $\dfrac{7}{0}$ 36. $\dfrac{12}{4}$

2.9 *In Problems 37 to 40, write each whole number in prime factor form.*

37. 30   38. 54   39. 126   40. 288

2.10 *Simplify each collection of symbols.*

41. $2 + 3 + \dfrac{9 + 5}{7} - 3(2)$ 42. $\dfrac{6 \cdot (5 - 2)}{9} + \dfrac{3 \cdot (6 + 2)}{6}$

Evaluate each expression.

43. $3 \cdot (x - 1) + \dfrac{3 \cdot x - 5}{2}$; $x = 3$

44. $x \cdot (2 \cdot x - 1) + \dfrac{5 \cdot x + 4}{x}$; $x = 2$

2.11 *Find the solution set of each equation or inequality by inspection. Assume that $x \in W$.*

45. $x + 6 = 9$ 46. $x - 1 = 9$ 47. $5 \cdot x = 23$

48. $\dfrac{x}{4} = 2$ 49. $x \leq 3$ 50. $x < 4$

B

51. Is $\{x \mid 1 \leq x < 5, x \in W\} \cap \{x \mid x > 9, x \in W\}$ equal to $\{0\}$? Why or why not?

52. Let $R = \{2, 4, 6\}$, $S = \{6, 7, 8, 9\}$, and $T = \{1, 3, 5, 7, 9\}$. State the cardinality of:

    a. $(R \cap S) \cup T$ b. $S \cup (R \cap T)$ c. $(T \cap \varnothing) \cup S$

53. List the elements in each set:

    a. $\{x \mid x < 3, x \in W\} \cap \{2, 4, 6, 8\}$.

    b. $\{y \mid 5 \leq y < 10, y \in W\} \cup \{y \mid y < 2, y \in W\}$.

54. Graph $\{0, 1, 2, 3\} \cup \{x \mid 1 < x \leq 7, x \in W\}$.

55. Graph the array of the Cartesian product $\{2, 4, 6, 8\} \times \{d, e, f\}$.

56. If $x \cdot y = 0$, and $x \neq 0$, what can you say about y?

57. If $x, y \in W$, what are the conditions on x and y for $y - x$ to represent a whole number?

58. Explain why 15 is a factor of $35 \cdot 36$.

Find the solution set of each equation or inequality by inspection. Assume that $x \in W$.

59. $4 \cdot x + 3 = 15$ 60. $6 \leq x < 11$

3
chapter

The Set of Integers

To this point we have been considering the set of whole numbers W. Elements of the set of natural numbers $N = \{1, 2, 3, \cdots\}$, a subset of the set of whole numbers, are used to represent physical quantities such as distance (2 miles), money (10 dollars), and temperature (50 degrees). In this way, they are used in a cardinal sense (how many?). Since the natural numbers do not always enable us to differentiate between distances in opposite directions from a starting point, gains or losses of money, or degrees above or below zero, the set of whole numbers is enlarged to allow us to make such differentiations. Thus, if the temperature at a certain time was $4°$ and it dropped $7°$, the new temperature can be represented by an element in the new set.

This new enlarged set of numbers, called the set of **integers,** enables us to give meaning to a difference such as $4 - 7$, or, in general, to $a - b$, for *every* pair of whole numbers, a, b, and, in fact, for *every* pair of elements a, b in the new set.

3.1 Some Properties of Integers

To consider integers, we return briefly to a number line which has been extended to the *left* of the origin.

For each *natural number* corresponding to a point to the *right* of the origin, we *assume* that there exists a number which can be assigned to a point located in the same relative position to the *left* of the origin. Each of these two numbers is the coordinate of a point located the same distance from but on opposite sides of the origin; each is the **negative** of the other. For example, if the points C and D in Figure 3.1 are located on a number

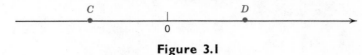

Figure 3.1

line the same distance from but on opposite sides of the origin, then the number associated with the point C is the negative of the number associated with the point D. Similarly, the number associated with the point D is the negative of the number associated with the point C.

We shall first represent the negative of a natural number by the same numeral but with a small raised dash or minus sign preceding it. Thus, ⁻2 denotes the negative of 2 and is read "negative two." ⁻3 denotes the negative of 3 and ⁻107 denotes the negative of 107. Numbers such as ⁻2, ⁻3, and ⁻107 are referred to as **negative numbers,** or, in particular, as **negative integers.** The association of these numbers with points on the number line is shown in Figure 3.2.

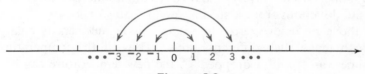

Figure 3.2

The term "positive number" is used to indicate any number other than 0 that is not a negative number. Any natural number is also referred to as a **positive integer.**

Sometimes the small symbol ⁺ is prefixed to a numeral representing a positive number. For example, ⁺3 and 3 both represent positive numbers and there should be no confusion in using either form.

The set that includes all the natural numbers and their negatives together with the number 0 is called the **set of integers.** This set is represented by the letter J. Thus,

$$J = \{\cdots \, ^-3, \, ^-2, \, ^-1, \, 0, \, 1, \, 2, \, 3, \, \cdots\}.$$

Using the number line, we observe that the negative of any number a is unique, and that, if a is a positive integer, then ^-a is a negative integer. On the other hand, if a is a negative integer, then ^-a is a positive integer. The negative of an integer is also called the **additive inverse** of the integer.

We now make the following assumption.

For every $a \in J$, there exists a unique integer ^-a, such that

$$a + {}^-a = 0 \quad and \quad {}^-a + a = 0.$$

This property is called the **additive inverse law.** The set J is the smallest set of numbers which contains the whole numbers and satisfies this property.

Examples. a. $7 + {}^-7 = 0$ b. $20 + {}^-20 = 0$ c. $^-32 + 32 = 0$

The fact that we are assuming that the additive inverse of any integer is unique (there is only one such element) implies that if

$$a + b = 0, \quad then \quad b = {}^-a \quad and \quad a = {}^-b,$$

and if

$$a = b, \quad then \quad {}^-a = {}^-b.$$

Examples. a. If $a + 3 = 0$, then $a = {}^-3$.   b. If $2 + x = 0$, then $x = {}^-2$.

When a set of numbers is extended to form an enlarged set, it is desirable that properties possessed by the original numbers carry over to the new numbers. Thus *we shall assume that the integers satisfy the laws of equality and order which were assumed to be satisfied by the whole numbers* (see pages 22 and 23).

The additive inverse law states that every integer has a negative. The following theorem, called the **double negative law,** is a consequence of the additive inverse law.

Theorem 3.1.   If $a \in J$, then $^-(^-a) = a$.

Examples. a. $^-(^-3) = 3$ b. $^-[^-(^-3)] = {}^-3$ c. $^-(^-x) = x$

We now refer to a symbol in the set $\{\cdots {}^-3, {}^-2, {}^-1, 0, 1, 2, 3, \cdots\}$ as a basic numeral for an integer. Thus, in the examples above, 3 is the basic numeral for $^-({}^-3)$, and $^-3$ is the basic numeral for $^-[{}^-({}^-3)]$.

Exercise 3.1

A

Locate the elements of each set on a number line.

Examples.   a. $\{4, {}^-2, 0, {}^-7\}$ b. $\{{}^-4, {}^-3, {}^-2, {}^-1, 0, 1, 2, \cdots\}$

Solutions

a.

b.

1. $\{2, {}^-5, 6, {}^-4\}$ 2. $\{3, {}^-9, {}^-7, 1\}$

3. $\{3, {}^-5, {}^-7, 6, {}^-1\}$ 4. $\{{}^-7, 0, 3, {}^-5, 1\}$

5. $\{{}^-25, 30, {}^-15, {}^-5, 20\}$ 6. $\{{}^-70, {}^-55, 50, 60, 75\}$

7. $\{{}^-2, {}^-1, 0, 1, 2\}$ 8. $\{{}^-13, {}^-12, {}^-11, {}^-10, {}^-9\}$

9. $\{\cdots {}^-3, {}^-2, {}^-1, 0, 1, 2\}$ 10. $\{{}^-2, {}^-1, 0, 1, 2, 3, \cdots\}$

Write the additive inverse (negative) of each integer. $x, y \in J$.

Examples.   a. 7   b. $^-3$

Solutions.   a. $^-7$ b. $^-({}^-3)$ or 3

11. 9   12. 14   13. $^-8$ 14. $^-64$

15. $^-33$ 16. 52   17. x 18. ^-y

19. If x is a positive integer, then ^-x is a_____integer.

20. If x is a negative integer, then ^-x is a_____integer.

State the axiom(s) or theorem(s) that justify each statement. $x, y \in J$.

21. $4 + ({}^-4) = 0$ 22. $^-5 + 5 = 0$

23. $^-(^-48) = 48$

24. $^-[^-(^-8)] = ^-8$

25. $x + (^-x) = 0$

26. $^-(^-y) = y$

B

27. Write each statement in symbolic form. Use a as the variable.
   a. Additive inverse law   b. Double negative law

28. Complete the statements in the proof of the theorem:
   If $a \in J$, then $^-(^-a) = a$.

Statements	**Reasons**
a. $a, ^-a, ^-(^-a) \in J$	a. Hypothesis; additive inverse law
b. $^-a + [^-(^-a)] = 0;\ ^-a + a = 0$	b. _____
c. $^-(^-a) = a$	c. _____

3.2 Equality and Order; Absolute Value

We observed in Section 2.3 that if $a < b$, the point corresponding to a on the number line will be to the left of the point corresponding to b. At this time, if we use the same criteria for this relation between two integers, we have that for two points on a number line the integer corresponding to the point on the left is less than the one corresponding to the point on the right.

Example. Locate the integers $^-5$ and $^-3$ on a number line and compare the integers using a symbol of order.

Solution

The point corresponding to $^-5$ is located to the left of the point corresponding to $^-3$; thus $^-5 < ^-3$.

Sometimes we are concerned with the distance between two points on a number line *and* the direction of one point in relation to the second point on the line. For example, the number 4 corresponds to the point 4 units to the right of the point corresponding to 0, and the number ⁻4 corresponds to the point 4 units to the left of the point corresponding to 0, as in Figure 3.3. Sometimes we are concerned *only* with the length of a line segment

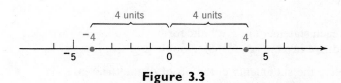

Figure 3.3

and we give a name to the positive number (or zero) associated with such a length. The positive number 4 associated with the *length* of the line segment joining the point at the origin to the point at 4 or the point at ⁻4 is called the **absolute value** of 4 or ⁻4 and is represented symbolically by placing the numeral between a pair of vertical bars. For example, $|4| = 4$ and $|{-4}| = 4$. In general, the absolute value of an integer, a, is defined as follows.

Definition 3.1.   For all $a \in J$, if a is positive or zero, then $|a| = a$;   if a is negative, then $|a| = {}^{-}a$.

Note that if a is negative, ⁻a is a positive integer. Thus *the absolute value of any integer is non-negative.*

Examples.

a. $|3| = 3$ b. $|{-3}| = {}^-(-3)$ c. $|0| = 0$
$\qquad\qquad\qquad\qquad\quad = 3$

Exercise 3.2

A

Consider the number line

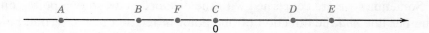

*where the coordinates of A, B, C, D, E, and F are the integers a, b, c, d, e, and f,
respectively. Replace the comma in each pair of letters with the proper symbol,
<, =, or >.*

1. *e, a*   2. *c, e*   3. *d, c*   4. *b, d*   5. *f, b*

6. *e, d*   7. *e, f*   8. *c, a*   9. *a, d*   10. *f, c*

*Replace the comma in each number pair with the proper symbol, <, =, >. (Hint:
Think of the number line.)*

11. 1, 7   12. 8, 2   13. ⁻5, ⁻9   14. ⁻6, ⁻4

15. ⁻4, 4   16. 15, ⁻15   17. ⁻3, 7   18. ⁻8, 10

19. ⁻21, 0   20. ⁻9, ⁻9   21. 12, ⁻(⁻12)   22. ⁻(⁻5), ⁻5

23. ⁻3 + 3, 0   24. 4 + (⁻4), 0   25. 6 + (⁻6), 3   26. ⁻2 + 2, 2

Write the absolute value of each integer as a basic numeral. $x, y, z \in J$.

Examples.   a. 3   b. ⁻3

Solutions.   a. $|3| = 3$ b. $|^-3| = 3$

27. 25   28. 16   29. ⁻7   30. ⁻42

31. ⁻(⁻5)   32. ⁻(⁻15)   33. ⁻[⁻(⁻6)]   34. ⁻[⁻(⁻58)]

*Replace the comma in each number pair with the proper symbol, <, =, >.
$x, y \in J$.*

35. $|^-3|, |3|$ 36. $|4|, |^-4|$ 37. $|^-5|, |2|$

38. $|2|, |^-3|$ 39. $|^-10|, |^-8|$ 40. $|16|, |18|$

41. $|^-1|, |0|$ 42. $|3|, |^-2|$ 43. $|^-2 + 2|, |0|$

44. $|^-5 + 5|, |^-1|$ 45. $|^-y|, |y|$ 46. $|^-x + x|, |x| \ (x \neq 0)$

3.3 Sums and Differences

We want the properties of the set of integers to be consistent with the
properties of the set of whole numbers. Thus we shall take for the set of
integers the laws pertaining to addition for the set W (see pages 29 and

30). *We assume that the set of integers is closed for addition and that addition is commutative and associative.*

The closure law of addition implies that the sum of two natural numbers is a natural number. Since a *positive integer* is a natural number, it follows that *the sum of two positive integers is a positive integer.* Thus both $3 + 4$ and $^+3 + {}^+4$ can be represented by the basic numeral, 7, or $^+7$.

How about the sum of two negative integers such as $^-3 + {}^-5$? Since $3 + {}^-3 = 0$ and $5 + {}^-5 = 0$, we can write

$$(3 + {}^-3) + (5 + {}^-5) = 0.$$

The associative and commutative laws of addition permit writing this statement as

$$(3 + 5) + ({}^-3 + {}^-5) = 0.$$

Hence, because the additive inverse is unique, $^-3 + {}^-5$ must be the additive inverse of $3 + 5$, which can be written as $^-(3 + 5)$.

The preceding discussion suggests the following theorem.

Theorem 3.2.   If $a, b \in J$, then $^-a + {}^-b = {}^-(a + b)$.

If we restrict a and b to be positive integers, then ^-a and ^-b are negative integers in the statement $^-a + {}^-b = {}^-(a + b)$, and the theorem states that *the sum of two negative integers is a negative integer.*

Examples.   a. $^-3 + {}^-8 = {}^-(3 + 8) = {}^-11$

b. $^-14 + {}^-7 = {}^-(14 + 7) = {}^-21$

The sum of two positive integers is a positive integer, and the sum of two negative integers is a negative integer. What is the nature of the sum of a positive integer and a negative integer? Consider the sum of 3 and $^-4$. The number represented by the basic numeral $^-4$ can also be written as $^-3 + {}^-1$. Thus,

$$3 + {}^-4 = 3 + ({}^-3 + {}^-1)$$

$$= (3 + {}^-3) + {}^-1 \qquad \text{Associative law of addition}$$

$$= 0 + {}^-1 \qquad \text{Additive inverse law}$$

$$= {}^-1 \qquad \text{Identity law of addition}$$

In this case, the sum of a positive integer and a negative integer is a negative integer. Now consider the sum $^-7 + 9$. This can be written as

$$^-7 + 9 = {}^-7 + ({}^+7 + {}^+2)$$
$$= ({}^-7 + {}^+7) + 2 \qquad \text{Associative law of addition}$$
$$= 0 + 2 \qquad\qquad \text{Additive inverse law}$$
$$= 2 \qquad\qquad\quad \text{Identity law of addition}$$

In this example, the sum of a positive integer and a negative integer is a positive integer. Notice that in the first example, $|^-4| > |3|$, and in the second, $|9| > |^-7|$. These examples illustrate the following theorem.

Theorem 3.3. If $a, b \in J$, and

if $a > b > 0$, then $a + (^-b) = {}^+(a - b)$;

if $b > a > 0$, then $a + (^-b) = {}^-(b - a)$.

The proof parallels the argument in the examples above.

Examples. Write each sum as a basic numeral.

a. $^+7 + (^-15)$ b. $^-3 + 9$ c. $^-12 + 2$

Solutions

a. $^-(15 - 7) = {}^-8$ b. $^+(9 - 3) = 6$ c. $^-(12 - 2) = {}^-10$

The numerals we have been using to represent integers, that is, numerals with raised $^-$ and $^+$ signs, such as $^-8$ and $^+3$, are not widely used. We have used such symbols for integers to emphasize the distinction between the symbol $^-$ used with the symbol for a negative number and the sign used to denote the operation of subtraction. The sign that is a part of the numeral is normally centered. That means that $^-8$ is generally written as -8, and $^+3$ as $+3$, or simply 3. If this symbolism is adopted, however, it becomes necessary to use parentheses in denoting sums, because where $^-8 + {}^+3$ is clear enough, $-8 + +3$ is not. Therefore, numerals for integers are frequently enclosed in parentheses to keep the sign portion of the numeral distinct from the symbol used to denote the operation. For example,

$^-8 + {}^+3$ might be written as $(-8) + (+3)$, or $(-8) + (3)$. Since, in this case, no ambiguity is possible with respect to -8, it can be written as $-8 + (3)$, or even as $-8 + 3$.

We can interpret Theorems 3.2 and 3.3 by using a number line if we associate the positive integers with the counting of marked points to the *right* on the number line and negative integers with counting to the *left*. For example, the sum $5 + 3$ can be thought of as the number associated with the point arrived at after counting off five of the chosen segments to the right of the origin and then, from this point, counting off three more segments to the right, as in Figure 3.4.

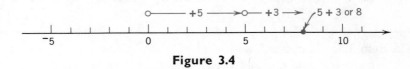

Figure 3.4

Similarly, the sum $-4 + (-3)$ appears on the number line as in Figure 3.5a, the sum $-4 + (+7)$ in Figure 3.5b, and the sum $5 + (-10)$ in Figure 3.5c.

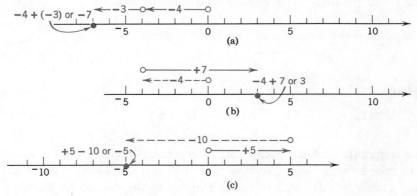

Figure 3.5

Recall that, in the discussion in Section 2.8, an expression such as $8 - 12$ did not represent a whole number and therefore was not defined for the set of whole numbers. The set of whole numbers is *not closed* with respect to the operation of subtraction. A representation of this difference on a number line appears in Figure 3.6, where the *subtraction* of the *positive number* 12 is associated with counting to the *left*.

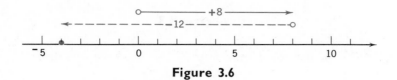

Figure 3.6

Observe that the point arrived at after counting 8 units to the right and 12 units to the left has the coordinate of −4. In the set of integers, the difference 8 − 12 equals −4. Similarly, every pair of integers has a difference, as stated in the following theorem.

Theorem 3.4. If $a, b \in J$, then $a - b = a + (-b)$.

Thus an expression such as 8 − 12 means $8 + (-12)$, for which the basic numeral is −4. Similarly, 12 − 8 means $12 + (-8)$, for which the basic numeral is clearly 4. It can be shown (though we will not do so) that Theorem 3.4 is consistent with our previous definition of $a - b$ for the set of whole numbers. We need not be overly concerned with the operation of subtraction in the system of integers since the difference of two integers can now be viewed as *the **sum** of the first integer and the additive inverse of the second.* Hence, since the set of integers is closed with respect to addition, it is also closed with respect to subtraction.

Examples. a. $7 - 11 = 7 + (-11) = -4$

   b. $7 - (-11) = 7 + [-(-11)] = 7 + 11 = 18$

   c. $-7 - 11 = -7 + (-11) = -18$

In Section 3.2, we discussed the meaning of $a < b$, where a and b are integers, by using the number line. We are now able to make a definition without relying on the number line.

Definition 3.2. For all $a, b \in J$,
   $a < b$ if there is a positive integer c such that $a + c = b$.

Example. $3 < 5$ because there is a positive integer 2, such that $3 + 2 = 5$.

Exercise 3.3

A

If a, b, $c \in J$, name the axiom or theorem that justifies each statement.

1. $a + b \in J$

2. $a + b = b + a$

3. $(a + b) + c = a + (b + c)$

4. $a + 0 = a$

5. If $a = b$, then $a + c = b + c$.

6. Define the difference $a - b$. a, $b \in J$.

Write each sum as a basic numeral.

Examples. a. $^-6 + {}^-2$ b. $8 + (-3)$ c. $-2 + 3 + (-4)$

Solutions.   a. $^-8$ b. 5 c. -3

7. $^+2 + {}^+4$ 8. $^+5 + {}^+8$ 9. $^+7 + {}^-10$ 10. $^+4 + {}^-3$

11. $^-6 + {}^-2$ 12. $^-9 + {}^-6$ 13. $^-15 + 0$ 14. $0 + {}^+6$

15. $8 + 4$ 16. $-3 + 4$

17. $-7 + (-7)$ 18. $-(-5) + (-5)$

19. $-9 + (-1) + (-7)$ 20. $-8 + 15 + (-7)$

21. $12 + (-3) + (-5)$ 22. $6 + (-3) + 7$

23. $9 + (-7) + (-5)$ 24. $-1 + 7 + (-4)$

25. $2 + (-2) + (-10)$ 26. $-9 + (-1) + (-4)$

Write each difference as a basic numeral.

Examples. a. $5 - 8$ b. $4 - (-2)$ c. $-3 - (-7)$

Solutions

a. $5 - 8 = 5 + (-8) = -3$

b. $4 - (-2) = 4 + [-(-2)] = 4 + 2 = 6$

c. $-3 - (-7) = -3 + [-(-7)] = -3 + 7 = 4$

27. $-6 - 2$ 28. $4 - 9$ 29. $8 - 1$ 30. $-4 - (-8)$

31. $-8 - (-4)$ 32. $3 - (-7)$ 33. $13 - 0$ 34. $-25 - 0$

35. $0 - 2$ 36. $0 - (-4)$

37. $[7 + (-2)] - 1$ 38. $[-1 + (-9)] - 6$

Complete each statement.

39. If x is a positive integer, then $x + x$ is a ————— integer.

40. If x is a negative integer, then $x + x$ is a ————— integer.

41. If x is a negative integer and y is a positive integer and $|x| = |y|$, then $x + y =$ —————.

42. If x is a negative integer and y is a positive integer and $|x| < |y|$, then $x + y$ is a ————— integer.

43. If x is a negative integer and y is a positive integer and $|x| > |y|$, then $x + y$ is a ————— integer.

B

In the following problems, $x, y \in J$.

44. Under what conditions is $x - y$ equal to 0?

45. Under what conditions is $x - y$ a positive integer?

46. Under what conditions is $x - y$ a negative integer?

47. Under what conditions is $x + y$ equal to 0?

48. Under what conditions is $x + y$ a positive integer?

49. Supply the missing reasons for the proof of the theorem:

If $a, b \in J$, then $^-a + {}^-b = {}^-(a + b)$.

Statements	**Reasons**
a. $0 + 0 = 0$	a. Additive identity law
b. $a + {}^-a = 0,\ b + {}^-b = 0$	b. —————
c. $(a + {}^-a) + (b + {}^-b) = 0$	c. —————
d. $(a + b) + ({}^-a + {}^-b) = 0$	d. Associative and commutative laws of addition
e. $^-a + {}^-b = {}^-(a + b)$	e. The additive inverse is unique. $[({}^-a + {}^-b)$ must be equal to the negative of $(a + b)$ if the statement in Step d is to be true.]

3.4 Products and Quotients

Again, we shall take for the set of integers the laws pertaining to multiplication for the set of whole numbers (see pages 34 and 35). *We assume that the set of integers is closed for multiplication, multiplication is commutative and associative, and the distributive law holds.*

The closure law of multiplication asserts that the product of two natural numbers is a natural number. Since a positive integer is a natural number, *the product of two positive integers is a positive integer.*

What is the nature of the product of a positive integer and a negative integer such as $4 \cdot (-3)$? Consider the following:

$$3 + (-3) = 0 \qquad \text{Additive inverse law}$$

$$4 \cdot [3 + (-3)] = 4 \cdot 0 \qquad \text{Multiplication law}$$

$$4 \cdot 3 + 4 \cdot (-3) = 4 \cdot 0 \qquad \text{Distributive law}$$

$$4 \cdot 3 + 4 \cdot (-3) = 0 \qquad \text{Zero factor law}$$

Because the additive inverse of an integer is unique, $4 \cdot (-3)$ must be the negative of $4 \cdot 3$ and therefore equals $-(4 \cdot 3)$, or -12. This example suggests the following theorem.

Theorem 3.5.   If $a, b \in J$, then $a \cdot (-b) = -(a \cdot b)$.

Using the commutative law of multiplication, we can also write

$$a \cdot (-b) = (-b) \cdot a$$

and then by the substitution law

$$(-b) \cdot a = -(b \cdot a) = -(a \cdot b).$$

Now, by restricting a and b to denote positive integers, we have from this theorem that *the product of a positive integer and a negative integer is a negative integer.*

Examples.   a. $2 \cdot (-3) = -6$ \qquad b. $-5 \cdot 6 = -30$

Using an argument similar to that employed for the product of a positive integer and a negative integer, we can determine the nature of the product of two negative integers such as $(-3) \cdot (-5)$. Consider the following:

$$5 + (-5) = 0 \qquad \text{Additive inverse law}$$
$$-3 \cdot [5 + (-5)] = -3 \cdot 0 \qquad \text{Multiplication law}$$
$$-3 \cdot 5 + (-3) \cdot (-5) = -3 \cdot 0 \qquad \text{Distributive law}$$
$$-3 \cdot 5 + (-3) \cdot (-5) = 0 \qquad \text{Zero factor law}$$

Because the additive inverse of an integer is unique, $(-3) \cdot (-5)$ must be the negative of $-3 \cdot 5$, and therefore equals $-(-3 \cdot 5)$, or $3 \cdot 5$, or 15. This example suggests the following theorem.

Theorem 3.6. If $a, b \in J$, then $(-a) \cdot (-b) = a \cdot b$.

If a and b are restricted to denote positive integers, this theorem states that *the product of two negative integers is a positive integer.*

Examples. a. $-2 \cdot (-3) = 6$ b. $-25 \cdot (-4) = 100$

Summarizing the preceding conclusions on the multiplication of integers, we have:

1. The product of any integer and 0 is 0.
2. The product of two positive integers is a positive integer.
3. The product of a positive integer and a negative integer is a negative integer.
4. The product of two negative integers is a positive integer.

Because we want any definition of division in the system of integers to be consistent with the definition of division in the system of whole numbers, we paraphrase Definition 2.4 (page 47) for a, $b \in J$ and obtain the following:

The quotient of two integers, a and b $(b \neq 0)$ denoted by $\dfrac{a}{b}$ or a/b is the integer q, if it exists, such that $b \cdot q = a$.

Examples. a. $\dfrac{10}{-2} = -5$, because $(-2) \cdot (-5) = 10$.

     b. $\dfrac{-3}{7}$ is not an integer, because no integer q exists such that

        $7 \cdot q = -3$.

Notice that if a and b are integers, $b \neq 0$, the quotient a/b is an integer if and only if b is an integral factor of a. Hence, the set of integers is *not closed* with respect to division. If a, b, $a/b \in J$, then the quotient a/b will be a member of the set of positive integers, a member of the set of negative integers, or 0. Consider the following possibilities.

1. If a and b are positive, a/b must also be positive since the product of two positive integers is positive.
2. If a is positive and b is negative, a/b must be negative because the product of two negative integers is positive.
3. If a is negative and b is positive, a/b must be negative because the product of a positive integer and a negative integer is negative.
4. If a is negative and b is negative, a/b must be positive because the product of a negative integer and a positive integer is negative.
5. If $a = 0$, a/b must be equal to 0 because the product of any integer and 0 is 0.

From the above, then, the quotient of two integers which are either both positive or both negative will be a positive integer, if such an integer exists. The quotient of two integers, one positive and the other negative, will be a negative integer, if such an integer exists. The quotient of zero divided by any non-zero integer is zero.

Examples

a. $\dfrac{15}{5} = 3$, because $5 \cdot 3 = 15$ b. $\dfrac{-24}{-6} = 4$, because $-6 \cdot 4 = -24$

c. $\dfrac{-6}{2} = -3$,   d. $\dfrac{8}{-4} = -2$,

    because $2 \cdot (-3) = -6$ because $(-4) \cdot (-2) = 8$

e. $\dfrac{0}{3} = 0$, because $3 \cdot 0 = 0$ f. $\dfrac{0}{-7} = 0$, because $-7 \cdot 0 = 0$

Two integers whose difference is 1 or ⁻1 are called **consecutive integers.** A number of the form $2 \cdot k$, $k \in J$ is called a *multiple* of 2. If an integer can be represented as $2 \cdot k$, $k \in J$, it is called an **even integer.** Any two even integers whose difference is 2, or ⁻2, are called **consecutive even integers.** Thus, elements of $\{\cdots {}^-6, {}^-4, {}^-2, 0, 2, 4, 6 \cdots\}$ are the even integers and pairs like ⁻10, ⁻8 or 6, 8 are consecutive even integers.

An integer which can be represented in the form $2 \cdot k + 1$, $k \in J$ is called an **odd integer.** Any two odd integers whose difference is 2, or $^-2$, are called **consecutive odd integers.** Thus, elements of $\{\cdots -5, -3, -1, 1,$ $3, 5 \cdots\}$ are the odd integers and pairs like $^-13, ^-11$ or 5, 7 are consecutive odd integers.

Exercise 3.4

A

If $a, b, c \in J$, name the axiom or theorem that justifies each statement.

1. $a \cdot b \in J$

2. $a \cdot b = b \cdot a$

3. $(a \cdot b) \cdot c = a \cdot (b \cdot c)$

4. $a \cdot (b + c) = a \cdot b + a \cdot c$

5. $a \cdot 1 = a$

6. $a \cdot 0 = 0$

7. If $a = b$, then $a \cdot c = b \cdot c$.

8. $a + (-a) = 0$

Write each product as a basic numeral.

Examples.   a. $-3 \cdot 2$ b. $-3 \cdot (-2)$ c. $[(-3) \cdot (-2)] \cdot (-2)$

Solutions   a. -6 b. 6   c. $(6) \cdot (-2) = -12$

9. $5 \cdot 9$

10. $3 \cdot 4$

11. $-6 \cdot 5$

12. $3 \cdot (-8)$

13. $-9 \cdot (-6)$

14. $-1 \cdot (-8)$

15. $-2 \cdot 0$

16. $0 \cdot (-7)$

17. $1 \cdot (-4) \cdot (2)$

18. $-8 \cdot 7 \cdot 5$

19. $5 \cdot (-4) \cdot (-2)$

20. $-6 \cdot 7 \cdot (-2)$

21. $2 \cdot 0 \cdot (-2)$

22. $-5 \cdot 8 \cdot 0$

23. $-3 \cdot (-7) \cdot (-2)$

24. $-6 \cdot (-2) \cdot (-9)$

25. $2 \cdot (-3) \cdot 8 \cdot (-6)$

26. $-4 \cdot 7 \cdot (-1) \cdot (-10)$

Write each quotient as a basic numeral if such an integer exists. If the quotient does not exist in the set of integers, so state.

Examples.   a. $\dfrac{-12}{3}$ b. $\dfrac{-25}{-5}$ c. $\dfrac{13}{4}$

Solutions.   a. -4 b. 5   c. Does not exist.

27. $\dfrac{8}{-4}$ 28. $\dfrac{24}{-3}$ 29. $\dfrac{-15}{3}$ 30. $\dfrac{-42}{6}$

31. $\dfrac{-14}{-2}$ 32. $\dfrac{-35}{-7}$ 33. $\dfrac{11}{3}$ 34. $\dfrac{-14}{5}$

35. $\dfrac{0}{-6}$ 36. $\dfrac{-19}{0}$ 37. $\dfrac{3 \cdot 4}{-2}$ 38. $\dfrac{-4 \cdot 3}{2 \cdot (-6)}$

Find two integral factors of the first number in each pair such that the sum of the factors is the second number.

Examples.   a. $-30, 1$ b. $-72, -21$

Solutions.   a. $(6)(-5); \; [6 + (-5) = 1]$ b. $-24 \cdot 3; \; (-24 + 3 = -21)$

39. $4, -4$ 40. $0, -5$ 41. $-12, 1$ 42. $20, -9$

43. $-8, 7$ 44. $-9, 0$ 45. $-18, 3$ 46. $-15, 2$

Complete each statement.

47. If x is a positive integer and y is a negative integer, then $x \cdot y$ is a _____ integer.

48. If x and y are both positive integers, then $x \cdot y$ is a _____ integer.

49. If x and y are both negative integers, then $x \cdot y$ is a _____ integer.

50. If x is a negative integer and y is a positive integer, then $x \cdot y$ is a _____ integer.

51. If x and y are both positive integers or both negative integers, then x/y is a _____ integer, if such an integer exists.

52. If x is a positive integer and y is a negative integer, then x/y is a _____ integer, if such an integer exists.

53. If x is a negative integer and y is a positive integer, then x/y is a _____ integer, if such an integer exists.

54. If x is an integer ($x \neq 0$), then $0/x = $ _____.

B

55. Specify the set whose elements are the first three positive multiples of 2.

56. Specify the set whose elements are the first five positive multiples of 3.

57. Specify the set whose elements are the first seven positive multiples of 4.

58. Specify the set whose elements are the first nine positive multiples of 5.

Example.   List the members in $\{2 \cdot k - 1 \mid k \in \{1, 2, 3, 4\}\}$.

Solution.   If $k = 1$, then $2 \cdot k - 1 = 2 \cdot 1 - 1 = 1$.

  If $k = 2$, then $2 \cdot k - 1 = 2 \cdot 2 - 1 = 3$.

  If $k = 3$, then $2 \cdot k - 1 = 2 \cdot 3 - 1 = 5$.

  If $k = 4$, then $2 \cdot k - 1 = 2 \cdot 4 - 1 = 7$.

Hence, the required set is $\{1, 3, 5, 7\}$.

59. List the members in $\{2 \cdot k + 1 \mid k \in \{1, 2, 3, 4\}\}$.

60. List the members in $\{2 \cdot k + 2 \mid k \in \{4, 5, 6, 7\}\}$.

61. List the members in $\{3 \cdot k - 1 \mid k \in \{1, 2, 3, 4\}\}$.

62. List the members in $\{3 \cdot k - 3 \mid k \in \{5, 6, 7, 8, 9\}\}$.

63. Supply the missing reasons for the proof of the theorem:

If $a, b \in J$, then $a \cdot (-b) = -(a \cdot b)$.

Statements	*Reasons*
a. $b + (-b) = 0$	a._____
b. $a \cdot [b + (-b)] = a \cdot 0$	b. Multiplication law of equality
c. $a \cdot b + a \cdot (-b) = a \cdot 0$	c._____
d. $a \cdot b + a \cdot (-b) = 0$	d._____
e. $a \cdot (-b) = -(a \cdot b)$	e. The additive inverse is unique. [There is only one number which, when added to $a \cdot b$ in Step d. equals 0, namely, $-(a \cdot b)$.]

64. Supply the missing reasons for the proof of the theorem:

If $a, b \in J$, then $(-a) \cdot (-b) = a \cdot b$.

Statements	*Reasons*
a. $b + (-b) = 0$	a._____
b. $-a \cdot [b + (-b)] = -a \cdot 0$	b._____

$Statements$	$Reasons$
c. $-a \cdot b + (-a) \cdot (-b) = -a \cdot 0$	c. Distributive law
d. $-a \cdot b + (-a) \cdot (-b) = 0$	d. Zero factor law
e. $-a \cdot b = -(a \cdot b)$	e. Theorem 3.5
f. $(-a) \cdot (-b) = a \cdot b$	f._____

3.5 Order of Operations; Numerical Evaluation

The agreement that we made in Section 2.10 concerning the order of operations in the set of whole numbers is also valid in the set of integers. Furthermore, expressions containing variables can now be evaluated for any integer replacements of the variables.

Exercise 3.5

A

Simplify each expression.

Examples.   a. $3 \cdot (-3) + (-2) \cdot (-5) - 4 \cdot (-2)$

b. $-1 \cdot (3 - 5) - \dfrac{2 - 11}{3}$

Solutions.   a. $3 \cdot (-3) + (-2) \cdot (-5) - 4 \cdot (-2) = -9 + 10 + 8$
$$= 9$$

b. $-1 \cdot (3 - 5) - \dfrac{2 - 11}{3} = -1 \cdot (-2) - \dfrac{-9}{3}$
$$= 2 - (-3)$$
$$= 2 + 3 = 5$$

1. $2 \cdot (-4) - (-3) \cdot (-5) + (-6) \cdot (-7)$

2. $7 \cdot (-2) - 3 \cdot (-4) - (-2) \cdot (-8)$

3. $4 \cdot (-5) - 3 \cdot (7) + (-5) \cdot (-2)$

4. $-2 \cdot (-4) + 5 \cdot (-3) - (-6) \cdot (-2)$

5. $5 \cdot (-6) - 4 \cdot (7) + (-18) \cdot (-3)$

6. $9 \cdot (-2) + 6 \cdot (8) - 5 \cdot (6)$

7. $-3 \cdot (-4) + (-6) \cdot (0) + 2 \cdot (-5)$

8. $-6 \cdot (-4) - (-7) \cdot (0) + 0 \cdot (-9)$

9. $6 \cdot (-3) - \dfrac{12}{3}$

10. $\dfrac{18}{2} - 2 \cdot (-3)$

11. $\dfrac{16}{-4} - \dfrac{12}{-3} + \dfrac{-15}{-5}$

12. $\dfrac{-28}{4} - \dfrac{27}{-3} + \dfrac{-35}{-7}$

13. $3 \cdot (2 - 5) - 2 \cdot (4 - 6)$

14. $6 \cdot (4 - 8) + 3 \cdot (-2 + 3)$

15. $-7 \cdot (-5 + 2) + 3 \cdot (1 - 5)$

16. $-3 \cdot (-14 + 8) + 5 \cdot (10 - 14)$

17. $6 \cdot (-8 + 10) - 4 \cdot (9 - 7)$

18. $9 \cdot (-3 + 5) - 7 \cdot (4 - 7)$

19. $3 \cdot (-2) - \dfrac{3 - 11}{2}$

20. $4 \cdot (-3) - \dfrac{4 - 13}{-3}$

21. $5 \cdot (2 - 4) + \dfrac{6 - 18}{-4}$

22. $3 \cdot (3 - 7) - \dfrac{4 - 1}{-3}$

23. $\dfrac{3 \cdot (5 - 9)}{6} - \dfrac{2 \cdot (3 - 17)}{7}$

24. $\dfrac{-2 \cdot (4 - 19)}{5} - \dfrac{-3 \cdot (5 - 13)}{6}$

Evaluate each expression for the given values of the variables.

Examples. a. $2 \cdot (x + 6) - 5 \cdot x + 3; \ x = -2$

b. $5 \cdot x + 2 \cdot x \cdot y - 3 \cdot y: \ x = -3, y = -2$

Solutions. a. $2 \cdot (x + 6) - 5 \cdot x + 3 = 2 \cdot [(-2) + 6] - 5 \cdot (-2) + 3$

$$= 2 \cdot (4) - 5 \cdot (-2) + 3$$

$$= 8 + 10 + 3 = 21.$$

b. $5 \cdot x + 2 \cdot x \cdot y - 3 \cdot y = 5 \cdot (-3) + 2 \cdot (-3) \cdot (-2) - 3 \cdot (-2)$

$$= -15 + 12 + 6 = 3.$$

25. $2 \cdot (x + 5) + x - 4; \ x = -3$

26. $3 \cdot (x + 1) + 2 \cdot x - 3; \ x = -4$

27. $4 \cdot (x - 1) - 5 \cdot x + 7; \ x = -5$

28. $2 \cdot (x - 2) - 7 \cdot x - 15; \ x = -2$

29. $5 \cdot x + x \cdot (x + 2)$; $x = -3$ 30. $2 \cdot x - x \cdot (x + 5)$; $x = -1$

31. $7 \cdot x - 2 \cdot x \cdot (2 - x)$; $x = 3$ 32. $5 \cdot x - 3 \cdot x \cdot (1 - x)$; $x = 4$

33. $\dfrac{x + 6 + 2}{x} - \dfrac{2 - x}{4 + x}$; $x = -2$ 34. $\dfrac{2 \cdot x + 4}{x} - \dfrac{3 \cdot x - 6}{2 \cdot x}$; $x = -2$

35. $3 \cdot x + x \cdot y - 2 \cdot y$; $x = -3, y = -2$

36. $2 \cdot x + 3 \cdot x \cdot y + 3 \cdot y$; $x = 2, y = -3$

37. $4 \cdot x - 2 \cdot x \cdot y + 5 \cdot y$; $x = -1, y = 4$

38. $3 \cdot x - 6 \cdot x \cdot y - 4 \cdot y$; $x = -2, y = -3$

3.6   Equations and Inequalities Involving Integers

Recall from Section 2.11 that some equations, or inequalities, do not have solutions in W. We now consider the set of integers as the replacement set for the variable.

Examples.   Find the solution set of:

a. $x + 5 = 3$ b. $4 - x = 5$ c. $-2 < x < 3$

Solutions

a. $\{-2\}$ because $-2 + 5 = 3$.   b. $\{-1\}$ because $4 - (-1) = 5$.

c. $\{-1, 0, 1, 2\}$ because -2 is less than every member of this set, every member of the set is less than 3, and the elements in the set are the only such integers.

Example c, above, concerns a sentence that places two conditions on the variable for the sentence to be true. Hence this sentence can also be written as $-2 < x$ *and* $x < 3$. Since any solution of the sentence must satisfy both conditions, the solution set is the intersection of these two sets and can be written as

$$\{x \mid -2 < x\} \cap \{x \mid x < 3\} = \{-1, 0, 1, 2\}.$$

The graph of each set is shown in Figure 3.7.

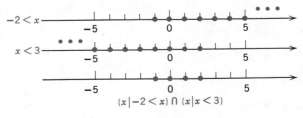

Figure 3.7

Exercise 3.6

A

Find the solution set of each equation over the set of integers.

Examples.   a. $x + 4 = 2$ b. $3 \cdot x = 2$

Solutions.   a. $\{-2\}$ because $-2 + 4 = 2$.

b. ∅ because there is no integer such that the product of the integer and 3 equals 2.

1. $x + 6 = 2$ 2. $x + 2 = 1$ 3. $x + 7 = -2$

4. $x + 4 = -1$ 5. $x = 9 - 4$ 6. $x = 6 - 8$

7. $x + 3 = 0$ 8. $x + 7 = 0$ 9. $x - 10 = -5$

10. $x - 5 = -4$ 11. $3 - x = 5$ 12. $6 - x = 4$

13. $8 - x = -7$ 14. $4 - x = -9$ 15. $2 \cdot x = 14$

16. $5 \cdot x = 15$ 17. $3 \cdot x = 12$ 18. $7 \cdot x = -21$

19. $-6 \cdot x = 25$ 20. $-4 \cdot x = 21$ 21. $-2 \cdot x = -8$

22. $-5 \cdot x = -35$ 23. $6 \cdot x = 0$ 24. $-4 \cdot x = 0$

25. $\dfrac{x}{3} = 4$ 26. $\dfrac{x}{5} = 7$ 27. $\dfrac{x}{1} = -2$

28. $\dfrac{x}{7} = -9$ 29. $\dfrac{x}{-3} = 2$ 30. $\dfrac{x}{-4} = 1$

31. $\dfrac{x}{-2} = -8$ 32. $\dfrac{x}{-9} = -5$ 33. $\dfrac{10}{x} = 3$

34. $\dfrac{15}{x} = 9$ 35. $\dfrac{-6}{x} = -2$ 36. $\dfrac{-8}{x} = -1$

Find the solution set of each inequality over the set of integers. Graph each solution set on a number line.

Example. $-4 < x \leq 2$

Solution on page 86.

Solution. $\{-3, -2, -1, 0, 1, 2\}$ because -4 is less than every element of the set and every element in the set is less than or equal to 2.

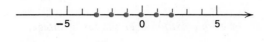

37. $-1 < x < 4$ 38. $-4 < x < -1$ 39. $0 < x \leq 3$

40. $3 < x < 6$ 41. $-2 \leq x \leq 2$ 42. $-3 \leq x \leq 1$

CHAPTER SUMMARY

3.1   To give meaning to the difference $a - b$ for *any* whole numbers a and b, additional members are added to the set of whole numbers to form the **set of integers,** $J = \{\cdots -3, -2, -1, 0, 1, 2, 3, \cdots\}$. The axioms and their consequences for the set of whole numbers are adopted for the set of integers. In addition, we assume that *for each* $a \in J$, *there exists a unique element* $-a$ *called the* **additive inverse** *or* **negative** *of a such that* $\mathbf{a + (-a) = 0}$. This assumption is called the **additive inverse law.**

3.2   The absolute value of a member in the set of integers is defined as follows:

$$|\mathbf{a}| = \mathbf{a}, \textit{ if } \mathbf{a} \geq \mathbf{0}, \quad \textit{and} \quad |\mathbf{a}| = -\mathbf{a}, \textit{ if } \mathbf{a} < \mathbf{0}.$$

3.3–3.4   The assumptions that the sum of two positive integers is a positive integer, that the product of two positive integers is a positive integer, and the additive inverse law lead to the following consequences:

a. *the sum of two negative integers is a negative integer;*

b. *the sum of a positive integer and a negative integer is positive or negative, depending on whether the integer with greater absolute value is positive or negative, respectively;*

c. *the product of a positive integer and a negative integer is a negative integer; and*

d. *the product of two negative integers is a positive integer.*

3.3   In the set of integers, the difference, $a - b$, is equal to $a + (-b)$. This view of a difference is consistent with the definition in the set of whole numbers.

3.4 The quotient of two nonzero integers is:
a. *positive if both integers are either positive or negative;*
b. *negative if one integer is positive and the other is negative.*

3.5 The order of operations agreed upon in the set of whole numbers is valid in the set of integers.

3.6 Equations and inequalities which do not have solutions in the set of whole numbers may have solutions in the set of integers.

CHAPTER REVIEW

A

3.1 *Write the additive inverse of each of the following.* $x, y \in J$.

1. 17
2. -14
3. $-y$
4. $-(-x)$

If $a, b, c \in J$, *name the axiom or theorem that justifies each statement.*

5. $a \cdot (b + c) = a \cdot b + a \cdot c$
6. If $a = b$, then $a \cdot c = b \cdot c$.
7. $(a + b) + c = a + (b + c)$
8. $a \cdot 0 = 0$
9. $a + (-a) = 0$
10. $(a \cdot b) \cdot c = a \cdot (b \cdot c)$
11. $a \cdot b \in J$
12. $a + b = b + a$
13. $a \cdot 1 = a$
14. $a + 0 = a$
15. $a \cdot b = b \cdot a$
16. $a + b \in J$

3.2 *Replace the comma in each number pair with the appropriate symbol,* $<, =,$ *or* $>$.

17. $-7, 3$
18. $-2, -5$
19. $-(-3), -3$
20. $0, -(-5)$
21. $|5|, |-5|$
22. $|-8|, |3|$
23. $|-3|, |0|$
24. $-|-4|, 4$

3.3 *Write each sum or difference as a basic numeral.*

25. $7 + 8$
26. $-8 + 2$
27. $4 - 7$
28. $-6 - 10$
29. $-8 + 12$
30. $4 - 4$
31. $6 + 3 - 8$
32. $9 - 7 - 6$

3.4 *Write each product or quotient as a basic numeral. If such a number does not exist, so state.*

33. $5 \cdot (-7)$ 34. $(-6) \cdot (-3)$ 35. $(-5) \cdot (3)$ 36. $(4) \cdot (5)$

37. $\dfrac{8}{-4}$ 38. $\dfrac{-12}{-3}$ 39. $\dfrac{0}{8}$ 40. $\dfrac{8}{0}$

3.5 *Simplify each expression.*

41. $-2 \cdot (-3) - 4 \cdot (-5) + \dfrac{6 - 15}{-3}$ 42. $2 \cdot (-3) + \dfrac{9 - 6}{-3}$

Evaluate each expression.

43. $5 \cdot (x - 2) + 15; \ x = -3$

44. $2 \cdot x - 7 \cdot x \cdot y - 3 \cdot y; \ x = -2, y = -1$

3.6 *Find the solution set of each equation or inequality over the set of integers.*

45. $x + 5 = -8$ 46. $x - 7 = 1$ 47. $6 \cdot x = 21$

48. $\dfrac{x}{-2} = 3$ 49. $-3 < x \le 0$ 50. $\dfrac{-5}{x} = 1$

B

In Problems 51 to 54, $x, y \in J$.

51. Under what conditions is $x - y = y - x$?

52. Under what conditions is $x - y > y - x$?

53. Under what conditions will $x/y < 0$?

54. Under what conditions will $x/y > 0$?

55. Specify the set whose elements are the first four positive multiples of 4.

56. Specify the set whose elements are the first four positive multiples of 7.

57. List the members in $\{2 \cdot k \mid k \in \{-3, -2, -1, 0\}\}$.

58. List the members in $\{3 \cdot k + 1 \mid k \in \{1, 2, 3, 4\}\}$.

4
chapter

The Set of Rational Numbers

The set of whole numbers was introduced in Chapter 2 and then extended in Chapter 3 by the addition of the negatives of the natural numbers to form the set of integers. We can associate all elements in the set of integers with points on the number line. Obviously, the set of integers does not include elements which are associated with points on a number line that lie between the graphs of the elements in the set of integers. For example, the point labeled A in Figure 4.1 lies between -3 and -2, and hence is not

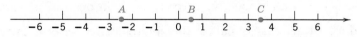

Figure 4.1

associated with an integer. Similarly, the points labeled B and C lie between the pair of integers 0, 1 and the pair 3, 4, respectively, and also are not associated with integers.

Thus, being limited to using only the set of integers does not allow us to deal with, for example, a temperature between -3 and -2 degrees, a distance between 3 and 4 miles, etc. Furthermore, the set of integers does not contain solutions to equations such as

$$3 \cdot x = 1 \quad \text{and} \quad 2 \cdot x = 5.$$

If the points A, B, and C are to be associated with numbers, what kind of numbers must they be? This chapter and the next chapter are concerned with the answer to this question and with the behavior of such numbers under the basic operations.

4.1 An Extension of the Set of Integers

As with the extension of the set of whole numbers, the set of integers is extended by adding some new elements. The new set, of course, still contains the integers as members, but, in addition, contains such new members as are necessary to insure that every pair of integers has a quotient. That is, every quotient a/b, where a and b are integers (provided that b does not equal 0), will represent one of the members of the new set; and conversely, every element of the new set can be viewed as the quotient of two integers. The extended set of numbers is called the **set of rational numbers** and will be denoted by Q.

We first extend the set of integers by making the following assumption:

For every integer a, except 0, there is a number $\dfrac{1}{a}$ such that

$$a \cdot \left(\frac{1}{a}\right) = 1 \quad \text{and} \quad \left(\frac{1}{a}\right) \cdot a = 1.$$

The number $\dfrac{1}{a}$ is called the **multiplicative inverse** of a, or the **reciprocal** of a. Thus, $\dfrac{1}{3}$ is the multiplicative inverse, or reciprocal, of 3. The above assumption, or axiom, is called the **multiplicative inverse law,** or the **reciprocal law.**

Recall that in the set of integers, a quotient such as $\dfrac{a}{b}$ has meaning if and only if a is a multiple of b, or b is an integral factor of a. If we are to extend the set of numbers so that $\dfrac{a}{b}$ has meaning for all a, $b \in J$, except $b = 0$, then the definition of a quotient must still hold. That is, the quotient $\dfrac{a}{b}$ equals the number q such that $b \cdot q = a$.

The definition of a quotient and the multiplicative inverse law permit us to rewrite quotients in another form that is very useful in many situations. For example, consider the quotient $\frac{5}{7}$. By the definition of a quotient we know that

$$\frac{5}{7} = q \quad \text{such that} \quad 7 \cdot q = 5. \tag{1}$$

Also, because $7 \neq 0$, the multiplicative inverse law assures us that the number $\frac{1}{7}$ exists. Then, multiplying both members of the second equation above by $\frac{1}{7}$, we have

$$\frac{1}{7} \cdot (7 \cdot q) = \frac{1}{7} \cdot 5.$$

By applying the associative law of multiplication to the left-hand member and the commutative law of multiplication to the right-hand member we obtain

$$\left(\frac{1}{7} \cdot 7\right) \cdot q = 5 \cdot \frac{1}{7}.$$

Now, from the multiplicative inverse law, $\frac{1}{7} \cdot 7 = 1$, and from the multiplicative identity law, $1 \cdot q = q$. Thus,

$$q = 5 \cdot \frac{1}{7}.$$

Finally, substituting $\frac{5}{7}$ for q from (1) above, we have

$$\frac{5}{7} = 5 \cdot \frac{1}{7}.$$

This example suggests the following theorem.

Theorem 4.1. If a, $b \in J$ ($b \neq 0$), then

$$\frac{a}{b} = a \cdot \frac{1}{b}.$$

Examples.   a. $\dfrac{3}{4} = 3 \cdot \dfrac{1}{4}$ b. $\dfrac{8}{113} = 8 \cdot \dfrac{1}{113}$

Extending the notion of a multiplicative inverse to every rational number we have the more general multiplicative inverse law.

> *For every nonzero rational number,* $\dfrac{a}{b}$, *there exists a unique rational number,*
>
> $\dfrac{1}{\dfrac{a}{b}}$, *with the property that* $\dfrac{a}{b} \cdot \dfrac{1}{\dfrac{a}{b}} = 1$ *and* $\dfrac{1}{\dfrac{a}{b}} \cdot \dfrac{a}{b} = 1.$

Notice that in the symbol for the reciprocal of a/b one bar is heavier than the other. This is done to emphasize that the numerator of the fraction is 1 and that the denominator is the rational number a/b. The fact that *the multiplicative inverse of a non-zero rational number is unique means that, for each rational number, there is one and only one multiplicative inverse.* Thus, if $a \cdot b = 1$, then $b = 1/a$ and $a = 1/b$.

A symbol for a quotient such as a/b is called a **fraction.** Although all rational numbers can be represented by fractions, we shall see later that *not all fractions represent rational numbers.*

Since the set of rational numbers contains the set of integers, the integers are also rational numbers and any integer, such as 4, is a rational number and can be represented by the symbol 4/1. That is, for $a \in Q$, a and $a/1$ represent the same rational number.

Since the set of integers is a subset of the set of rational numbers, we want the axioms adopted for the integers to be applicable also to the elements of the enlarged set. Thus *we adopt these same axioms for the rational numbers. Furthermore, the theorems which depend only on these axioms hold for all rational numbers.*

Exercise 4.1

A

What is the multiplicative inverse or reciprocal of each of the following? All variables are elements of J.

Examples.   a. -6 b. x $(x \neq 0)$ c. $x + y$ $(x \neq -y)$

Solutions. a. $\dfrac{1}{-6}$ b. $\dfrac{1}{x}$ c. $\dfrac{1}{x+y}$

1. 2   2. 5   3. -7 4. -3

5. 0   6. $\dfrac{2}{5}$ 7. $\dfrac{4}{3}$ 8. $\dfrac{-3}{8}$

9. y $(y \neq 0)$ 10. $x+4$ $(x \neq -4)$ 11. $y-7$ $(y \neq 7)$

12. $x-y$ $(x \neq y)$ 13. $a+b$ $(a \neq -b)$ 14. $m-n$ $(m \neq n)$

Write each quotient in the form $a \cdot \dfrac{1}{b}$. All variables are elements of J.

Examples. a. $\dfrac{3}{4}$ b. $\dfrac{2 \cdot x + y}{y}$ $(y \neq 0)$

Solutions. a. $3 \cdot \dfrac{1}{4}$ b. $(2 \cdot x + y) \cdot \dfrac{1}{y}$

15. $\dfrac{4}{7}$ 16. $\dfrac{3}{4}$ 17. $\dfrac{9}{5}$ 18. $\dfrac{3}{7}$

19. $\dfrac{x}{6}$ 20. $\dfrac{y}{7}$ 21. $\dfrac{x-y}{10}$ 22. $\dfrac{2 \cdot x + y}{4}$

23. $\dfrac{x}{y}$ $(y \neq 0)$ 24. $\dfrac{4 \cdot y}{3 \cdot x}$ $(x \neq 0)$

Name the axiom that justifies each statement. $a, b, c, d, e, f \in J$. Assume no denominator equals 0.

25. $\dfrac{a}{b} = \dfrac{a}{b}$ 26. If $\dfrac{a}{b} = \dfrac{c}{d}$, then $\dfrac{c}{d} = \dfrac{a}{b}$.

27. If $\dfrac{a}{b} = \dfrac{c}{d}$ and $\dfrac{c}{d} = \dfrac{e}{f}$, then $\dfrac{a}{b} = \dfrac{e}{f}$.   28. $\dfrac{a}{b} + \dfrac{c}{d} \in Q$

29. $\dfrac{a}{b} \cdot \dfrac{c}{d} \in Q$ 30. $\dfrac{a}{b} + \dfrac{c}{d} = \dfrac{c}{d} + \dfrac{a}{b}$

31. $\dfrac{a}{b} \cdot \dfrac{c}{d} = \dfrac{c}{d} \cdot \dfrac{a}{b}$ 32. $\left(\dfrac{a}{b} \cdot \dfrac{c}{d}\right) \cdot \dfrac{e}{f} = \dfrac{a}{b} \cdot \left(\dfrac{c}{d} \cdot \dfrac{e}{f}\right)$

33. $\left(\dfrac{a}{b} + \dfrac{c}{d}\right) + \dfrac{e}{f} = \dfrac{a}{b} + \left(\dfrac{c}{d} + \dfrac{e}{f}\right)$

34. $\dfrac{e}{f} \cdot \left(\dfrac{a}{b} + \dfrac{c}{d}\right) = \dfrac{e}{f} \cdot \dfrac{a}{b} + \dfrac{e}{f} \cdot \dfrac{c}{d}$

35. $\dfrac{a}{b} + 0 = \dfrac{a}{b}$

36. $\dfrac{a}{b} \cdot 1 = \dfrac{a}{b}$

37. $\dfrac{a}{b} + \left(-\dfrac{a}{b}\right) = 0$

38. $\dfrac{a}{b} \cdot \dfrac{1}{\dfrac{a}{b}} = 1$

In Problems 39 to 42, name the theorem that justifies each statement. $a, b, c, d, e,$ $f \in J$. Assume no denominator equals 0.

39. $\dfrac{a}{b} \cdot 0 = 0$

40. If $\dfrac{a}{b} = \dfrac{c}{d}$, then $\dfrac{a}{b} + \dfrac{e}{f} = \dfrac{c}{d} + \dfrac{e}{f}$.

41. If $\dfrac{a}{b} = \dfrac{c}{d}$, then $\dfrac{a}{b} \cdot \dfrac{e}{f} = \dfrac{c}{d} \cdot \dfrac{e}{f}$.

42. If $\dfrac{a}{b} = \dfrac{c}{d}$, then $\dfrac{a}{b} + \left(-\dfrac{e}{f}\right) = \dfrac{c}{d} + \left(-\dfrac{e}{f}\right)$.

Consider the set $A = \left\{\dfrac{1}{2},\ -3, 0, 1,\ -\dfrac{2}{3}, \dfrac{7}{4},\ -8,\ -\dfrac{5}{9}\right\}.$

43. Specify the subset of A which contains all whole numbers in A.

44. Specify the subset of A which contains all integers in A.

45. Specify the subset of A which contains all positive rational numbers in A.

46. Specify the subset of A which contains all negative integers in A.

Consider the set $B = \left\{-2, \dfrac{1}{3},\ -1, \dfrac{3}{2},\ -\dfrac{4}{7}, 6, \dfrac{1}{8}, 0, 4\right\}.$

47. Specify the subset of B which contains all natural numbers in B.

48. Specify the subset of B which contains all positive integers in B.

49. Specify the subset of B that contains all non-negative rational numbers in B.

50. Specify the subset of B which contains all negative rational numbers in B.

B

51. Explain why x cannot be 0 if $1/x \in Q$.

52. Explain why x cannot equal $-y$ if $1/(x + y) \in Q$.

53. For what value of x is $\dfrac{x}{5}$ equal to 0?

54. For what value of x is $\dfrac{5}{x}$ undefined?

55. Write the multiplicative inverse law in symbolic form.

56. Construct a Venn diagram showing the relationship between the sets of natural numbers, whole numbers, integers, and rational numbers.

4.2 Some Properties of Rational Numbers

Since rational numbers can be expressed in the form $\dfrac{a}{b}$, and since we expect the new set to have the property of closure for both addition and multiplication, we would expect that the sum of any two rational numbers, for example $\dfrac{2}{3} + \dfrac{3}{4}$, can be expressed as a single fraction. Before we can discuss such procedures we must be able to determine when two or more fractions represent the same rational number. Actually, any rational number can be represented by infinitely many fractions. For example,

$$\frac{4}{4} = 1, \text{ because } 4 \cdot 1 = 4 \quad \text{and} \quad \frac{2}{2} = 1, \text{ because } 2 \cdot 1 = 2,$$

and for any non-zero integer, b,

$$\frac{b}{b} = 1, \text{ because } b \cdot 1 = b.$$

Similarly,

$$\frac{12}{4} = 3, \text{ because } 4 \cdot 3 = 12, \quad \text{and} \quad \frac{18}{6} = 3, \text{ because } 6 \cdot 3 = 18.$$

Before this notion is extended to every rational number, we shall look at the representation of rational numbers on the number line. If the fraction 3/3 is used to represent the number 1, and the point 3 units of measure to the right of the origin is represented by the number 1, the number line would appear as in Figure 4.2.

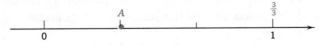

Figure 4.2

Since $\frac{3}{3} = 3 \cdot \frac{1}{3}$, then $\frac{3}{3}$ can be considered as 3 one-thirds and the right-hand endpoint of the first segment to the right of the origin, labeled A, corresponds to the rational number 1/3. An extension of this idea permits the representation of the origin as 0/3, and the right-hand end point of the second segment as 2/3. The same procedure can be used for locating points on a number line corresponding to every rational number.

The fact that an integer can be represented by different fractions makes it possible to label some points on a number line as shown in Figure 4.3. An

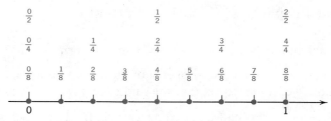

Figure 4.3

examination of this number line reveals that the point corresponding to 4/8 is also labeled 2/4 or 1/2. Therefore,

$$\frac{4}{8} = \frac{2}{4} = \frac{1}{2},$$

since all are associated with the same point and name the same number. Similarly,

$$\frac{2}{8} = \frac{1}{4} \quad \text{and} \quad \frac{6}{8} = \frac{3}{4}.$$

While the number line can be used to demonstrate the equality of some fractions, it is desirable to have a more general way of determining when two fractions are equal; that is, when two fractions denote the same rational number.

If several pairs of equal fractions on the number line in Figure 4.3 are examined, it will be noted that the product of the numerator of the first and the denominator of the second is equal to the product of the denominator of the first and the numerator of the second. For example

$$\frac{4}{8} = \frac{2}{4} \quad \text{and} \quad 4 \cdot 4 = 8 \cdot 2.$$

This suggests the following theorem.

Theorem 4.2. If $a, b, c, d \in J$ $(b, d \neq 0)$, and

$$\text{if } \frac{a}{b} = \frac{c}{d}, \text{ then } a \cdot d = b \cdot c;$$

$$\text{if } a \cdot d = b \cdot c, \text{ then } \frac{a}{b} = \frac{c}{d}.$$

Examples.   a. $\dfrac{2}{3} = \dfrac{12}{18}$, because $2 \cdot 18 = 3 \cdot 12$

b. $\dfrac{3}{7} \neq \dfrac{7}{15}$, because $3 \cdot 15 \neq 7 \cdot 7$

Now we have a general method for determining when two fractions represent the same rational number. Next, we shall find ways of generating fractions which represent the same rational number. The following example illustrates one such method. By the reflexive law,

$$2 \cdot (3 \cdot 4) = 2 \cdot (3 \cdot 4).$$

By the commutative law and associative law of multiplication

$$(2 \cdot 4) \cdot 3 = (3 \cdot 4) \cdot 2,$$

and from Theorem 4.2 above,

$$\frac{2 \cdot 4}{3 \cdot 4} = \frac{2}{3}.$$

This example suggests the following theorem.

Theorem 4.3. If $a, b, c \in J$ $(b, c \neq 0)$, then $\dfrac{a \cdot c}{b \cdot c} = \dfrac{a}{b}$.

For convenience, this theorem will be referred to as the **fundamental principle of fractions.** The symmetric property of equality permits writing this statement as

$$\frac{a}{b} = \frac{a \cdot c}{b \cdot c} \quad (b, c \neq 0).$$

The two forms of Theorem 4.3 can be interpreted to mean:

If the numerator and denominator of a fraction are multiplied or divided by the same non-zero integer, the resulting fraction denotes the same rational number as does the original fraction.

Examples.   a. $\dfrac{2 \cdot 3}{5 \cdot 3} = \dfrac{2}{5}$ b. $\dfrac{4}{7} = \dfrac{4 \cdot 5}{7 \cdot 5}$ c. $\dfrac{x + 2}{3} = \dfrac{(x + 2) \cdot 2}{3 \cdot 2}$

The numerator and denominator of a fraction are frequently referred to as the **terms** of the fraction. If a fraction $\dfrac{a}{b}$ is written as $\dfrac{a \cdot c}{b \cdot c}$, the resulting fraction is in **higher terms.** If a fraction $\dfrac{a \cdot c}{b \cdot c}$ is written as $\dfrac{a}{b}$, the resulting fraction is in **lower terms,** and is said to be **reduced.** A fraction is said to be in **lowest terms** when its numerator and denominator are **relatively prime;** that is, they have no common integral factor except 1 or -1. For convenience, we call such a fraction a **basic fraction.**

A fraction whose numerator and denominator are not relatively prime can be changed to a basic fraction by first writing the numerator and denominator in a factored form in which common factors exist, and then applying the fundamental principle of fractions.

Examples.   a. $\dfrac{12}{20} = \dfrac{3 \cdot 4}{5 \cdot 4} = \dfrac{3}{5}$ b. $\dfrac{6 \cdot x}{9 \cdot x} = \dfrac{2 \cdot 3 \cdot x}{3 \cdot 3 \cdot x} = \dfrac{2}{3}$ $(x \neq 0)$

Diagonal lines are sometimes used to show this procedure. Example a. would then appear as

$$\frac{12}{20} = \frac{\overset{3}{\cancel{12}}}{\underset{5}{\cancel{20}}} = \frac{3}{5}.$$

If you use the format with the diagonal lines, you should simply consider this as an alternate form of the fundamental principle of fractions. Changing fractions to basic fractions should be accomplished mentally wherever possible.

Exercise 4.2

A

Replace the comma in each number pair with the proper symbol, = or ≠.

Examples. a. $\dfrac{3}{4}, \dfrac{6}{8}$　　　　　b. $\dfrac{5}{6}, \dfrac{7}{8}$

Solutions.

a. Because $3 \cdot 8 = 4 \cdot 6$, then $\dfrac{3}{4} = \dfrac{6}{8}$.　b. Because $5 \cdot 8 \neq 6 \cdot 7$, then $\dfrac{5}{6} \neq \dfrac{7}{8}$.

1. $\dfrac{7}{9}, \dfrac{14}{18}$　　2. $\dfrac{4}{7}, \dfrac{12}{21}$　　3. $\dfrac{7}{8}, \dfrac{14}{17}$　　4. $\dfrac{2}{5}, \dfrac{22}{50}$

5. $\dfrac{0}{6}, \dfrac{0}{7}$　　6. $\dfrac{0}{4}, \dfrac{0}{2}$　　7. $\dfrac{1}{3}, \dfrac{0}{2}$　　8. $\dfrac{6}{5}, \dfrac{0}{4}$

9. $\dfrac{2 \cdot 5}{3}, \dfrac{4 \cdot 5}{6}$　　10. $\dfrac{8}{3}, \dfrac{2 \cdot 8}{6}$　　11. $\dfrac{-9}{2}, \dfrac{18}{-4}$　　12. $\dfrac{3}{-10}, \dfrac{-9}{30}$

Apply the fundamental principle of fractions and write each of the following as a basic fraction. $x, y \in J$ $(x, y \neq 0)$.

Examples. a. $\dfrac{3 \cdot 5}{4 \cdot 5}$　　　b. $\dfrac{14}{21}$　　　c. $\dfrac{3x}{15y}$

Solutions. a. $\dfrac{3}{4}$　　　b. $\dfrac{2 \cdot 7}{3 \cdot 7} = \dfrac{2}{3}$　　　c. $\dfrac{x \cdot 3}{5y \cdot 3} = \dfrac{x}{5y}$

13. $\dfrac{3 \cdot 4}{4 \cdot 4}$　　14. $\dfrac{5 \cdot 7}{3 \cdot 7}$　　15. $\dfrac{6 \cdot 2}{5 \cdot 2}$　　16. $\dfrac{7 \cdot 2}{7 \cdot 7}$

17. $\dfrac{9}{15}$　　18. $\dfrac{35}{42}$　　19. $\dfrac{22}{33}$　　20. $\dfrac{72}{84}$

21. $\dfrac{12 \cdot x}{15 \cdot x}$　　22. $\dfrac{65 \cdot y}{117 \cdot y}$　　23. $\dfrac{16 \cdot x}{36 \cdot y}$　　24. $\dfrac{26 \cdot x}{143 \cdot y}$

Write the missing numerator or denominator so that each second fraction will be equal to the first. $x, y \in J$ $(x, y \neq 0)$.

Examples.   a. $\dfrac{2}{3} = \dfrac{?}{6}$ b. $\dfrac{5}{13} = \dfrac{10}{?}$

Solutions.   a. $\dfrac{2}{3} = \dfrac{2 \cdot 2}{3 \cdot 2} = \dfrac{4}{6}$ b. $\dfrac{5}{13} = \dfrac{5 \cdot 2}{13 \cdot 2} = \dfrac{10}{26}$

25. $\dfrac{4}{5} = \dfrac{?}{10}$ 26. $\dfrac{7}{8} = \dfrac{?}{24}$ 27. $\dfrac{0}{2} = \dfrac{?}{36}$ 28. $\dfrac{3}{x} = \dfrac{?}{2 \cdot x}$

29. $\dfrac{3 \cdot x}{7 \cdot y} = \dfrac{9 \cdot x}{?}$ 30. $\dfrac{x}{2} = \dfrac{?}{4 \cdot y}$ 31. $\dfrac{3}{4} = \dfrac{3 \cdot y}{?}$ 32. $\dfrac{5 \cdot x}{4 \cdot y} = \dfrac{?}{12 \cdot y}$

B

33. Which of the following rational numbers are equal to $\dfrac{7}{8}$?

a. $\dfrac{7 + 1}{8 + 1}$ b. $\dfrac{7 - 5}{8 - 5}$ c. $\dfrac{7 \cdot 3}{8 \cdot 3}$

d. $\dfrac{7 \div 4}{8 \div 4}$ e. $\dfrac{7 \cdot (-6)}{8 \cdot (-6)}$ f. $\dfrac{7 \div 7}{8 \div 8}$

34. Which of the following rational numbers are equal to $\dfrac{30}{40}$?

a. $\dfrac{30 \cdot 5}{40 \cdot 5}$ b. $\dfrac{30 - 7}{40 - 7}$ c. $\dfrac{30 + 9}{40 + 9}$

d. $\dfrac{30 \cdot (-2)}{40 \cdot (-2)}$ e. $\dfrac{30 \div 3}{40 \div 4}$ f. $\dfrac{30 \div 7}{40 \div 7}$

35. Supply the missing reasons for the proof of the theorem:

If $a, b, c, d \in J$ $(b, d \neq 0)$ and if $a \cdot d = b \cdot c$, then $\dfrac{a}{b} = \dfrac{c}{d}$.

Statements	Reasons
a. $a \cdot d = b \cdot c$	a. Hypothesis
b. $\dfrac{1}{b} \in Q; \; \dfrac{1}{d} \in Q$	b. _____
c. $\dfrac{1}{b} \cdot \dfrac{1}{d} \in Q$	c. _____

d. $\left(\dfrac{1}{b}\cdot\dfrac{1}{d}\right)\cdot(a\cdot d)=\left(\dfrac{1}{b}\cdot\dfrac{1}{d}\right)\cdot(b\cdot c)$ d. Multiplication law of equality

e. $\left(a\cdot\dfrac{1}{b}\right)\cdot\left(d\cdot\dfrac{1}{d}\right)=\left(c\cdot\dfrac{1}{d}\right)\cdot\left(b\cdot\dfrac{1}{b}\right)$ e. Commutative and associative laws of multiplication

f. $\left(a\cdot\dfrac{1}{b}\right)\cdot 1=\left(c\cdot\dfrac{1}{d}\right)\cdot 1$ f. Multiplicative inverse law

g. $a\cdot\dfrac{1}{b}=c\cdot\dfrac{1}{d}$ g. _____

h. $\dfrac{a}{b}=\dfrac{c}{d}$ h. _____

36. Supply the missing reasons for the proof of the theorem:

If $a,\,b,\,c,\,d\in J$ $(b,\,d\neq 0)$, and if $\dfrac{a}{b}=\dfrac{c}{d}$, then $a\cdot d=b\cdot c$.

Statements	**Reasons**
a. $\dfrac{a}{b}=\dfrac{c}{d}$	a. Hypothesis
b. $a\cdot\dfrac{1}{b}=c\cdot\dfrac{1}{d}$	b. Theorem 4.1
c. $(b\cdot d)\cdot\left(a\cdot\dfrac{1}{b}\right)=(b\cdot d)\cdot\left(c\cdot\dfrac{1}{d}\right)$	c. _____
d. $(a\cdot d)\cdot\left(b\cdot\dfrac{1}{b}\right)=(b\cdot c)\cdot\left(d\cdot\dfrac{1}{d}\right)$	d. Commutative and associative laws of multiplication
e. $(a\cdot d)\cdot 1=(b\cdot c)\cdot 1$	e. _____
f. $a\cdot d=b\cdot c$	f. _____

37. Show that Theorem 4.1 follows from Definition 2.4.

38. Prove the theorem:

If $a,\,b,\,c\in J$ $(b,\,c\neq 0)$, then $\dfrac{a\cdot c}{b\cdot c}=\dfrac{a}{b}$.

4.3 Sums

Consider the sum $\dfrac{2}{7} + \dfrac{3}{7}$. By Theorem 4.1, $\dfrac{2}{7} = 2 \cdot \dfrac{1}{7}$ and $\dfrac{3}{7} = 3 \cdot \dfrac{1}{7}$. Therefore,

$$\frac{2}{7} + \frac{3}{7} = 2 \cdot \frac{1}{7} + 3 \cdot \frac{1}{7}.$$

By applying the right-hand distributive law, we obtain

$$2 \cdot \frac{1}{7} + 3 \cdot \frac{1}{7} = (2 + 3) \cdot \frac{1}{7},$$

and again, from Theorem 4.1

$$(2 + 3) \cdot \frac{1}{7} = \frac{2 + 3}{7} = \frac{5}{7}.$$

This example suggests the following theorem.

Theorem 4.4.   If $a, b, c \in J$ $(c \neq 0)$, then $\dfrac{a}{c} + \dfrac{b}{c} = \dfrac{a + b}{c}$.

A sum such as $\dfrac{2}{3} + \dfrac{3}{4}$ cannot be written directly as a basic fraction by Theorem 4.4 since 1/3 and 1/4 do not have the same denominator. How, then, can we find the basic fraction which represents the sum of two rational numbers whose fractions do not have the same denominator? One or both of the fractions can be replaced by fractions in higher or lower terms so that the resulting fractions have the same denominator. The basic fraction can then be written directly from Theorem 4.4.

It is desirable to raise the terms of the fractions so that the common denominator will be the **least common multiple** of the denominators, that is, the **least common denominator** of the fractions. The least common multiple of two or more natural numbers is the smallest natural number that is exactly divisible (can be divided with no remainder) by each of the given numbers. Thus, 6 is the least common multiple of 2 and 3 because 6 is the smallest natural number exactly divisible by both 2 and 3. Very often the least common multiple of a set of natural numbers can be found by inspection. When inspection fails us, however, we can find the least common multiple as follows:

1. Express each number in prime factor form.
2. Write as factors of a product each different prime factor occurring in any of the numbers, including each factor the greatest number of times it occurs in any of the given numbers.

Examples. Find the least common multiple of:

a. 12 and 15.   b. 8, 18, and 24.

Solutions.   a. The numbers $\longrightarrow$ 12   and   15

appear in prime factor form as $\longrightarrow$ $2 \cdot 2 \cdot 3$ and $3 \cdot 5.$

The least common multiple
contains the factors $\longrightarrow$ $2 \cdot 2 \cdot 3 \cdot 5$
for which the basic numeral is $\longrightarrow$ 60.

Thus, 60 is the least common multiple of 12 and 15 because 60 is the smallest number that is exactly divisible by 12 and 15.

b. The numbers   8,   18,   and   24

contain the factors $\longrightarrow$ $2 \cdot 2 \cdot 2$, $2 \cdot 3 \cdot 3$,   and $2 \cdot 2 \cdot 2 \cdot 3$

The least common multiple
contains the factors $\longrightarrow$ $2 \cdot 2 \cdot 2 \cdot 3 \cdot 3$
for which the basic numeral is $\longrightarrow$ 72.

Thus, the least common multiple of 8, 18, and 24 is 72 because 72 is the smallest number exactly divisible by each of these numbers.

Two (or more) fractions with different denominators can be rewritten as a single fraction by first finding the least common denominator and using the fundamental principle of fractions (Theorem 4.3) to write each fraction as an equivalent fraction with this denominator. We then use Theorem 4.4 to complete the process

Examples.   a. $\dfrac{2}{3} + \dfrac{3}{4}$ b. $\dfrac{7}{12} + \dfrac{2}{15}$

Solutions

a. By inspection we determine the least common denominator to be 12. Therefore,

b. From the example above, we observe that the least common denominator is 60. Therefore,

Solutions continued.

$$\frac{2}{3} + \frac{3}{4} = \frac{2 \cdot (4)}{3 \cdot (4)} + \frac{3 \cdot (3)}{4 \cdot (3)}$$

$$= \frac{8}{12} + \frac{9}{12} = \frac{17}{12}.$$

$$\frac{7}{12} + \frac{2}{15} = \frac{7 \cdot (5)}{12 \cdot (5)} + \frac{2 \cdot (4)}{15 \cdot (4)}$$

$$= \frac{35}{60} + \frac{8}{60} = \frac{43}{60}.$$

Exercise 4.3

A

Find the least common multiple of the elements in each set. $x, y, z \in J.$

Example. $\{8, 12\}$

Solution. In factored form $8 = 2 \cdot 2 \cdot 2$ and $12 = 2 \cdot 2 \cdot 3$. The least common multiple contains the factors 2, 2, 2, and 3. The least common multiple is $2 \cdot 2 \cdot 2 \cdot 3$ or 24.

1. $\{9, 15\}$
2. $\{10, 25\}$
3. $\{15, 35\}$
4. $\{16, 24\}$
5. $\{42, 49\}$
6. $\{33, 88\}$
7. $\{14, 21, 35\}$
8. $\{20, 24, 32\}$
9. $\{2 \cdot x, 3 \cdot y\}$
10. $\{4 \cdot x, 12 \cdot y\}$
11. $\{3 \cdot x, 6 \cdot y, 9 \cdot x \cdot y\}$
12. $\{5 \cdot x, 10 \cdot y, 3 \cdot x \cdot x\}$

Write each sum as a basic fraction or basic numeral.

Example. $\frac{1}{3} + \frac{1}{4} + \frac{5}{12}$

Solution. By inspection, the least common denominator equals 12 and we have

$$\frac{1}{3} + \frac{1}{4} + \frac{5}{12} = \frac{1 \cdot 4}{3 \cdot 4} + \frac{1 \cdot 3}{4 \cdot 3} + \frac{5}{12} = \frac{4 + 3 + 5}{12} = \frac{12}{12} = 1.$$

13. $\frac{3}{5} + \frac{1}{5}$
14. $\frac{1}{8} + \frac{5}{8}$
15. $\frac{5}{7} + \frac{2}{7}$
16. $\frac{2}{9} + \frac{7}{9}$
17. $\frac{2}{3} + \frac{7}{3}$
18. $\frac{3}{4} + \frac{5}{4}$
19. $\frac{2}{5} + \frac{4}{5} + \frac{3}{5}$
20. $\frac{1}{9} + \frac{5}{9} + \frac{7}{9}$
21. $\frac{1}{7} + \frac{3}{14}$

22. $\dfrac{2}{5} + \dfrac{2}{15}$

23. $\dfrac{2}{3} + \dfrac{3}{5}$

24. $\dfrac{8}{5} + \dfrac{4}{7}$

25. $\dfrac{1}{5} + \dfrac{2}{3} + \dfrac{4}{15}$

26. $\dfrac{1}{3} + \dfrac{3}{7} + \dfrac{6}{21}$

27. $\dfrac{3}{4} + \dfrac{1}{8} + \dfrac{7}{12}$

28. $\dfrac{2}{5} + \dfrac{7}{10} + \dfrac{1}{15}$

29. $\dfrac{3}{7} + \dfrac{1}{6} + \dfrac{2}{5}$

30. $\dfrac{2}{11} + \dfrac{1}{3} + \dfrac{4}{5}$

Replace each comma with the proper order symbol $<, =, >$. (*Hint: Change each pair of fractions so that they have the same denominators.*)

31. $\dfrac{8}{5}, \dfrac{17}{10}$

32. $\dfrac{4}{7}, \dfrac{16}{28}$

33. $\dfrac{1}{2}, \dfrac{6}{13}$

34. $\dfrac{7}{3}, \dfrac{9}{4}$

35. $\dfrac{0}{5}, \dfrac{1}{6}$

36. $\dfrac{2}{5}, \dfrac{4}{9}$

37. $\dfrac{3}{11}, \dfrac{4}{15}$

38. $\dfrac{7}{13}, \dfrac{8}{15}$

B

39. Supply the missing reasons for the proof of the theorem:

If $a, b, c \in J$ ($c \neq 0$), then $\dfrac{a}{c} + \dfrac{b}{c} = \dfrac{a+b}{c}$.

Statements	**Reasons**
a. $\dfrac{a}{c} + \dfrac{b}{c} = \dfrac{a}{c} + \dfrac{b}{c}$	a. ————————————
b. $\dfrac{a}{c} = a \cdot \dfrac{1}{c}$; $\dfrac{b}{c} = b \cdot \dfrac{1}{c}$	b. Theorem 4.1
c. $\dfrac{a}{c} + \dfrac{b}{c} = a \cdot \dfrac{1}{c} + b \cdot \dfrac{1}{c}$	c. Substitution law
d. $\dfrac{a}{c} + \dfrac{b}{c} = (a + b) \cdot \dfrac{1}{c}$	d. ————————————
e. $\dfrac{a}{c} + \dfrac{b}{c} = \dfrac{a+b}{c}$	e. ————————————

4.4 Differences

A rational number is the quotient of two integers and can be written as a fraction. Hence the numerator could be either positive or negative (or zero) and the denominator could be either positive or negative. The four possible cases (other than the numerator being equal to zero) lead to two cases; either the signs are alike or they are different. According to the laws pertaining to the signs, discussed in Section 3.4, if the signs are alike the quotient is positive, and if they are unlike the quotient is negative. For example,

$$\frac{+3}{+4} = \frac{-3}{-4} = \frac{3}{4} \quad \text{and} \quad \frac{-3}{4} = \frac{3}{-4} = -\frac{3}{4}.$$

Thus there are three signs associated with the fractional representation of a rational number, namely, the sign of the numerator, the sign of the denominator, and the sign of the fraction itself. The example above suggests the following theorem.

Theorem 4.5.   If $a, b \in W$ $(b \neq 0)$, then

$$\frac{a}{b} = -\frac{-a}{b} = -\frac{a}{-b} = \frac{-a}{-b},$$

and

$$-\frac{a}{b} = \frac{-a}{b} = \frac{a}{-b} = -\frac{-a}{-b}.$$

Theorem 4.5 implies that if any two of the three signs associated with a fraction are changed, the resulting fraction represents the same rational number as the original fraction. Thus, if a fraction has exactly two minus signs, it may be written in the form $\frac{a}{b}$. If a fraction has one or three minus signs, it may be written as $\frac{-a}{b}$. For convenience we will call $\frac{a}{b}$ and $\frac{-a}{b}$ **standard forms of fractions.** It is customary to write the answer to a problem in one of these forms.

As with the integers, the difference of rational numbers $\frac{a}{b} - \frac{c}{d}$ is viewed as the sum $\frac{a}{b} + \left(-\frac{c}{d}\right)$, where the negative of a rational number obeys the additive inverse axiom, $\frac{c}{d} + \left(-\frac{c}{d}\right) = 0$. The following theorem follows directly from the definition of a difference.

Theorem 4.6. If $a, b, c \in J$ ($c \neq 0$), then

$$\frac{a}{c} - \frac{b}{c} = \frac{a - b}{c}.$$

Examples. a. $\dfrac{7}{9} - \dfrac{2}{9} = \dfrac{7 - 2}{9} = \dfrac{5}{9}$

b. $\dfrac{3}{5} - \dfrac{1}{2}$

The least common denominator is 10. Hence,

$$\frac{3}{5} - \frac{1}{2} = \frac{3 \cdot 2}{5 \cdot 2} - \frac{1 \cdot 5}{2 \cdot 5} = \frac{6 - 5}{10} = \frac{1}{10}.$$

Exercise 4.4

A

Write each difference as a basic fraction or basic numeral.

Example. $\dfrac{2}{5} - \dfrac{1}{3}$

Solution. By inspection, the least common denominator is 15 and we have

$$\frac{2}{5} - \frac{1}{3} = \frac{2 \cdot 3}{5 \cdot 3} - \frac{1 \cdot 5}{3 \cdot 5} = \frac{6 - 5}{15} = \frac{1}{15}.$$

1. $\dfrac{5}{7} - \dfrac{2}{7}$

2. $\dfrac{7}{9} - \dfrac{2}{9}$

3. $\dfrac{7}{6} - \dfrac{1}{6}$

4. $\dfrac{15}{11} - \dfrac{4}{11}$

5. $\dfrac{1}{8} - \dfrac{5}{8} + \dfrac{7}{8}$

6. $\dfrac{2}{5} - \dfrac{4}{5} + \dfrac{3}{5}$

7. $\dfrac{1}{6} - \dfrac{5}{12}$

8. $\dfrac{3}{5} - \dfrac{4}{15}$

9. $\dfrac{2}{7} - \dfrac{4}{3}$

10. $\dfrac{7}{2} - \dfrac{4}{5}$

11. $\dfrac{3}{4} + \dfrac{5}{8} - \dfrac{1}{32}$

12. $\dfrac{2}{3} - \dfrac{11}{7} + \dfrac{1}{21}$

13. $\dfrac{3}{5} - \dfrac{9}{10} + \dfrac{4}{15}$ 14. $\dfrac{1}{4} - \dfrac{5}{8} + \dfrac{11}{12}$ 15. $\dfrac{3}{2} + \dfrac{1}{6} - \dfrac{7}{18}$

16. $\dfrac{5}{3} + \dfrac{2}{9} - \dfrac{7}{27}$ 17. $\dfrac{1}{11} - \dfrac{2}{3} - \dfrac{9}{5}$ 18. $\dfrac{5}{7} - \dfrac{1}{6} - \dfrac{8}{5}$

Replace each comma with the proper symbol, = or ≠, to make a true statement. Assume no numerator or denominator equals 0.

19. $\dfrac{-7}{-3}, -\dfrac{7}{3}$ 20. $\dfrac{-3}{4}, -\dfrac{3}{4}$ 21. $\dfrac{-5}{8}, \dfrac{-5}{-8}$

22. $\dfrac{x}{-3}, -\dfrac{x}{-3}$ 23. $\dfrac{(x+y)}{-z}, \dfrac{-(x+y)}{-z}$ 24. $\dfrac{3 \cdot x}{5 \cdot y}, \dfrac{-3 \cdot (-x)}{-5 \cdot y}$

B

Justify each statement without using Theorem 4.5. $a, b \in W$ ($b \neq 0$).

25. $\dfrac{-a}{-b} = \dfrac{a}{b}$ 26. $-\dfrac{-a}{-b} = -\dfrac{a}{b}$ 27. $\dfrac{-a}{b} = -\dfrac{a}{b}$

28. $\dfrac{a}{-b} = -\dfrac{a}{b}$ 29. $-\dfrac{-a}{b} = \dfrac{a}{b}$ 30. $-\dfrac{a}{-b} = \dfrac{a}{b}$

31. Supply the missing reasons for the proof of the theorem:

If $a, b, c \in J$ ($c \neq 0$), then $\dfrac{a}{c} - \dfrac{b}{c} = \dfrac{a - b}{c}$.

	Statements		**Reasons**
a.	$\dfrac{a}{c} - \dfrac{b}{c} = \dfrac{a}{c} + \left(-\dfrac{b}{c}\right)$	a.	Theorem 3.4
b.	$\dfrac{a}{c} - \dfrac{b}{c} = \dfrac{a}{c} + \left(\dfrac{-b}{c}\right)$	b.	Theorem 4.5
c.	$\dfrac{a}{c} - \dfrac{b}{c} = \dfrac{a + (-b)}{c}$	c. _____
d.	$\dfrac{a}{c} - \dfrac{b}{c} = \dfrac{a - b}{c}$	d. _____

4.5 Products and Quotients

We now consider the operations of multiplication and division on elements in the set of rational numbers and inquire how a product or a quotient can be written as a basic fraction.

For example, consider the product $\frac{2}{3} \cdot \frac{4}{5}$. Using Theorem 4.1, the product $\frac{2}{3} \cdot \frac{4}{5}$ can be rewritten as

$$\left(2 \cdot \frac{1}{3}\right) \cdot \left(4 \cdot \frac{1}{5}\right).$$

Since $(3 \cdot 5) \cdot \frac{1}{3 \cdot 5} = 1$, we can write

$$\frac{2}{3} \cdot \frac{4}{5} = \left(2 \cdot \frac{1}{3}\right) \cdot \left(4 \cdot \frac{1}{5}\right) \cdot \left[(3 \cdot 5) \cdot \frac{1}{3 \cdot 5}\right]$$

We can now use the associative and commutative laws of multiplication to rewrite the right-hand member and obtain

$$\frac{2}{3} \cdot \frac{4}{5} = (2 \cdot 4) \cdot \frac{1}{3 \cdot 5} \cdot \left(3 \cdot \frac{1}{3}\right) \cdot \left(5 \cdot \frac{1}{5}\right).$$

Since by the multiplicative inverse law, $3 \cdot \frac{1}{3}$ and $5 \cdot \frac{1}{5}$ each equals 1, we have

$$\frac{2}{3} \cdot \frac{4}{5} = (2 \cdot 4) \cdot \frac{1}{3 \cdot 5} \cdot 1 \cdot 1,$$

and by the identity law of multiplication,

$$\frac{2}{3} \cdot \frac{4}{5} = 2 \cdot 4 \cdot \frac{1}{3 \cdot 5}. \tag{1}$$

Finally, by Theorem 4.1, $2 \cdot 4 \cdot \frac{1}{3 \cdot 5} = \frac{2 \cdot 4}{3 \cdot 5}$, and by substituting in the right member of (1) we have

$$\frac{2}{3} \cdot \frac{4}{5} = \frac{2 \cdot 4}{3 \cdot 5} = \frac{8}{15}.$$

The above example suggests that if the product $\frac{a}{b} \cdot \frac{c}{d}$ $(b, d \neq 0)$ is to be consistent with previous definitions, axioms, and theorems, we must have the following theorem.

Theorem 4.7.   If $\dfrac{a}{b}, \dfrac{c}{d} \in Q$ $(b, d \neq 0)$, then $\dfrac{a}{b} \cdot \dfrac{c}{d} = \dfrac{a \cdot c}{b \cdot d}$.

Examples.

a. $\dfrac{2}{3} \cdot \dfrac{5}{7} = \dfrac{2 \cdot 5}{3 \cdot 7} = \dfrac{10}{21}$ b. $\dfrac{x}{7} \cdot \dfrac{y}{3} = \dfrac{x \cdot y}{7 \cdot 3} = \dfrac{x \cdot y}{21}$ $(x, y \in J)$

The fraction resulting from rewriting

$$\dfrac{a}{b} \cdot \dfrac{c}{d} \quad \text{as} \quad \dfrac{a \cdot c}{b \cdot d}$$

is not a basic fraction if there are any factors common to either numerator and either denominator in the original expression. For example,

$$\dfrac{3}{4} \cdot \dfrac{2}{5} = \dfrac{3 \cdot 2}{4 \cdot 5} = \dfrac{6}{20} .$$

The fraction 6/20 is not a basic fraction since both 6 and 20 contain the common factor 2. It can be written as a basic fraction by factoring the numerator and denominator and invoking the fundamental principle of fractions. Thus,

$$\dfrac{6}{20} = \dfrac{3 \cdot 2}{10 \cdot 2} = \dfrac{3}{10} .$$

   An alternate approach to obtaining a basic fraction for a product of two fractions is to apply the fundamental principle before multiplying.

Example.   Express $\dfrac{15}{8} \cdot \dfrac{4}{55}$ as a basic fraction.

Solution.   We first factor each numerator and denominator to obtain

$$\dfrac{3 \cdot 5}{2 \cdot 2 \cdot 2} \cdot \dfrac{2 \cdot 2}{5 \cdot 11} .$$

We now observe that 5 is a common factor of both a numerator and a denominator, and that $2 \cdot 2$ is also a common factor. Thus, by applying the fundamental principle, we can write the product as

$$\dfrac{3}{2} \cdot \dfrac{1}{11} ,$$

and then as

$$\frac{3 \cdot 1}{2 \cdot 11} = \frac{3}{22}.$$

Now let us consider a quotient of two rational numbers. First we consider a quotient such as $\dfrac{1}{\frac{5}{7}}$ (or $1 \div \frac{5}{7}$) in which 1 is divided by a rational number. How can we rewrite such a quotient as a basic fraction? Observe that $\dfrac{5}{7} \cdot \dfrac{7}{5} = \dfrac{35}{35} = 1$, and hence, $\dfrac{7}{5}$ is the multiplicative inverse of $\dfrac{5}{7}$. Since $\dfrac{1}{\frac{5}{7}}$ is also the multiplicative inverse of $\dfrac{5}{7}$, it must be true that

$$\frac{1}{\frac{5}{7}} = \frac{7}{5},$$

because the multiplicative inverse of a number is unique. This numerical example suggests the following theorem.

Theorem 4.8.   If $a, b \in J$ $(a, b \neq 0)$, then $\dfrac{1}{\frac{a}{b}} = \dfrac{b}{a}$.

Examples.   a. $\dfrac{1}{\frac{2}{3}} = \dfrac{3}{2}$ b. $\dfrac{1}{\frac{5}{4}} = \dfrac{4}{5}$ c. $1 \div \dfrac{9}{13} = \dfrac{13}{9}$

We can now proceed to quotients of rational numbers in which the dividend (numerator) is not 1. For example, consider the quotient $(2/3)/(7/5)$. By applying the fundamental principle of fractions and multiplying the numerator and denominator by 5/7, we have,

$$\frac{\frac{2}{3}}{\frac{7}{5}} = \frac{\frac{2}{3} \cdot \frac{5}{7}}{\frac{7}{5} \cdot \frac{5}{7}} = \frac{\frac{2}{3} \cdot \frac{5}{7}}{1} = \frac{2}{3} \cdot \frac{5}{7}.$$

The preceding example suggests the following theorem.

Theorem 4.9. If $\dfrac{a}{b}, \dfrac{c}{d} \in Q \left(\dfrac{c}{d} \neq 0 \right)$, then $\dfrac{\dfrac{a}{b}}{\dfrac{c}{d}} = \dfrac{a}{b} \cdot \dfrac{d}{c}$.

Examples.   a. $\dfrac{\dfrac{7}{8}}{\dfrac{3}{5}} = \dfrac{7}{8} \cdot \dfrac{5}{3}$ b. $\dfrac{1}{2} \div \dfrac{1}{4} = \dfrac{1}{2} \cdot \dfrac{4}{1}$

$$= \dfrac{7 \cdot 5}{8 \cdot 3} = \dfrac{35}{24} \qquad\qquad = \dfrac{1 \cdot 4}{2 \cdot 1} = \dfrac{4}{2} = \dfrac{2 \cdot 2}{2} = 2$$

Because the set of rational numbers is closed with respect to multiplication, and because the quotient of two rational numbers,

$$\dfrac{a}{b} \div \dfrac{c}{d} = \dfrac{a}{b} \cdot \dfrac{d}{c} \qquad \left(\dfrac{c}{d} \neq 0 \right),$$

the product of two rational numbers, *the set of rational numbers is closed with respect to division, with the exception that elements in the set cannot be divided by 0.*

Note that Theorem 4.9 is consistent with the definition of a quotient in the set of integers, since by Theorem 4.1, if $a, b \in J$ ($b \neq 0$), the quotient

$$\dfrac{a}{b} = a \cdot \dfrac{1}{b},$$

the product of the numerator and the reciprocal of the denominator.

Exercise 4.5

A

Express each product as a basic fraction. $x, y \in J.$

Examples.   a. $\dfrac{3}{4} \cdot \dfrac{5}{7}$ b. $\dfrac{2}{3} \cdot \dfrac{6}{7}$

Solutions. a. $\dfrac{3}{4} \cdot \dfrac{5}{7} = \dfrac{3 \cdot 5}{4 \cdot 7} = \dfrac{15}{28}$ b. $\dfrac{2}{3} \cdot \dfrac{6}{7} = \dfrac{2 \cdot 6}{3 \cdot 7} = \dfrac{2 \cdot 2 \cdot 3}{7 \cdot 3} = \dfrac{4}{7}$

1. $\dfrac{4}{7} \cdot \dfrac{1}{3}$ 2. $\dfrac{5}{8} \cdot \dfrac{2}{9}$ 3. $\dfrac{3}{5} \cdot \dfrac{6}{7}$ 4. $\dfrac{2}{3} \cdot \dfrac{11}{12}$

5. $\dfrac{x}{5} \cdot \dfrac{2}{3}$ 6. $\dfrac{4}{5} \cdot \dfrac{y}{7}$ 7. $\dfrac{x}{6} \cdot \dfrac{y}{6}$ 8. $\dfrac{x}{3} \cdot \dfrac{y}{8}$

9. $\dfrac{7}{13} \cdot \left(\dfrac{-5}{9}\right)$ 10. $\dfrac{-21}{17} \cdot \dfrac{13}{6}$ 11. $\dfrac{-4}{5} \cdot \left(\dfrac{-3}{2}\right)$ 12. $\dfrac{-2}{7} \cdot \left(\dfrac{-6}{5}\right)$

13. $\dfrac{-3}{5} \cdot \dfrac{x}{4}$ 14. $\dfrac{-x}{3} \cdot \dfrac{5}{2}$ 15. $\dfrac{-x}{7} \cdot \left(\dfrac{-1}{4}\right)$ 16. $\dfrac{-3}{8} \cdot \left(\dfrac{-x}{5}\right)$

17. $\left(\dfrac{1}{2} \cdot \dfrac{1}{2}\right) \cdot \dfrac{1}{2}$ 18. $\dfrac{3}{4} \cdot \left(\dfrac{3}{4} \cdot \dfrac{3}{4}\right)$ 19. $\dfrac{-2}{5} \cdot \left(\dfrac{2}{5} \cdot \dfrac{1}{5}\right)$ 20. $\dfrac{-4}{3} \cdot \left(\dfrac{4}{3} \cdot \dfrac{1}{3}\right)$

Express the reciprocal of each rational number as a basic fraction. If none exists, so state. $x, y \in J.$

Examples. a. 5 b. $\dfrac{2}{3}$ c. $\dfrac{x}{2}$

Solutions. a. $\dfrac{1}{5}$ b. $\dfrac{3}{2}$ c. $\dfrac{2}{x}$ $(x \neq 0)$

21. 4 22. -7 23. $\dfrac{-3}{5}$ 24. $\dfrac{9}{2}$

25. $\dfrac{3}{4}$ 26. $\dfrac{5}{2}$ 27. $\dfrac{y}{5}$ 28. $\dfrac{6}{x}$ $(x \neq 0)$

29. $\dfrac{x}{y}$ $(y \neq 0)$ 30. $\dfrac{-y}{x}$ $(x \neq 0)$ 31. $x + 2$ 32. $\dfrac{1}{y-3}$ $(y \neq 3)$

Write each quotient as a basic fraction. $x, y \in J.$ *State any restrictions on the variables.*

Examples. a. $\dfrac{5}{8} \div \dfrac{2}{3} = \dfrac{5}{8} \cdot \dfrac{3}{2} = \dfrac{15}{16}$ b. $\dfrac{x}{2} \div \dfrac{y}{3} = \dfrac{x}{2} \cdot \dfrac{3}{y} = \dfrac{3 \cdot x}{2 \cdot y}$ $(y \neq 0)$

33. $\dfrac{5}{8} \div \dfrac{1}{3}$ 34. $\dfrac{4}{7} \div \dfrac{5}{9}$ 35. $\dfrac{2}{3} \div \dfrac{3}{5}$ 36. $\dfrac{3}{7} \div \dfrac{11}{12}$

37. $\dfrac{x}{6} \div \dfrac{5}{9}$ 38. $\dfrac{y}{4} \div \dfrac{7}{8}$ 39. $\dfrac{x}{3} \div \dfrac{y}{4}$ 40. $\dfrac{3 \cdot x}{8} \div \dfrac{y}{5}$

41. $\dfrac{9}{13} \div \dfrac{5 \cdot y}{7}$ 42. $\dfrac{6}{17} \div \dfrac{13 \cdot x}{21}$ 43. $\dfrac{x}{11} \div \dfrac{y}{4}$ 44. $\dfrac{y}{6} \div \dfrac{x}{7}$

Factor each expression as shown. $x, y \in J$. *State any restrictions on the variables.*

Example. $\dfrac{1}{2} \cdot x + \dfrac{3}{2} \cdot y = \dfrac{1}{2} \cdot (? + ?)$

Solution. Using Theorem 4.1 and the distributive law, we obtain

$$\dfrac{1}{2} \cdot x + \dfrac{3}{2} \cdot y = \dfrac{1}{2} \cdot x + \dfrac{1}{2} \cdot 3 \cdot y$$

$$= \dfrac{1}{2} \cdot (x + 3 \cdot y).$$

45. $\dfrac{2}{3} \cdot x + \dfrac{2}{3} \cdot y = \dfrac{2}{3} \cdot (? + ?)$ 46. $\dfrac{3}{5} \cdot x + \dfrac{3}{5} \cdot y = \dfrac{3}{5} \cdot (? + ?)$

47. $\dfrac{5}{8} \cdot x - \dfrac{1}{8} \cdot y = \dfrac{1}{8} \cdot (? - ?)$ 48. $\dfrac{4}{7} \cdot x - \dfrac{6}{7} \cdot y = \dfrac{1}{7} \cdot (? - ?)$

49. $\dfrac{3}{4} \cdot x + \dfrac{5}{12} \cdot y = \dfrac{1}{12} \cdot (? + ?)$ 50. $\dfrac{3}{2} \cdot x + \dfrac{7}{10} \cdot y = \dfrac{1}{10} \cdot (? + ?)$

B

51. What restriction must be placed on the elements of Q in order to assure that a quotient of two elements in the set exists?
52. What property (resulting from the quotient of two elements) does Q have that is not a property of either W or J?

4.6 Equations Involving Rational Numbers

Some equations that do not have solutions in the set of integers may have solutions in the set of rational numbers. Consider the equation $3x = 2$. There is no integer replacement for x such that the product of 3 and the integer is equal to 2. However, observe that

$$3 \cdot \dfrac{2}{3} = 2.$$

Hence, the solution of $3x = 2$ is the rational number $2/3$.

Exercise 4.6

A

Write the solution set of each equation by inspection.

Examples. a. The solution set of $3 \cdot x = 1$ is $\left\{\dfrac{1}{3}\right\}$ because $3 \cdot \left(\dfrac{1}{3}\right) = 1$

b. The solution set of $3 \cdot x = -5$ is $\left\{\dfrac{-5}{3}\right\}$ because $3 \cdot \left(\dfrac{-5}{3}\right) = -5$.

1. $2 \cdot x = 2$ 2. $5 \cdot x = 5$ 3. $7 \cdot x = 2$ 4. $3 \cdot x = 8$

5. $4 \cdot x = -3$ 6. $6 \cdot x = -9$ 7. $-7 \cdot x = 4$ 8. $-8 \cdot x = 13$

Example. The solution set of $x + \dfrac{1}{5} = \dfrac{3}{5}$ is $\left\{\dfrac{2}{5}\right\}$ because $\dfrac{2}{5} + \dfrac{1}{5} = \dfrac{3}{5}$.

9. $x + \dfrac{1}{3} = \dfrac{5}{3}$ 10. $x + \dfrac{2}{5} = \dfrac{6}{5}$ 11. $x + \dfrac{3}{4} = \dfrac{8}{4}$

12. $x + \dfrac{4}{7} = \dfrac{9}{7}$ 13. $x - \dfrac{1}{2} = \dfrac{4}{2}$ 14. $x - \dfrac{2}{3} = \dfrac{4}{3}$

15. $x - \dfrac{5}{4} = \dfrac{1}{4}$ 16. $x - \dfrac{3}{7} = \dfrac{5}{7}$ 17. $x + \dfrac{3}{2} = 0$

18. $x + \dfrac{3}{4} = 0$ 19. $x - \dfrac{5}{8} = 0$ 20. $x - \dfrac{7}{4} = 0$

21. $x + \dfrac{1}{2} = -\dfrac{1}{2}$ 22. $x + \dfrac{2}{3} = -\dfrac{2}{3}$ 23. $x - \dfrac{3}{4} = -\dfrac{2}{4}$

24. $x - \dfrac{8}{7} = -\dfrac{3}{7}$ 25. $\dfrac{7}{5} - x = \dfrac{1}{5}$ 26. $\dfrac{9}{7} - x = \dfrac{3}{7}$

B

Example. The solution set of $x + \dfrac{1}{3} = \dfrac{5}{4}$ can be obtained by first writing the equation equivalently as

$$x + \dfrac{4}{12} = \dfrac{15}{12}.$$

By inspection, the solution set is $\left\{\dfrac{11}{12}\right\}$ because $\dfrac{11}{12} + \dfrac{4}{12} = \dfrac{15}{12}$.

27. $x + \dfrac{2}{3} = \dfrac{5}{6}$ 　　　　28. $x + \dfrac{3}{4} = \dfrac{11}{12}$ 　　　　29. $x + \dfrac{1}{2} = \dfrac{4}{3}$

30. $x + \dfrac{4}{5} = \dfrac{5}{4}$ 　　　　31. $x - \dfrac{1}{4} = \dfrac{7}{12}$ 　　　　32. $x - \dfrac{3}{2} = \dfrac{5}{6}$

4.7 Decimal Notation

You learned in arithmetic that a **decimal fraction** is another way of writing a fraction in which the denominator is a number such as 10, 100, 1000, etc. Thus $0.3 = \dfrac{3}{10}$, $0.25 = \dfrac{25}{100}$, and $0.705 = \dfrac{705}{1000}$. Since each of the right-hand members is the quotient of two integers, we see that at least some decimal fractions are representations of rational numbers.

Each of the decimal fractions, or more simply, **decimals,** has a finite number of digits to the right of the decimal point. Decimals of this kind are called **terminating decimals.** *A terminating decimal is one way of representing certain rational numbers.*

Now consider the rational number $\dfrac{3}{11}$. If we attempt to convert this to decimal form, we obtain

$$
\begin{array}{r}
0.2727 \\
11\overline{)3.0000} \\
2\,2 \\
\hline
80 \\
77 \\
\hline
30 \\
22 \\
\hline
80 \\
77 \\
\hline
3
\end{array}
$$

Notice that the only remainders are 8 and 3 and that the division is never exact so that the group of digits, 27, will continue to repeat itself. A decimal whose digits are repeated continuously in some particular pattern of grouping is called a **repeating decimal,** or **periodic decimal.** *A repeating decimal is one way of representing certain rational numbers.* Since the digits repeat themselves continuously, a repeating decimal is a **nonterminating decimal.**

Since a repeating decimal is nonterminating, we need some symbol to indicate that the digits do repeat. One way of designating that a decimal is repeating is to place a bar over the repeating group of digits.

Example. $0.27\overline{27}$, $4.132\overline{32}$, and $1.234567\overline{567}$ are repeating decimals.

Some decimals are neither nonterminating nor repeating. While we do not do so here, it can be shown that such decimals do not represent rational numbers.

Any fraction of the form $\frac{a}{b}$ $(b \neq 0)$ can be written as a decimal by performing the indicated division.

Examples.   a. $\frac{7}{20}$ b. $\frac{2}{11}$

Solutions

$$\begin{array}{r} 0.35 \\ \text{a. } 20\overline{)7.00} \\ 6\,0 \\ \hline 1\,00 \\ 1\,00 \\ \hline 0 \end{array}$$

Hence, $\frac{7}{20} = 0.35$.

$$\begin{array}{r} 0.1818\cdots \\ \text{b. } 11\overline{)2.0000} \\ 1\,1 \\ \hline 90 \\ 88 \\ \hline 20 \\ 11 \\ \hline 90 \\ 88 \\ \hline 2 \end{array}$$

Hence, $\frac{2}{11} = 0.18\overline{18}$.

If a remainder of 0 is obtained, the decimal fraction will terminate as in Example a above. When a remainder appears for a second time in the division process, digits in the quotient will start to repeat as in Example b above. Observe that 9 and 2 appear as remainders for a second time.

It can be shown (although we shall not do so) that every terminating decimal and every periodic decimal is a rational number and that every rational number can be represented as a terminating decimal or as a periodic decimal.

A *terminating* decimal can be written directly as a fraction in the form a/b where a and b denote integers.

Examples.   a. 0.123   b. 0.3   c. 0.0003

Solutions.   a. $\dfrac{123}{1000}$ b. $\dfrac{3}{10}$ c. $\dfrac{3}{10000}$

A repeating decimal can also be written as a fraction of the form a/b, but since the procedure involves techniques of equation solving that we have not considered, we do not discuss it here.

   You have seen that rational numbers can be associated with points on the number line. For example, the number 1/2 is associated with the point half-way between the points associated with 0 and 1; the number 1/4 is associated with the point half-way between the points associated with 0 and 1/2; etc. If this process is continued indefinitely, it is obvious that the graphs of some rational numbers can be placed very close together. In fact, it can be shown that there is at least one rational number between any two given rational numbers no matter how close together they may be. This property of the rational numbers is called **denseness.** However, as you will see in Chapter 5, the points associated with the rational numbers *do not* completely "fill" the number line.

Exercise 4.7

A

Express each rational number in decimal notation.

Examples.   a. $\dfrac{1}{2}$ b. $\dfrac{1}{3}$ c. $\dfrac{23}{8}$

Solutions.   a. $2\overline{|1.0}$ with 0.5 above   b. $3\overline{|1.00}\cdots$ with $0.3\overline{3}$ above   c. $8\overline{|23.000}$ with 2.875 above

          $\dfrac{1}{2} = 0.5$ $\dfrac{1}{3} = 0.3\overline{3}$ $\dfrac{23}{8} = 2.875$

1. $\dfrac{1}{4}$ 2. $\dfrac{1}{5}$ 3. $\dfrac{3}{4}$ 4. $\dfrac{4}{5}$

5. $\dfrac{5}{8}$ 6. $\dfrac{3}{16}$ 7. $\dfrac{21}{8}$ 8. $\dfrac{17}{16}$

9. $\dfrac{4}{9}$ 10. $\dfrac{5}{6}$ 11. $\dfrac{13}{3}$ 12. $\dfrac{22}{7}$

13. $\dfrac{52}{11}$ 14. $\dfrac{63}{13}$ 15. $\dfrac{111}{25}$ 16. $\dfrac{217}{42}$

Express each rational number in the form $\dfrac{a}{b}$, $a, b \in J.$

Examples.   a. $0.321 = \dfrac{321}{1000}$ b. $0.75 = \dfrac{75}{100} = \dfrac{3}{4}$

17. 0.7 18. 0.77 19. 0.2 20. 0.25

21. 0.125 22. 0.875 23. 0.0625 24. 0.1875

CHAPTER SUMMARY

4.1   An extension of the set of integers, which includes every quotient a/b, $a, b \in J$ $(b \neq 0)$ is called the **set of rational numbers, Q.** The axioms for the set of integers are adopted for the set of rational numbers. In addition the **multiplicative inverse law** is adopted:

For $\dfrac{a}{b} \in Q \left(\dfrac{a}{b} \neq 0 \right)$, *there exists a rational number,* $\dfrac{1}{\frac{a}{b}}$, *such that*

$$\frac{a}{b} \cdot \frac{1}{\frac{a}{b}} = 1.$$

In the set of rational numbers, the quotient $\dfrac{a}{b}$, $a, b \in J$ $(b \neq 0)$ is equal to $a \cdot \dfrac{1}{b}$.

4.2–4.5   The following theorems are applicable to the set of rational numbers. The variables $a, b, c, d \in Q$ and no denominator equals 0.

$\dfrac{a}{b} = \dfrac{c}{d}$ if and only if $a \cdot d = b \cdot c$ Theorem 4.2

$\dfrac{a \cdot c}{b \cdot c} = \dfrac{a}{b}$ or $\dfrac{a}{b} = \dfrac{a \cdot c}{b \cdot c}$ Theorem 4.3

$$\frac{a}{c} + \frac{b}{c} = \frac{a+b}{c}$$ Theorem 4.4

$$\frac{a}{b} = -\frac{-a}{b} = -\frac{a}{-b} = \frac{-a}{-b} ; \quad -\frac{a}{b} = \frac{-a}{b} = \frac{a}{-b} = -\frac{-a}{-b}$$

Theorem 4.5

$$\frac{a}{c} - \frac{b}{c} = \frac{a-b}{c}$$ Theorem 4.6

$$\frac{a}{b} \cdot \frac{c}{d} = \frac{a \cdot c}{b \cdot d}$$ Theorem 4.7

$$\frac{1}{\frac{a}{b}} = \frac{b}{a}$$ Theorem 4.8

$$\frac{a}{b} \div \frac{c}{d} = \frac{a}{b} \cdot \frac{d}{c}$$ Theorem 4.9

4.6 Some equations that do not have solutions in the set of integers do have solutions in the set of rational numbers.

4.7 Rational numbers can be represented by fractions in the form $\frac{a}{b}$, $a, b \in J$, or by **terminating** or **periodic decimal fractions**.

CHAPTER REVIEW

A

4.1 *Find the multiplicative inverse of each of the following.*

1. 5

2. $\frac{1}{8}$

3. $x + y \quad (x \neq -y)$

4. $\frac{1}{x+y} \quad (x \neq -y)$.

4.2 *Replace the comma in each number pair with the proper symbol,* $=$ *or* $\neq$. $x, y \in J \quad (x, y \neq 0)$.

5. $\frac{4}{7}, \frac{20}{35}$

6. $\frac{3 \cdot 5}{8}, \frac{6 \cdot 5}{2 \cdot 8}$

Write the basic fraction for each quotient. $x, y \in J$ $(x, y \neq 0)$.

7. $\dfrac{12}{30}$

8. $\dfrac{30 \cdot x}{36 \cdot y}$

Write the missing numerator so that the second fraction is equal to the first. $x, y \in J$ $(x, y \neq 0)$.

9. $\dfrac{3}{7} = \dfrac{?}{42}$

10. $\dfrac{2 \cdot x}{3 \cdot y} = \dfrac{?}{12 \cdot y}$

4.3 *Find the least common multiple of each set of integers* $(x, y \neq 0)$.

11. $\{12, 30, 40\}$

12. $\{4 \cdot x, \; 10 \cdot y, \; 15 \cdot y\}$

4.4–4.5 *Write each expression as a basic fraction or basic numeral.* $x, y \in J$ $(x, y, \neq 0)$.

13. $\dfrac{5}{7} + \dfrac{1}{7}$

14. $\dfrac{3}{11} + \dfrac{x}{11}$

15. $\dfrac{5}{6} - \dfrac{3}{4} + \dfrac{1}{5}$

16. $\dfrac{1}{2} + \dfrac{x}{7} - \dfrac{y}{21}$

17. $\dfrac{3}{7} \cdot \dfrac{4}{5}$

18. $\dfrac{-4}{x} \cdot \dfrac{y}{7}$

19. $\dfrac{4}{3} \div \dfrac{2}{3}$

20. $\dfrac{-x}{5} \div \dfrac{y}{7}$

4.6 *Find the solution set of each equation.*

21. $2 \cdot x = 3$

22. $3 \cdot x = -4$

23. $x + \dfrac{7}{8} = \dfrac{9}{8}$

24. $x - \dfrac{3}{5} = \dfrac{1}{5}$

25. $x - \dfrac{9}{2} = 0$

26. $x - \dfrac{1}{7} = \dfrac{-5}{7}$

4.7 *Express each rational number in decimal notation.*

27. $\dfrac{5}{8}$

28. $\dfrac{64}{9}$

Express each rational number in the form a/b, $a, b \in J$.

29. 0.85

30. 0.423

B

31. Which of the following rational numbers equal $\dfrac{13}{17}$?

a. $\dfrac{13 \cdot 2}{17 \cdot 2}$

b. $\dfrac{13 - 3}{17 - 3}$

c. $\dfrac{13 + 4}{17 + 4}$

d. $\dfrac{13 \cdot (-5)}{17 \cdot (-5)}$

e. $\dfrac{13 \div 6}{17 \div 6}$

f. $\dfrac{13 \div (-7)}{17 \div (-7)}$

32. Which of the following rational numbers have the same graph?

$$\frac{3}{2}, \frac{6}{4}, \frac{12}{10}, \frac{12}{8}, \frac{18}{12}, \frac{18}{14}, \frac{24}{16}$$

Find the solution set of each equation.

33. $x + \dfrac{3}{2} = \dfrac{6}{5}$

34. $x - \dfrac{4}{5} = \dfrac{-1}{12}$

5
chapter

The Set of Real Numbers

In Section 4.7 we noted that the set of rational numbers is dense, but that the graphs of the elements in the set do not completely fill the number line—there are points on the number line that do not correspond to rational numbers.

In this chapter we first consider a new way to represent rational numbers, and then extend this notation to the representation of numbers that can be associated with points on the number line and that are not rational.

5.1 Radical Notation

A product in which all of the factors are the same number is called a **power** of that number. Some of these powers have special names. For example, the product of a number multiplied by itself is called a **square.**

Examples.   a. $2 \cdot 2$ is the square of 2.

b. $(-3)(-3)$ is the square of -3.

c. $x \cdot x$ is the square of x.

Sometimes we need to find a number whose square is a given number. In this case each of the two equal factors whose product is the given number is called a **square root** of the given number. For example, 4 is a square root of 16 because $4 \cdot 4$ is 16.

123

This concept can be extended to a number expressed as the product of any number of equal factors. Because $(-3) \cdot (-3) \cdot (-3) = -27$, we say that $(-3) \cdot (-3) \cdot (-3)$ or -27 is the **cube** of -3, and -3 is a **cube root** of -27; because $2 \cdot 2 \cdot 2 \cdot 2 = 16$, we say that 16 is the **fourth power** of 2, and 2 is a **fourth root** of 16, and so on. In this chapter we shall limit our discussion to square roots of positive numbers.

Because a square root of a number is one of two *equal* factors, the two factors must be both positive or both negative. Since the product of two positive numbers and the product of two negative numbers is always positive, a positive number will have two square roots, one positive and one negative. For example, because $4 \cdot 4 = 16$ and $(-4) \cdot (-4) = 16$, both 4 and -4 are square roots of 16.

Symbolically, the *positive* square root of a number, such as 16, is represented by $\sqrt{16}$. An expression such as this is called a **radical expression.** The symbol "$\sqrt{}$" is called a **radical sign** and the number under the radical sign is called the **radicand.** In general we have the following definition.

Definition 5.1.   For all $a \in Q$, $a \geq 0$,

$\sqrt{a}$ is the non-negative number such that $\sqrt{a} \cdot \sqrt{a} = a$.

To indicate the *negative* square root of a number, a minus sign $(-)$ is placed in front of the radical sign.

Examples.   a. $\sqrt{16} = 4$ b. $\sqrt{25} = 5$

c. $-\sqrt{16} = -4$ d. $-\sqrt{25} = -5$

We now have another symbol for a rational number. Thus 4 and $\sqrt{16}$ are different symbols for the same rational number. Similarly, $\sqrt{4/25}$ and $2/5$ are different symbols for the same rational number. How can we determine if a square root of a number is a rational number? This is not always easy to do. You probably know that the square roots of 1, 4, 9, 16, 25, 36, 49, 81, and 100 are rational numbers and that the square roots of all other natural numbers less than 100 are not rational numbers. The perfect squares equal to and less than 100 and their square roots are shown in color inside the back cover.

Exercise 5.1

A

Write each square root as a basic numeral or as a basic fraction.

Examples. a. $\sqrt{144} = 12$ b. $-\sqrt{64} = -8$

1. $\sqrt{9}$ 2. $\sqrt{25}$ 3. $\sqrt{49}$ 4. $\sqrt{64}$

5. $-\sqrt{4}$ 6. $-\sqrt{36}$ 7. $-\sqrt{81}$ 8. $-\sqrt{121}$

9. $\sqrt{\dfrac{4}{9}}$ 10. $\sqrt{\dfrac{16}{25}}$ 11. $-\sqrt{\dfrac{49}{81}}$ 12. $-\sqrt{\dfrac{36}{100}}$

Express each rational number as the square root of a rational number.

Examples. a. 5   b. -3 c. $\dfrac{3}{4}$

Solutions. a. $5 = \sqrt{25}$ b. $-3 = -\sqrt{9}$ c. $\dfrac{3}{4} = \sqrt{\dfrac{9}{16}}$

13. 9   14. 11   15. 14   16. 18

17. -5 18. -7 19. -10 20. -14

21. $\dfrac{3}{8}$ 22. $\dfrac{4}{7}$ 23. $-\dfrac{5}{9}$ 24. $-\dfrac{8}{15}$

Rewrite each expression as a basic numeral or as a basic fraction.

Examples. a. $\sqrt{25} - \sqrt{16} = 5 - 4 = 1$

b. $\sqrt{\dfrac{9}{4}} + \sqrt{\dfrac{25}{16}} = \dfrac{3}{2} + \dfrac{5}{4} = \dfrac{6}{4} + \dfrac{5}{4} = \dfrac{11}{4}$

25. $\sqrt{25} + \sqrt{9}$ 26. $\sqrt{81} + \sqrt{16}$ 27. $\sqrt{64} - \sqrt{100}$

28. $\sqrt{36} - \sqrt{4}$ 29. $\dfrac{\sqrt{121}}{\sqrt{36}}$ 30. $\dfrac{\sqrt{49}}{\sqrt{81}}$

31. $\dfrac{\sqrt{16} + \sqrt{49}}{22}$ 32. $\dfrac{\sqrt{25} - \sqrt{64}}{9}$ 33. $\sqrt{\dfrac{4}{9}} + \sqrt{\dfrac{81}{16}}$

34. $\sqrt{\dfrac{100}{9}} - \sqrt{\dfrac{25}{36}}$ 35. $\sqrt{\dfrac{25}{4}} \cdot \sqrt{\dfrac{49}{9}}$ 36. $\sqrt{\dfrac{121}{36}} \div \sqrt{\dfrac{100}{25}}$

B

Specify the replacements for the variable so that each radical expression represents a real number.

Examples. a. $\sqrt{x-2}$ b. $\sqrt{x+4}$

Solutions. a. $x \geq 2$ b. $x \geq -4$

37. $\sqrt{x-5}$ 38. $\sqrt{x-4}$ 39. $\sqrt{x+3}$ 40. $\sqrt{x+7}$

41. The product $(1.4) \cdot (1.4) = 1.96$ and the product $(1.5) \cdot (1.5) = 2.25$. Can you find a rational number x between 1.4 and 1.5 such that $x \cdot x = 2$?

42. Can you find a rational number x between 1.7 and 1.8 such that $x \cdot x = 3$?

5.2   The Set of Irrational Numbers

On page 123 we noted that some points on the number line do not correspond to rational numbers. However, such points do correspond to numbers, even though they are not rational numbers.

In problem 41, Exercise 5.1, you were asked to find a rational number x such that $x \cdot x = 2$. That is, is $\sqrt{2}$, the positive square root of 2, a rational number? We now discuss this problem in more detail.

In Section 5.1, we defined the symbol $\sqrt{a}$, $a \geq 0$, to be the non-negative number such that when it is multiplied by $\sqrt{a}$, the product is a. Since

$$(1.4) \cdot (1.4) = 1.96 \quad \text{and} \quad (1.5) \cdot (1.5) = 2.25,$$

you can see that $\sqrt{2}$ lies somewhere between 1.4 and 1.5. Now consider the numbers 1.41 and 1.42, each of which is between 1.4 and 1.5. Since

$$(1.41) \cdot (1.41) = 1.9881 \quad \text{and} \quad (1.42) \cdot (1.42) = 2.0664,$$

it is evident that $\sqrt{2}$ lies between 1.41 and 1.42. We can repeat this "squeezing process" several times to obtain the sequence of rational numbers associated with points on the line as in Figure 5.1. Only that part of the number line showing numbers between 0 and 1.5 is indicated and the broken line indicates the scale has been changed to permit more detail in the figure.

By selecting numbers which are the coordinates of points closer and closer to the point P, we appear to be approaching a number such that, when it is multiplied by itself, the product equals 2. At least, from the illustration below, we see that

$$1.4142 < \sqrt{2} < 1.4143.$$

This use of the number line has not really answered the question whether there is a rational number represented by a terminating or periodic decimal fraction, or fraction in the form a/b, such that $(a/b)^2 = 2$. It can be shown, although we shall not do so here, that *there is no such rational number.* However, in Figure 5.1, the line segment from 0 to P with length $\sqrt{2}$ does

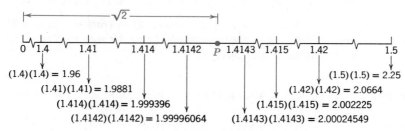

Figure 5.1

exist. Thus, the number $\sqrt{2}$ can be associated with a point on the number line.

There are many numbers such as $\sqrt{2}$, $\sqrt{3}$, $\sqrt[3]{7}$, and π, which cannot be written as a fraction a/b, where a and b are integers, but which can be associated with points on a number line. Such numbers are called **irrational numbers.** The set of irrational numbers is infinite; it will be designated by the capital letter, H. At this time, we shall limit ourselves to irrational numbers such as $\sqrt{2}$, $\sqrt{3}$, and $\sqrt{5}$, which when multiplied by themselves yield the rational numbers 2, 3, and 5, respectively.

Although an irrational number cannot be written as a fraction a/b where a and b are integers, and therefore cannot be represented by a terminating or periodic decimal fraction, we can certainly use a rational number to *approximate* the irrational number as closely as we like. For example, in the illustration above, you observed that

$$\sqrt{2} \approx 1.4142,$$

where the symbol $\approx$ means "is approximately equal to." The table inside the back cover shows approximations of some irrational numbers, namely, the square roots of certain integers between 1 and 100. As noted in Section 5.1, the numbers shown in color have square roots that are rational numbers. The square roots of the other numbers are irrational and the corresponding rational number in decimal form is only an approximation to the irrational number.

Exercise 5.2

A

Let $U = \{-\sqrt{2}, 1, \frac{2}{3}, \sqrt{36}, \sqrt{37}, -\sqrt{9}, \sqrt{31}, 0, -\sqrt{17}, \sqrt{81}, -\sqrt{16}\}.$

List the members in each subset of U, *where* $Q = \{rational\ numbers\}$ *and* $H = \{irrational\ numbers\}$. *Use the table inside the back cover, to obtain approximations for irrational numbers, if necessary.*

Examples.   a. $\{x \mid 0 < x < 10, x \in Q\}$ b. $\{x \mid 0 < x < 10, x \in H\}$

Solutions.   a. $\left\{1, \frac{2}{3}, \sqrt{36}, \sqrt{81}\right\}$ b. $\{\sqrt{37}, \sqrt{31}\}$

1. $\{x \mid x \in Q\}$ 2. $\{x \mid x \in H\}$

3. $\{x \mid x > 0, x \in H\}$ 4. $\{x \mid x \leq 0, x \in Q\}$

5. $\{x \mid -10 \leq x < 0, x \in Q\}$ 6. $\{x \mid 5 < x < 6, x \in H\}$

7. $\{x \mid x < -3, x \in Q\}$ 8. $\{x \mid x > 5, x \in Q\}$

9. $\{x \mid -10 < x < -6, x \in H\}$ 10. $\{x \mid 5 \leq x \leq 8, x \in Q\}$

11. $\{x \mid x \in H\} \cup \{x \mid x \in Q\}$ 12. $\{x \mid x \in H\} \cap \{x \mid x \in Q\}$

Graph each set of numbers. In general, the graphs will have to be approximated. Use the table to obtain approximations for irrational numbers.

Example. $\{-\sqrt{15}, -2, \sqrt{17}\}$

Solution.

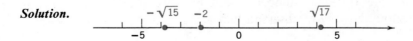

13. $\{-\sqrt{10}, 0, \sqrt{5}\}$

14. $\{-\sqrt{23}, -3, \sqrt{2}\}$

15. $\{-6, -\sqrt{3}, \sqrt{34}\}$

16. $\{-\sqrt{80}, -\sqrt{26}, \sqrt{61}\}$

17. $\{-\sqrt{7}, \sqrt{5}, \sqrt{30}\}$

18. $\{-\sqrt{54}, -\sqrt{17}, 5\}$

19. $\{-\sqrt{25}, 2, \sqrt{21}\}$

20. $\{-\sqrt{72}, -3, \sqrt{16}\}$

21. Construct a Venn diagram showing the relationships between the sets of natural numbers N, whole numbers W, integers J, rational numbers Q, and irrational numbers H.

22. Which pairs of the sets N, W, J, Q, and H are disjoint?

Which of the following statements are true for every value of the variable in its replacement set?

23. If $x \in W$, then $x \in N$.

24. If $y \in J$, then $y \in H$.

25. If $z \in N$, then $z \in Q$.

26. If $a \in Q$, then $a \in W$.

27. If $b \in H$, then $b \in W$.

28. If $c \in W$, then $c \in J$.

29. If $x \in W$, then $x \in Q$.

30. If $y \in N$, then $y \in J$.

Rewrite (to the nearest hundredth) using values from the table.

Example. $\sqrt{15} + \sqrt{7} \approx 3.873 + 2.646 \approx 6.52$

31. $\sqrt{17} + \sqrt{8}$ 32. $\sqrt{52} - \sqrt{19}$ 33. $\sqrt{61} \cdot \sqrt{13}$

34. $\sqrt{98} \div \sqrt{23}$ 35. $3(\sqrt{21} - \sqrt{14})$ 36. $2(\sqrt{37} + \sqrt{44})$

5.3 Some Properties of Real Numbers

You have seen that a rational number can be represented by a terminating or periodic decimal. An irrational number can be represented by a *non*-terminating, *non*-periodic decimal. The set formed by the union of the set of rational numbers Q and the set of irrational numbers H is called **the set of real numbers** and is denoted by R. Every real number has a decimal representation and every decimal represents a real number.

The sets Q and H are *disjoint*, but the elements of both sets have a common characteristic—they can be associated with points on a number line. It can be shown, though this is beyond the scope of this book, that there is a one-to-one correspondence between the set of real numbers and the set of points on a number line. This means that *every* real number can be represented as a point on a number line and that *every* point on a number line has a coordinate which is a real number. Thus we say a number line is completely "filled" with the graphs of all real numbers and express this fact by saying that the set of real numbers has the **property of completeness.**

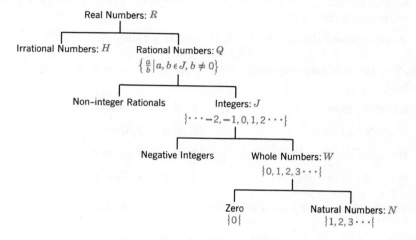

Figure 5.2

Figure 5.2 summarizes the relationship between the different sets of numbers which have been discussed thus far. In particular, you should be familiar with the following sets:

1. The set of natural numbers, N, symbolized by $\{1, 2, 3, \cdots\}$.

2. The set of whole numbers, W, symbolized by $\{0, 1, 2, 3, \cdots\}$.

3. The set of integers, J, symbolized by $\{\cdots -3, -2, -1, 0, 1, 2, 3, \cdots\}$.

4. The set of rational numbers, Q, whose elements are all those numbers that can be represented in the form a/b, where $a, b \in J$ ($b \neq 0$) or as terminating or periodic decimals. Examples are $-5/6, 21/27, 3, -7, 0.43$, and $0.6\bar{6}$.

5. The set of irrational numbers, H, whose elements are those numbers whose representations are non-terminating, non-periodic decimals. Examples are $\sqrt{2}, \pi, -\sqrt{7}$, and $\sqrt[3]{15}$. An irrational number cannot be represented in the form a/b, where a and b are integers.

6. The set of real numbers, R, which is the union of the set of rational numbers and the set of irrational numbers.

It is with the set of real numbers, or subsets of the real numbers, that we shall be most concerned in the remainder of this book.

Since the set of rational numbers Q is a subset of the set of real numbers R, we want the real numbers to have the same properties that are applicable to the rational numbers. Hence, we take the set of axioms previously adopted for the rational numbers for the set of real numbers. The theorems previously established are therefore also applicable to the set of real numbers. Thus we now have:

1. The set of real numbers.

2. Two basic operations on the elements in the set (addition and multiplication).

3. Some assumptions (axioms) concerning the elements and governing the operations. These are listed in the appendix on page 387.

4. Some logical consequences (theorems) of these assumptions. These are listed in the appendix on page 389.

The four features above constitute the **real number system.** We have not considered all possible theorems in the system. We have at this time considered only those theorems that are useful for developments in the following chapters.

Another useful fact concerning the set of real numbers is that it contains solutions of some equations and inequalities which do not have solutions in the set of rational numbers.

Definition 5.1 restricts a in $\sqrt{a}$ to be a non-negative number. This means that if the radicand contains a variable, the replacement set for the variable must be restricted to those replacements for which the radicand represents a non-negative number. For example, $\sqrt{x-3}$ represents a real number if and only if $x \geq 3$. If $x < 3$, then $x - 3$ is negative and $\sqrt{x-3}$ is not defined. Furthermore, Definition 5.1 specifies that $\sqrt{a}$ is a non-negative number for $a \geq 0$. Hence, an equation of the form $\sqrt{x} = c$, $c \in R$, has a solution if and only if $c \geq 0$.

Examples. Find the solution set in R for each equation.

a. $\sqrt{x} = 3$ 　　　　 b. $\sqrt{x} = -5$ 　　　　 c. $\sqrt{x} + 2 = 4$

Solutions on page 132.

Solutions

a. {9}; because $\sqrt{9} = 3$.   b. ∅ ; because $\sqrt{x}$ must be non-negative.

c. {4}; because $\sqrt{4} + 2 = 2 + 2 = 4$.

Exercise 5.3

A

Let N = {*natural numbers*}, Q = {*rational numbers*}, W = {*whole numbers*}, H = {*irrational numbers*}, J = {*integers*}, *and* R = {*real numbers*}. *Consider* R *to be the universe. State whether each statement is true or false.*

Examples.   a. $N \subset W$ b. $J \cap W = J$

Solutions.   a. True   b. False

1. $J \subset Q$ 2. $J \subset W$ 3. $Q \cup J = R$ 4. $N \cap W = J$

5. $H \subset R$ 6. $Q \cup H = R$ 7. $Q \cap H = R$ 8. $N \cup W = W$

9. $R \cap H = H$ 10. $Q' = H$ 11. $Q \cap Q' = \emptyset$ 12. $H' \cup J = R$

Graph each set.

Examples.   a. $\{x \mid 0 < x \le 3, x \in J\}$ b. $\{x \mid 0 < x \le 3, x \in R\}$

Solutions.   a.

b.

The solid line indicates an infinite set of points; zero is not in the set.

13. $\{x \mid -3 \le x \le 2, x \in J\}$ 14. $\{z \mid 4 < z \le 10, z \in J\}$

15. $\{y \mid -2 \le y < 2, y \in R\}$ 16. $\{x \mid -12 \le x \le -5, x \in R\}$

17. $\{y \mid y \in N\}$ 18. $\{x \mid x \in W\}$

19. $\{x \mid x \in R\}$ 20. $\{x \mid x \in J\}$

21. $\{z \mid z > 0, z \in J\}$ 22. $\{y \mid y \le 0, y \in W\}$

23. $\{z \mid z < 0, z \in R\}$ 24. $\{x \mid x \ge 0, x \in N\}$

Name an axiom of the real number system that justifies each statement. Assume
$x, y, z \in R$, *all variables in the radicands to be positive numbers, and no denominator*
to be equal to 0.

Examples.   a. $(x \cdot y) \cdot z = x \cdot (y \cdot z)$ b. $x + 3 = 3 + x$

Solutions

a. Associative law of multiplication   b. Commutative law of addition

25. $2 \cdot \sqrt{3} \in R$

26. If $x + \sqrt{5} = 9$, then $9 = x + \sqrt{5}$.

27. $-\sqrt{5} + \sqrt{5} = 0$

28. If $x = y$ and $y = \sqrt{2}$, then $x = \sqrt{2}$.

29. $-2 \cdot (x + y) = -2 \cdot x - 2 \cdot y$ 30. If $y > x$ and $x > 0$, then $y > 0$.

31. $2 + (x - z) = (x - z) + 2$ 32. $(x + y) \cdot 2 = (y + x) \cdot 2$

33. $(x + y) + 0 = (x + y)$ 34. $x < y$, $x = y$, or $x > y$

35. $(x - y) \cdot 1 = x - y$ 36. $(x + y - z) \cdot \dfrac{1}{(x + y - z)} = 1$

37. $\dfrac{x}{y} = \dfrac{x}{y}$ 38. $x + \sqrt{y} \in R$

39. $3 \cdot x \cdot \dfrac{1}{3 \cdot x} = 1$ 40. $\sqrt{x} \cdot \dfrac{1}{\sqrt{y}} \in R$

41. $x \cdot (y + \sqrt{z}) = x \cdot y + x \cdot \sqrt{z}$ 42. $(x + \sqrt{y}) + z = x + (\sqrt{y} + z)$

43. $(3 \cdot \sqrt{x}) \cdot \dfrac{1}{\sqrt{z}} = 3 \cdot \left(\sqrt{x} \cdot \dfrac{1}{\sqrt{z}} \right)$ 44. If $x = y \cdot \sqrt{2}$, then $y \cdot \sqrt{2} = x$.

State a theorem by name to justify each statement. Assume $x, y, z \in R$, *all*
variables in the radicands to be positive numbers, and no denominator to be equal
to 0.

45. $\dfrac{1}{x \cdot y} \cdot 0 = 0$ 46. $\dfrac{1}{\sqrt{x}} = \dfrac{\sqrt{x}}{\sqrt{x}\sqrt{x}}$

47. If $x - 3 = 2 \cdot y$, then $x - 3 + 3 = 2 \cdot y + 3$.

48. If $x = y + 2$, then $6 \cdot x = 6 \cdot (y + 2)$.

49. If $\dfrac{x+y}{z} = \dfrac{2 \cdot y}{z}$, then $z \cdot \left(\dfrac{x+y}{z}\right) = z \cdot \left(\dfrac{2 \cdot y}{z}\right)$.

50. If $x = y$, then $x + \sqrt{2} = y + \sqrt{2}$.

51. If $\dfrac{x+y}{x-y} = \dfrac{3}{4}$, then $[4 \cdot (x-y)] \cdot \dfrac{(x+y)}{x-y} = [4 \cdot (x-y)] \cdot \dfrac{3}{4}$.

52. If $3 \cdot (x-y) = z$, then $\dfrac{1}{3} \cdot [3 \cdot (x-y)] = \dfrac{1}{3} \cdot z$.

53. $\dfrac{\sqrt{x} \cdot \sqrt{3}}{\sqrt{y} \cdot \sqrt{3}} = \dfrac{\sqrt{x}}{\sqrt{y}}$

54. $0 \cdot \dfrac{\sqrt{x}}{y} = 0$

Specify the replacements for the variable so that each radical expression represents a real number.

Examples. a. $\sqrt{x-3}$ b. $\sqrt{x+5}$

Solutions.   a. $x \geq 3$ b. $x \geq -5$

55. $\sqrt{x-6}$ 56. $\sqrt{x+2}$ 57. $\sqrt{8+x}$ 58. $\sqrt{2-x}$

Find the solution set in R for each equation.

Examples. a. $\sqrt{x} + 3 = 7$ b. $\sqrt{x} + 5 = 3$

Solutions.   a. $\{16\}$, because $\sqrt{16} = 4$ and $4 + 3 = 7$.

b. $\varnothing$, because $\sqrt{x}$ is always non-negative and there is no such real number whose sum with 5 equals 3.

59. $\sqrt{x} = 2$ 60. $\sqrt{x} = 5$ 61. $\sqrt{x} = -4$

62. $\sqrt{x} = -1$ 63. $\sqrt{x} - 3 = 0$ 64. $\sqrt{x} - 4 = 0$

65. $\sqrt{x} + 3 = 3$ 66. $\sqrt{x} + 4 = 4$ 67. $2 - \sqrt{x} = 1$

68. $5 - \sqrt{x} = 2$ 69. $3 + \sqrt{x} = 2$ 70. $4 + \sqrt{x} = 3$

71. $\sqrt{x} - 4 = 1$ 72. $\sqrt{x} - 3 = 3$ 73. $x - \sqrt{4} = 3$

74. $x + \sqrt{9} = 7$ 75. $x \cdot x = 2$ 76. $x \cdot x = 5$

77. $-(x \cdot x) = -3$ 78. $-(x \cdot x) = -7$

B

Graph each set.

79. $\{x \mid 1 < x \leq 5, x \in R\} \cap \{x \mid 3 \leq x < 27, x \in R\}$

80. $\{y \mid y \leq 3, y \in R\} \cup \{y \mid y < 0, y \in R\}$

81. $\{x \mid x \in W\} \cap \{x \mid x \in Q\}$

82. $\{z \mid z \in R\} \cap \{z \mid z < 0, z \in J\}$

CHAPTER SUMMARY

5.1 Some rational numbers have square roots that are rational numbers and can be represented in **radical form.** If $a \in Q$, $a \geq 0$, then $\sqrt{a}$ is the non-negative number such that $\sqrt{a} \cdot \sqrt{a} = a$.

5.2 The elements in the set of **irrational numbers** cannot be expressed as a quotient, $\frac{a}{b}$, a, $b \in J$. They can be *approximated* by rational numbers to any degree of accuracy that is desired.

5.3 The union of the set of rational numbers and the set of irrational numbers constitutes the set of **real numbers.** There is a *one-to-one correspondence* between the elements of the set of real numbers and the points on a number line. To each real number there corresponds a point on the line and to each point there corresponds a real number. Definitions, axioms, and theorems applicable to the set of rational numbers are also applicable to the set of real numbers. (See Appendix II, page 394, for a summary of the definitions, axioms, and theorems that are applicable to the real number system.) Some open sentences which do not have solutions in the set of rational numbers have solutions in the set of real numbers.

CHAPTER REVIEW

5.1 *Write each square root as a basic numeral or as a basic fraction.*

1. $-\sqrt{169}$

2. $\sqrt{\dfrac{4}{81}}$

Express each rational number as the square root of a rational number.

3. 7   4. -8 5. $\dfrac{3}{5}$ 6. $-\dfrac{2}{3}$

5.2 *Let* $U = \left\{-\sqrt{5}, 7, \dfrac{3}{4}, \sqrt{9}, \sqrt{10}, 0, 2, -3\right\}$. *List the members in each subset of U, as indicated.*

7. $\{x \mid x < 0, x \in Q\}$ 8. $\{x \mid x \geq 0, x \in Q\}$

9. $\{x \mid x < 0, x \in H\}$ 10. $\{x \mid x \geq 0, x \in H\}$

Graph each set. Approximate the positions of the graphs of non-integral elements.

11. $\{-7, 1, \sqrt{9}\}$ 12. $\{-\sqrt{6}, 0, \sqrt{3}\}$

13. $\{x \mid 0 < x \leq 7, x \in J\}$ 14. $\{x \mid 0 < x \leq 7, x \in R\}$

15. $\{x \mid x > 3, x \in R\}$ 16. $\{x \mid x \leq 4, x \in J\}$

State whether each statement is true or false.

17. If $a \in Q$, then $a \in N$.   18. If $a \in H$, then $a \in R$.

19. $(W \cup N) \cap R = R$ 20. $(W \cap N) \cup R = R$

5.3 *State whether each statement is true or false.*

21. $Q \subset H$ 22. $Q \subset R$ 23. $Q \cup R = Q$

24. $H \cup J = R$ 25. $Q \cap H = R$ 26. $R \cap Q = Q$

Name an axiom or theorem that justifies each statement. $x, y, z \in R.$

27. If $x + \sqrt{3} = 9$, then $x + \sqrt{3} + (-\sqrt{3}) = 9 + (-\sqrt{3})$.

28. If $x \cdot y = 12$ and $x = \sqrt{3}$, then $\sqrt{3} \cdot y = 12$.

29. If $x < \sqrt{3}$ and $\sqrt{3} < y$, then $x < y$.

30. If $3 \cdot x = y$, then $\dfrac{1}{3} \cdot 3 \cdot x = \dfrac{1}{3} \cdot y$.

Find the solution set in R for each equation.

31. $\sqrt{x} = 6$ 32. $\sqrt{x} + 1 = 0$ 33. $3 - \sqrt{x} = 3$

34. $6 + \sqrt{x} = 8$ 35. $x - \sqrt{16} = 2$ 36. $x \cdot x = 11$

37. What four features constitute the real number system?

38. Graph $\{x \mid x \in Q\} \cup \{x \mid x \in H\}$

39. Graph $\{x \mid x \in Q\} \cap \{x \mid x \in H\}$

40. Describe the graph of $\{x \mid x \in Q\}$.

chapters 1-5
Cumulative Review

CHAPTER 1

1. A set may be identified by listing the names of the members or elements or by a _____.
2. The intersection of two sets, A and B, contains all elements that are common to both A and B. The _____ of the two sets contains all elements that belong either to A or B or to both.
3. The _____ of a set A in a universal set contains all elements of the universal set that are not contained in A.
4. _____ diagrams are geometric representations which exhibit relationships between sets.
5. The _____ product of two sets A and B, where $a \in A$ and $b \in B$, contains all ordered pairs (a, b) that can be formed.

In Problems 6 to 9, consider $U = \{0, 1, 2, 3\}$, $A = \{1, 2\}$, *and* $B = \{2, 3\}$.

6. $A \cup B =$ _____   7. $A \cap B =$ _____

8. $A' =$ _____   9. $A \times B =$ _____

CHAPTER 2

10. Whole numbers when associated with "how many" elements there are in a set are said to be used in a _____ sense.
11. The set represented symbolically as $\{1, 2, 3, \cdots\}$ is the set of _____ numbers.
12. A symbol, such as a, b, or c, used to represent an unspecified element in a specified set is called a(n) _____.
13. In mathematics, a formally stated assumption is called a postulate or a(n) _____.
14. A number corresponding to a point on a number line is called the _____ of the point.

15. If A and B represent disjoint sets, then the number associated with $A \cup B$ is called the _____ of the numbers associated with A and B.
16. When the result of performing an operation on two elements of a given set is always another element of the *same* set, the set is said to be _____ with respect to the operation.
17. A symbol in the set $\{0, 1, 2, 3, \cdot \cdot \cdot\}$ can be used to name a whole number. These symbols are called basic _____.
18. The cardinality of a Cartesian product of two sets whose cardinalities are a and b is called the _____ of a and b.
19. The graph of a Cartesian product is called an array or a(n) _____.
20. The _____ law is symbolized as follows: If a, b, $c \in W$, then $a \cdot (b + c) = a \cdot b + a \cdot c$.
21. Statements which are logical consequences of definitions and axioms are called _____.
22. If a, $b \in W$ and if $a < b$, then $a - b$ is (always/sometimes/never) a whole number.
23. To simplify an expression that involves several operations, first perform those operations above or below a _____ _____.
24. To evaluate an expression that contains a variable, replace the variable with its given value and then perform the indicated operations in the proper _____.
25. A solution of an equation or an inequality is a replacement for the variable that makes the equation or inequality a _____ _____.
26. The set of all solutions of an equation or inequality is called the _____ _____.

CHAPTER 3

27. The set of natural numbers and their negatives, together with zero, is called the set of _____.
28. If $a \in J$ ($a \neq 0$), then a and $-a$ are called the negatives or the _____ _____ of each other.
29. If $a \in J$, then $-a$ (always/sometimes/never) denotes a negative number.
30. The relative order of $-4, 6, -8$ can be shown by writing __ < __ < __.
31. The axioms and theorems for the system of whole numbers are (always/sometimes/never) applicable to the system of integers.
32. $|-5|$ is read "the _____ value of negative five."
33. The sum of two negative integers is (always/sometimes/never) a negative integer.

34. The sum of a positive integer and a negative integer is (always/sometimes/never) a negative integer.
35. If a and b are integers, then $a - b$ is (always/sometimes/never) an integer.
36. If a and b are integers, then $a - b$ is (always/sometimes/never) equal to $a + (-b)$.
37. Write $(-5) + (-7) + (15)$ as a basic numeral.
38. The product of a positive integer and a negative integer is (always/sometimes/never) a positive integer.
39. The product of two negative integers is (always/sometimes/never) a positive integer.
40. Write $(-5) \cdot (-3) \cdot (-2)$ as a basic numeral.
41. The product of any integer a and 0 is _____.
42. If $a \in J$, $a \neq 0$, then the quotient of 0 divided by a is zero and a divided by 0 is _____.

43. Simplify $3(2 + 4) - \dfrac{2(4 - 7)}{6}$.

44. Simplify $2(-5) + \dfrac{4 + 2 \cdot 5}{7}$.

45. Evaluate $7x - 5$ for $x = -2$.
46. Evaluate $2x - 3xy - 4y$ for $x = 2$ and $y = -2$.

CHAPTER 4

47. $\{a/b \mid a, b \in J, b \neq 0\}$ is called the set of _____ numbers.

48. If $a/b \in Q$ $(a/b \neq 0)$, then a/b and $\dfrac{1}{a/b}$ are called reciprocals or _____
_____ of each other.

49. The reciprocal of 7 is _____ and the reciprocal of 1/5 is _____.
50. The quotient a/b can be written as the product _____.
51. Either statement

$$\frac{a \cdot c}{b \cdot c} = \frac{a}{b} \quad \text{or} \quad \frac{a}{b} = \frac{a \cdot c}{b \cdot c}$$

is called the _____ _____ of fractions.
52. The least common multiple of the elements in the set $\{15, 18, 24\}$ is _____.

53. Write $\dfrac{3}{4} + \dfrac{4}{5}$ as a basic fraction.

54. The fractions $-\dfrac{-a}{b}$, $-\dfrac{a}{-b}$, and $\dfrac{-a}{-b}$ can also be written as _____.

55. The fractions $\dfrac{-a}{b}$, $\dfrac{a}{-b}$, and $-\dfrac{-a}{-b}$ can also be written as _____.

56. Write $\dfrac{13}{9} - \dfrac{2}{5}$ as a basic fraction.

57. Write $\dfrac{2}{3} \cdot \dfrac{6}{7}$ as a basic fraction.

58. Write $\dfrac{2}{3} \div \dfrac{6}{7}$ as a basic fraction.

59. If a/b, $c/d \in Q$, $c/d \neq 0$, then the quotient, a/b divided by c/d, is (always/sometimes/never) a rational number.

60. Change 0.97 to the form a/b.

61. Change 0.775 to the form a/b.

CHAPTER 5

62. Irrational numbers can be approximated to any desired degree of accuracy by _____ numbers.

63. The set of _____ numbers completely "fills" the number line.

In Problems 64 to 66, consider

$$U = \left\{ -\sqrt{64}, -\sqrt{48}, -\sqrt{5}, 0, \frac{3}{2}, 2, \sqrt{19}, \sqrt{49}, \sqrt{50} \right\}.$$

64. List the members in $\{x \mid -7 < x < 2, x \in R\}$.

65. List the members in $\{x \mid x \in J\} \cup \{x \mid x \in Q\}$.

66. List the members in $\{x \mid x \in J\} \cap \{x \mid x \in Q\}$.

67. The axioms for the set of rational numbers are (always/sometimes/never) applicable to the set of real numbers.

68. Graph $\{x \mid -2 < x \leq 2, x \in J\}$.

69. Graph $\{x \mid -2 < x \leq 2, x \in R\}$.

Replace each comma with the proper symbol: $\in$, $\subset$, $\cup$, *or* $\cap$.

70. N, J 71. $Q, H = R$ 72. $-5, J$ 73. Q, R

74. $Q, H = \varnothing$ 75. $4, Q$ 76. $\sqrt{5}, R$ 77. $R, J = R$

78. $\{$prime numbers$\}$, $\{$composite numbers$\} = \varnothing$.

Find the solution set in R for each equation.

79. $x + 2 = 5$ 80. $x + 1 = -4$ 81. $\dfrac{2}{3} + x = 4$

82. $x + 1 = \dfrac{3}{5}$ 83. $\dfrac{x}{5} = 2$ 84. $6 \cdot x = -5$

85. $\sqrt{x} - 3 = 7$ 86. $x \cdot x = 12$

6
chapter

Polynomials

Much of mathematics is concerned with writing mathematical models for word phrases or sentences. You have already learned about symbols (numerals and variables) which represent numbers and symbols for operations and grouping such as $+, -, \cdot, \div, (\ ), [\ ]$, etc. Any collection of symbols such as

$$5, \quad x, \quad x + y, \quad x \cdot (y + z) \quad \text{and} \quad \frac{x \cdot p + y \cdot p}{p \cdot q}$$

which represents a number is called an **algebraic expression.**

In this chapter you will learn more about using collections of symbols which represent numbers by working with a particular kind of an algebraic expression called a **polynomial.**

6.1 Forms of Products

Recall that when two numbers are paired by the operation of multiplication, the result is called a product. The two numbers which are paired are called the factors of the product. Since $4 \cdot 3 = 12$, the symbol 12 is the basic numeral for the product of 4 and 3; the expression $4 \cdot 3$ is called a factored form of 12.

The factored form of some products can be written without a multiplication sign, $\cdot$ or $\times$. When the intent is clear, we shall do so. Thus the factored form of 10 can be written as $2 \cdot 5$, $(2)(5)$, or $2(5)$. The product of two variables, such as a and b, will generally be shown as ab, or ba.* If a product contains a numerical factor and a variable, the numerical factor is generally written first.

Recall that any natural number greater than 1 which is exactly divisible only by itself and 1 is called a prime number and those which are not prime numbers are designated as composite numbers. When the factored form of a product contains only prime numbers as factors, it is called the **prime factor form** or the **completely factored form**.

Examples.   Write each number in prime factor form.

a. 35   b. 12   c. 135

Solutions

a. $35 = 5 \cdot 7$ b. $12 = 4 \cdot 3 = 2 \cdot 2 \cdot 3$ c. $135 = 5 \cdot 27 = 5 \cdot 3 \cdot 3 \cdot 3$

If a number appears more than once as a factor in a product, the form can be shortened by using a special notation.

Definition 6.1.   For all $a \in R$ and $m \in N$,

$$a^m = a \cdot a \cdot a \cdots a. \qquad (m \text{ factors})$$

In this notation, $2 \cdot 2 \cdot 2$ is written as 2^3 and $x \cdot x \cdot x \cdot x$, or $xxxx$, is written as x^4. The numbers 3 and 4, which are written to the right and a little above the 2 and the x, respectively, are called **exponents** and indicate how many times the respective factor appears in the product. The factor itself is called the **base.** Symbols such as 2^3 and x^4 are called **powers.** Some powers have special names. For example if the exponent is 2, we noted in Section 5.1 that the power is called a *square* and if the exponent is 3, the power is called a *cube*. Other powers are named according to the number used as an exponent.

Examples.   a. 5^2 is read "five squared," or "the square of five."

---

* In general, we shall not show alternate forms as answers. If your answer to a problem is simply an alternate form of the answer given, consider your result to be a correct one.

b. $(y + z)^3$ is read "the quantity $y + z$ cubed", or "the cube of the quantity $y + z$."

c. x^4 is read "x to the fourth power", "the fourth power of x," or more simply "x to the fourth."

When a product in completely factored form is shortened by using exponents, the resulting form is known as the **exponential form** of the product. For example,

$$2 \cdot 2 \cdot 2 \cdot 3 \cdot 3 \cdot 5 \cdot xxyyyz$$

is the completely factored form of a product, and its exponential form is

$$2^3 \cdot 3^2 \cdot 5x^2y^3z.$$

Note that, as in the cases of the numeral 5 and the variable z, when a numeral or a variable appears only once as a factor, the exponent "1" is not written.

As you will see in later chapters, if products are to be used in the solution of word problems, the mathematical models for word phrases must first be constructed.

Examples.   Construct mathematical models for each of the following.

a. The product of 3 and the square of y.

b. Four times the quantity $2 + y$ cubed.

Solutions.   a. $3y^2$ b. $4(2 + y)^3$

In Exercise 6.1 assume that all variables are elements of the set of real numbers. This assumption is to hold for all exercise sets in the remainder of the text unless otherwise stated.

Exercise 6.1

A

Write each product as a basic numeral.

Examples.   a. $2^2 \cdot 3^2$ b. $(-1)^3(-2)^2$ *(Solutions on p. 146)*

Solutions.   a. $2 \cdot 2 \cdot 3 \cdot 3 = 36$ b. $(-1)(-1)(-1)(-2)(-2) = -4$

1. 4^3 2. 3^5 3. $2 \cdot 5^3$

4. $2^3 \cdot 5$ 5. $(-4)^3$ 6. $(-3)^4$

7. $2^2 \cdot 3^3 \cdot 5$ 8. $2^3 \cdot 3^2 \cdot 5^2$ 9. $(-2)^2(-3)^2$

10. $(-2)^3(-5)^2$ 11. $(-1)^2(-2)^3(-3)^2$ 12. $(-1)^3(-2)^2(-3)^3$

Write each expression in completely factored form.

Examples.   a. $12p^2q^4$ b. $9r^3t$

Solutions.   a. $2 \cdot 2 \cdot 3 \cdot ppqqqq$ b. $3 \cdot 3rrrt$

13. 20   14. 36   15. $2^4 \cdot 3^2$ 16. $3^3 \cdot 5^3$

17. x^3 18. x^4 19. x^2y^2 20. x^3y^4

21. $24x^2yz^2$ 22. $48k^2m^2n$ 23. $(15)^2x^3y$ 24. $(28)^2xy^2$

Write each product in exponential form.

Examples.   a. $2 \cdot 2 \cdot 2xxy$ b. *rssttt*

Solutions.   a. 2^3x^2y b. rs^2t^3

25. $2 \cdot 2 \cdot 2$ 26. $2 \cdot 3 \cdot 3$ 27. *xxx*

28. *3mmmnn*   29. $2 \cdot 3 \cdot 3yyz$ 30. *6xyzz*

31. *5rsst*   32. *mmmnnnn*   33. *2rrs*

34. *rtttt*   35. *yyyyy*   36. *xxyyyz*

Write each word phrase as an algebraic expression.

Examples.   a. Five times x b. Three times the sum of five and x

Solutions.   a. $5x$ b. $3(5 + x)$

37. Two times six   38. Product of five and y

39. p multiplied by q 40. Three times the cost (c)

41. Four times the reciprocal of x 42. A number (n) multiplied by three

43. Five times the sum of three and x

44. Four times the product of the square of x and the cube of y

Justify each statement by an axiom, definition, or theorem.

45. $3(x^2y) = (3x^2)y$

46. $7pq^2 = 7q^2p$

47. $9x^2yz \cdot 1 = 9x^2yz$

48. $3r^2s^3 \cdot 0 = 0$

49. $5s^2t^3 \in R$

50. If $x^2y = z$, then $z = x^2y$.

51. If $q = x$ and $x = rs^3$, then $q = rs^3$.

52. If $x^2 = 6$ and $x = 2y$, then $(2y)^2 = 6$.

6.2 Definitions

For our purposes we define a **polynomial in one variable** as an algebraic expression such as

$$2x + 1, \qquad 3x^3 - 5x + 7, \qquad \text{and} \qquad 5,$$

where $x \in R$ and all exponents are natural numbers. Note that a constant such as 5 is also a polynomial. Expressions such as $2x/(3 + x)$, in which the variable appears in the denominator, or $3x^2 - 2\sqrt{x}$, in which the variable appears under a radical sign, are not polynomials.

In an expression written as the sum of several quantities, each of these quantities is called a **term** of the expression. For example, the terms of the polynomial $3x^2 - 5x + 7$, or $3x^2 + (-5x) + 7$, are $3x^2$, $-5x$, and 7.

Each factor in a term is called a **coefficient** of the product of the remaining factors. The numerical factor is called the **numerical coefficient** of the term. The numerical coefficient is generally called simply the "coefficient" of the term. If no numerical coefficient appears in a term, it is understood to be 1.

Examples. a. In the term $5xyz$, $5x$ is the coefficient of yz, y is the coefficient of $5xz$, 5 is the coefficient (or numerical coefficient) of xyz.

b. $3x^2 - x$ is a polynomial of two terms, $3x^2$ and $-x$. The coefficient of x^2 is 3 and the coefficient of x is -1.

Polynomials are named according to the number of terms they contain. A polynomial of one term is called a **monomial;** a polynomial of two terms is called a **binomial;** and one of three terms is called a **trinomial.** No special names are given to polynomials with more than three terms.

Examples.   a. 7, $4x^2$, and $8y^3$ are monomials.

b. $3r^2 - r$, $4y^2 - 7y$, and $2x^3 + x^2$ are binomials.

c. $3y^3 + 2y^2 - 6y$ and $2x^2 - 9x + 15$ are trinomials.

d. $1/x + y^2$, $\sqrt{x + 1}$, and $x^{-3} + y^{2/3}$ are not polynomials.

The greatest exponent of any variable in a polynomial determines the **degree of the polynomial** in that variable. However, if no variable occurs, as in a constant polynomial, the degree is zero.

Examples.   a. $x^2 + 2x - 5$ is a polynomial of second degree in x.

b. $3x^4 - 2x^2$ is a polynomial of fourth degree in x.

c. 5 is a polynomial of zero degree.

Polynomials may contain more than one variable, but it is beyond the scope of this discussion to consider the degree of such polynomials.

Recall that in Section 2.6 we agreed that the collection of symbols $a + b \cdot c$ is equal to the sum of the term a and the term which is the product of b and c. This agreement and the fact that the signs of grouping (parentheses, brackets, fraction bars, etc.) mean that the symbols so grouped are to be considered a single number lead to the following order in which the operations indicated in an algebraic expression should be performed. Note that powers are evaluated before other operations are performed.

1. Operations within parentheses.

2. Operations above and/or below a fraction bar.

3. Evaluation of powers.

4. Multiplications or divisions in order from left to right.

5. Additions or subtractions in order from left to right.

Examples

a. $5 + \dfrac{3(6-3)}{9} - 2$

$= 5 + \dfrac{3 \cdot 3}{9} - 2$

$= 5 + 1 - 2 = 4$

b. $\dfrac{2(3)^2 - 4}{7} + \dfrac{(-3)^2 - 2^2}{5}$

$= \dfrac{2 \cdot 9 - 4}{7} + \dfrac{9 - 4}{5}$

$= \dfrac{18 - 4}{7} + \dfrac{9 - 4}{5}$

$= \dfrac{14}{7} + \dfrac{5}{5} = 2 + 1 = 3$

The phrase "polynomial in x" (or some other variable) is used quite frequently in conjunction with symbols such as $P(x)$, $N(x)$, $D(x)$, and $Q(x)$. For example $P(x)$ names a polynomial in x, $P(y)$ names a polynomial in y, and so on. $P(a)$ is the value of a polynomial in which the variable has been replaced by the real number a.

Example. If

$$P(x) = 2x^2 + x,$$

then

$$P(4) = 2 \cdot (4)^2 + (4) = 32 + 4 = 36,$$

$$P(0) = 2 \cdot (0)^2 + (0) = 0,$$

and

$$P(-2) = 2 \cdot (-2)^2 + (-2) = 2(4) - 2 = 8 - 2 = 6.$$

The laws of closure for addition and multiplication of real numbers assure us that a polynomial represents a unique real number for each real number replacement of the variable in the polynomial. This process of finding the number represented by a polynomial when the real number replacement of its variable is specified is called **numerical evaluation.** Any number which is a replacement for the variable, and for which the polynomial is zero, is called a **zero of the polynomial.**

Example. If

$$P(x) = x + 3$$

then,

$$P(-3) = (-3) + 3 = 0$$

and -3 is a zero of the polynomial $x + 3$.

Exercise 6.2

A

Give the name and degree of each polynomial.

Examples.   a. $4x^3$ b. $2x^2 - x$ c. $3x^4 - 7x^2 + 2x$

Solutions.   a. monomial;   b. binomial;   c. trinomial;
                    degree 3   degree 2   degree 4

1. $5x$ 2. $3x^2$ 3. $4x^3 + 2$

4. $2x^2 - 5$ 5. $4x^2 - x$ 6. $7x^4 + 3x$

7. $5x^2 + 4x + 3$ 8. $2x^2 - 7x + 1$ 9. $3x^3 - 4x - 2$

10. $x^3 + 6x^2 + 4$ 11. $9x^3 + x^2 - x + 2$ 12. $11x^3 - 2x^2 + 3x + 7$

Write the numerical coefficient and exponent on the variable of each term.

Examples.   a. $3x^2 - 7x$ b. $x^3 + 3x^2$

Solutions.   a. 1st term: 3; 2   b. 1st term: 1; 3
                    2nd term: -7; 1   2nd term: 3; 2

13. $2x^3$ 14. $3x^5$ 15. $4x^4 - x^2$

16. $5x^3 - x^4$ 17. $4x^2 + x^3 - 2x^4$ 18. $7x^2 - 2x^3 + x^5$

Write the basic numeral for each expression.

Examples

a. $2 + 3 \cdot 4$ b. $\dfrac{1}{2} \cdot 2 + 2$ c. $\dfrac{2 + 4}{3} - 2$

Solutions

a. $2 + 3 \cdot 4 = 2 + 12$ b. $\dfrac{1}{2} \cdot 2 + 2 = 1 + 2$ c. $\dfrac{2 + 4}{3} - 2 = \dfrac{6}{3} - 2$

$= 14$ $= 3$ $= 2 - 2 = 0$

19. $5 \cdot 2 + 4$ 20. $3 + 6 \cdot 5$ 21. $-4 \cdot 7 + 8$

22. $-3 + 5 \cdot 8$ 23. $0 \cdot 3 + 2$ 24. $2 + 9 \cdot 0$

25. $0 + \dfrac{1}{3} \cdot 3$

26. $7 \cdot \dfrac{1}{7} - 1$

27. $2(3 - 1)^4$

28. $3(5 - 1)^3$

29. $4 + \dfrac{10 + 8}{3}$

30. $5 + \dfrac{9 + 7}{4}$

31. $\dfrac{4^2 - 1}{5} + \dfrac{3^3}{9}$

32. $\dfrac{8^2 - 7^2}{3(5)} + \dfrac{1 + 3^2}{2}$

If $P(x) = x^2 + 2x + 1$, *find the value for each of the following.*

Examples. a. $P(2)$ b. $P(-2)$

Solutions.   a. $P(2) = (2)^2 + 2(2) + 1$ b. $P(-2) = (-2)^2 + 2(-2) + 1$

$\qquad\qquad\qquad = 4 + 4 + 1$ $= 4 - 4 + 1$

$\qquad\qquad\qquad = 9$ $= 1$

33. $P(1)$ 34. $P(3)$ 35. $P(-1)$ 36. $P(0)$

If $P(x) = x^2 - 5x + 6$, *find the value for each of the following.*

37. $P(2)$ 38. $P(-3)$ 39. $P\left(\dfrac{1}{2}\right)$ 40. $P(0)$

If $P(y) = y^3 + 2y + 1$, *find the value for each of the following.*

41. $P(-1)$ 42. $P(2)$ 43. $P(0)$ 44. $P(3)$

If $P(t) = (t - 1)^2 + 3t$, *find the value for each of the following.*

45. $P(1)$ 46. $P(0)$ 47. $P(-1)$ 48. $P(-2)$

By inspection, find the zeros for each expression.

Examples. a. $3x^2$ b. $x - 2$ c. $x(x - 2)$

Solutions.   a. 0; because   b. 2; because   c. 0 and 2; because

$\qquad\qquad\quad 3 \cdot 0^2 = 0$ $2 - 2 = 0$ $0 \cdot (0 - 2) = 0$ and

$\qquad\qquad\qquad\qquad\qquad\qquad\qquad\qquad\qquad 2 \cdot (2 - 2) = 0$

49. $4x^3$ 50. $3y^2$ 51. $y + 4$

52. $2x - 6$ 53. $x(x - 1)$ 54. $y^2(y + 2)$

55. $(y - 1)(y + 2)$ 56. $y(y - 3)(y + 4)$

B

Justify each statement in Problems 57 to 62 with an axiom, definition, or theorem.

57. $3x^2y \in R$

58. $5x^2y + 7xy^2 = 7xy^2 + 5x^2y$

59. $x(y^2 + z^3) = xy^2 + xz^3$

60. $x^4(p^2 - q^2) = (p^2 - q^2)x^4$

61. If $x^2y + 1 = z$, then $6(x^2y + 1) = 6z$.

62. If $4x^2 + 7x - 3 = 10$, then $4x^2 + 7x - 3 + 3 = 10 + 3$.

63. If $P(x) = x^2 - 6x + 1$, find $P(1) \cdot P(2)$.

64. If $P(x) = x - 1$ and $Q(x) = x + 3$, find $P(3) - Q(2)$.

6.3 Sums and Differences

Algebraic expressions such as those we shall use as mathematical models for word problems in later sections are often quite complicated. However, since such expressions are to represent real numbers for real number replacements of the variables, the axioms, definitions, and theorems of the real number system enable us to rewrite these expressions in forms which are simpler to use. You should always keep in mind the fact that a polynomial represents a single *number* for specific replacements of its variables and hence, any alternate form must represent the *same* number for the *same* replacement. Such alternate forms are said to be **equivalent.**

Examples. a. $\dfrac{a}{b}$ is equivalent to $a \cdot \dfrac{1}{b}$ $(b \neq 0)$.

b. $a - b$ is equivalent to $a + (-b)$.

If two terms have the same variable in the same degree as a factor, as x in $2x + 3x$, they are called **like terms** or **similar terms.** The distributive law permits us to find a "simpler" equivalent form for the sum. By the distributive law,

$$2x + 3x = (2 + 3)x = 5x.$$

Consider the sum

$$(2x^2 + 3x + 5) + (3x^2 + 6x + 8).$$

The commutative and associative laws of addition permit us to write this sum as

$$(2x^2 + 3x^2) + (3x + 6x) + (5 + 8),$$

which, by the distributive law, can be written as

$$(2 + 3)x^2 + (3 + 6)x + (5 + 8),$$

and then as

$$5x^2 + 9x + 13.$$

Thus we write the sum of two polynomials in a simpler equivalent form by applying the definitions and axioms applicable to real numbers. This process can usually be accomplished mentally.

Examples

a. $(3x^2 + 4x + 7) + (2x^2 + 3x + 2) = 5x^2 + 7x + 9$

b. $(2x^2 + 3xy + 4y^2) + (x^2 + xy + 2y^2) = 3x^2 + 4xy + 6y^2$

Since the difference $a - b$ $(a, b, \in R)$ is the same as $a + (-b)$ and since each polynomial represents a real number, the difference of two polynomials can also be written as a sum. First note that the negative of a polynomial $P(x)$ can be obtained by changing the sign of each term of the polynomial. For example, if

$$P(x) = 2x^2 - 5x + 7$$

the negative of $P(x)$

$$-P(x) = -(2x^2 - 5x + 7) = -2x^2 + 5x - 7$$

because

$$(2x^2 - 5x + 7) + (-2x^2 + 5x - 7) = 0.$$

Hence, a difference such as

$$(7x^2 - 2x + 7) - (2x^2 - 8x + 6)$$

can be written equivalently as

$$(7x^2 - 2x + 7) + [-(2x^2 - 8x + 6)],$$

or

$$(7x^2 - 2x + 7) + (-2x^2 + 8x - 6),$$

and then as

$$5x^2 + 6x + 1.$$

Again, this process can usually be accomplished mentally.

Examples

a. $(7x^2 + 4x - 3) - (2x^2 + x - 4) = 7x^2 + 4x - 3 - 2x^2 - x + 4$

$$= 5x^2 + 3x + 1$$

b. $(6xy^2 + 3xy - 2y^3) - (5xy - 4y^3) = 6xy^2 + 3xy - 2y^3 - 5xy + 4y^3$

$$= 6xy^2 - 2xy + 2y^3$$

   In the foregoing examples we say that *like terms* have been combined and the expression has been simplified. Admittedly, what constitutes a simpler form for an algebraic expression is a matter of opinion and generally is determined by how the expression is to be used. In working with polynomials, we use the phrase "simplest form" to refer to the form of the polynomial in which no two terms are alike. The important idea, no matter what we call it, is that when like terms are combined, we are applying the distributive law, and that the algebraic expressions both before and after combining like terms represent the same real number for the same replacement of the variables by real numbers. The expressions are *equivalent*.

   Although it is not necessary to invoke formally an axiom, definition, or theorem of our number system each time an expression is rewritten equivalently, all such steps can be so justified. If you do not think in terms of the justifications, there is the danger that you will soon be manipulating symbols in a mechanical manner which may result in your equating expressions that are not equivalent, that is, which do not represent the same real numbers for the same real number replacements of the variables.

Exercise 6.3

A

Name the axiom, definition, or theorem that justifies each statement.

Example. $6x + (3y + 5x)$

a.	$= 6x + (5x + 3y)$	Commutative law of addition
b.	$= (6x + 5x) + 3y$	Associative law of addition
c.	$= (6 + 5)x + 3y$	Distributive law
d.	$= 11x + 3y$	Basic numeral for $6 + 5$

1. $3y + 2y$

    a. $= (3 + 2)y$

    b. $= 5y$

2. $7x + 3x$

    a. $= (7 + 3)x$

    b. $= 10x$

3. $(2x + 5x) + (y - 3y)$

    a. $= (2 + 5)x + (1 - 3)y$

    b. $= (2 + 5)x + [1 + (-3)]y$

    c. $= 7x - 2y$

4. $(6x - 4x) + (2y - 3y)$

    a. $= (6 - 4)x + (2 - 3)y$

    b. $= [6 + (-4)]x + [2 + (-3)]y$

    c. $= 2x - y$

5. $4x + (2y + x)$

    a. $= 4x + (x + 2y)$

    b. $= (4x + x) + 2y$

    c. $= (4 + 1)x + 2y$

    d. $= 5x + 2y$

6. $5x^2 + (3y - 2x^2)$

    a. $= 5x^2 + (-2x^2 + 3y)$

    b. $= [5x^2 + (-2x^2)] + 3y$

    c. $= [5 + (-2)]x^2 + 3y$

    d. $= 3x^2 + 3y$

Reduce the number of terms in each polynomial by combining similar terms. Perform as many steps as possible mentally.

Examples

a. $5x^2 - 2x + x^2 + 6x = 5x^2 + x^2 - 2x + 6x$

$$= 6x^2 + 4x$$

b. $3x^2y - 7xy + 2x + 4xy^2 + 2xy - 5x + 2x^2y$

$$= 3x^2y + 2x^2y - 7xy + 2xy + 2x - 5x + 4xy^2$$
$$= 5x^2y - 5xy - 3x + 4xy^2$$

7. $4x^2 - 3x + 2x^2 - 5x$

8. $6y - 2y^2 - 5y + 4y^2$

9. $7p^2 + 2q + 2p^2 - 3q$

10. $4h^2 - 3k + h^2 + 9k$

11. $2x^2 + 5xy + 2y^2 - 4xy$

12. $7xy - x^2 - 4xy + 2y^2$

13. $9rs - 5t + 12u - rs + 5t - 7u$

14. $2x^2y + 6y - 4x + 3x^2y - 7y + 3x$

Write each sum as a polynomial in simplest form.

Examples

a. $(4x^2 - 2x) + (x^2 + 1) = 4x^2 + x^2 - 2x + 1 = 5x^2 - 2x + 1$

b. $(3x^2 + 2x + 1) + (7x^2 - 3x - 2) = 3x^2 + 7x^2 + 2x - 3x + 1 - 2$

$$= 10x^2 - x - 1$$

15. $(3x^2 + 2x - 1) + (4x^2 + 3)$
16. $(7x^2 + 2x - 3) + (x^2 + 2x)$

17. $(2y^2 - y + 1) + (2y + 1)$
18. $(4t^2 - 2t - 1) + (3t^2 - 1)$

19. $(2x^3 - 7x + 1) + (5x^2 - 3)$
20. $(x^4 + x^2 + 1) + (x^3 - 1)$

21. $(x^2y - xy + xy^2) + (2x^2y + 3xy)$
22. $(r^2s^2 + 2rs + 1) + (3 - rs)$

23. $(5x^2 - 3x + 6) + (-3x^2 - x - 4)$

24. $(3y^2 + 5y + 7) + (-7y^2 - 5y - 6)$

25. $(9t^2 + 2t - 7) + (7t^2 - t - 2)$

26. $(7z^2 - 6z + 8) + (5z^2 - 3z + 9)$

27. $(4x^2y + 8xy - 3xy^2) + (-6x^2y + 4xy - 2xy^2)$

28. $(2 + 2st - 4s^2t^2) + (-3s^2t^2 - 6st - 3)$

Write each difference as a polynomial in simplest form.

Example. $(2x^2 - 3xy + 5y^2) - (3x^2 + 2xy - y^2)$

$$= (2x^2 - 3xy + 5y^2) + [-(3x^2 + 2xy - y^2)]$$

$$= (2x^2 - 3xy + 5y^2) + (-3x^2 - 2xy + y^2)$$

$$= 2x^2 - 3x^2 - 3xy - 2xy + 5y^2 + y^2$$

$$= -x^2 - 5xy + 6y^2$$

29. $(z^3 - 3z) - (2z^2 + z)$
30. $(y^2 + 4) - (y^2 - 3)$

31. $(2p^2 - 3p + 1) - (2p^2 + 1)$
32. $(x^2 - 3x + 1) - (2x^2 - 6x)$

33. $(3x^2 - 2x - 1) - (-2x - 3)$
34. $(7r^2 + 2r - 3) - (r^2 + 2r)$

35. $(xy^2 + xy - x^2y) - (2x^2y - 3xy)$
36. $(2xy^2 + 3xy - x) - (2xy + x)$

37. $(7x^2 - x + 4) - (-3x^2 + 5x - 4)$

38. $(6y^2 - 4y + 3) - (-2y^2 + 3y - 5)$

39. $(8t^2 - 5t - 1) - (-6t^2 - 7t + 2)$

40. $(6z^2 + 2z - 5) - (-3z^2 - 5z + 4)$

41. $(-3x^2y + 4xy + 6xy^2) - (9x^2y + 4xy - 6xy^2)$

42. $(2 - 7rs - 7r^2s^2) - (7 - rs - 8r^2s^2)$

B

Example. $2 - [2 - (x + 1)] + (x + 1) = 2 - [2 + (-x - 1)] + (x + 1)$

$$= 2 - (2 - x - 1) + (x + 1)$$
$$= 2 + (-(2 - x - 1)) + (x + 1)$$
$$= 2 + (-2 + x + 1) + (x + 1)$$
$$= 2 - 2 + x + 1 + x + 1$$
$$= 2x + 2$$

43. $5 - [5 - (x - 4)] + (x - 4)$ 44. $-3 - [-6 - (x + 2)] + (x - 2)$

45. $(x - 1) - [(x + 4) - 3] + 1$ 46. $(x + 6) - [(x - 3) + 6] + 4$

47. $-7 - [(4 - x) - 2] - (-2 + x)$ 48. $2 - [(-8 + x) - 5] + (1 + x)$

49. $(x + 7) - [(x - 6) - (x - 5)] + 4$

50. $(x - 3) - [(x + 1) - (x + 1)] + 9$

6.4 Products and Quotients of Monomials

Polynomials represent real numbers when the replacement sets for the variables are composed of real numbers. Hence the axioms, theorems, and definitions applicable to the set of real numbers enable us to express certain products and quotients in more useful equivalent forms. Consider the product $4x \cdot 3y$. The commutative and associative laws of multiplication enable us to write this as the product of the numerical coefficients times the product of the variable factors. Thus,

$$4x \cdot 3y = (4 \cdot 3) \cdot (x \cdot y) = 12xy.$$

Now, consider the product $x^2 \cdot x^3$. Since x^2 means $x \cdot x$ and x^3 means $x \cdot x \cdot x$, we have

$$x^2 \cdot x^3 = xx \cdot xxx = x^5.$$

This example suggests that the product of two monomials in one and the same variable can be written equivalently as a power of that variable with an exponent which is the sum of the exponents in the two monomials. More generally, we have

$$a^m \cdot a^n = \overbrace{(a \cdot a \cdot a \cdots a)}^{m \text{ factors}} \cdot \overbrace{(a \cdot a \cdot a \cdots a)}^{n \text{ factors}}$$

$$= \overbrace{(a \cdot a \cdot a \cdots a)}^{m + n \text{ factors}}.$$

The result of this argument can be stated as follows.

Theorem 6.1.   If $a \in R$ and $m,\ n \in N$, then

$$a^m \cdot a^n = a^{m+n}.$$

When a product is rewritten using this theorem, the expression is said to be simplified.

Examples.   a. $2x^2 \cdot 3x^5 = 2 \cdot 3 \cdot x^{2+5}$ b. $2xy \cdot 5y^2z = 2 \cdot 5 \cdot xy^{1+2}z$

$$= 6x^7 \qquad\qquad\qquad = 10xy^3z$$

Quotients of monomials can also be written in several equivalent forms. First let us recall that division by zero is undefined. Unless otherwise specified, we assume that in the following discussion and exercises *variables do not have replacements that would result in a zero divisor.*

Consider the quotient x^7/x^4. By Theorem 4.1 we can write

$$\frac{x^7}{x^4} = x^7 \cdot \frac{1}{x^4}$$

$$= (x^3 \cdot x^4) \cdot \frac{1}{x^4}$$

$$= x^3 \cdot \left(x^4 \cdot \frac{1}{x^4}\right) = x^3 \cdot 1 = x^3.$$

More generally, we have that if $a \in R$ $(a \neq 0)$, $m, n \in N$, and $m > n$,

$$\frac{a^m}{a^n} = a^m \cdot \frac{1}{a^n} = \overbrace{(a \cdot a \cdot a \cdots a}^{m-n \text{ factors}} \cdot \overbrace{a \cdot a \cdot a \cdots a}^{n \text{ factors}}) \cdot \frac{1}{\underbrace{a \cdot a \cdot a \cdots a}_{n \text{ factors}}}$$

$$= \overbrace{a \cdot a \cdot a \cdots a}^{m-n \text{ factors}} \cdot \left(\overbrace{a \cdot a \cdot a \cdots a}^{n \text{ factors}} \cdot \frac{1}{\underbrace{a \cdot a \cdot a \cdots a}_{n \text{ factors}}} \right)$$

$$= a^{m-n} \cdot 1$$

$$= a^{m-n}.$$

The result of this argument can be stated as follows.

Theorem 6.2. If $a \in R$ $(a \neq 0)$, $m, n \in N$, and $m > n$, then

$$\frac{a^m}{a^n} = a^{m-n}.$$

When a quotient is rewritten using this theorem, the expression is said to be simplified.

Examples. a. $\dfrac{y^5}{y^2} = y^{5-2}$ b. $\dfrac{12x^3y^4}{4xy^3} = \dfrac{12}{4} \cdot \dfrac{x^3}{x} \cdot \dfrac{y^4}{y^3}$

$\qquad\qquad\quad = y^3$ $\qquad\qquad\qquad\quad = 3 \cdot x^{3-1} \cdot y^{4-3}$

$\qquad\qquad\qquad\qquad\qquad\qquad\qquad = 3x^2y$

Note that in Theorem 6.2, the exponents m and n were restricted so that $m > n$. The cases where $m = n$ or $m < n$ are discussed in the next section.

Exercise 6.4

A

Justify each statement by an appropriate axiom, definition, or theorem.

Example. $(2x^2)(xy)$

    a. $= (2x^2 \cdot x)(y)$ Associative law of multiplication

    b. $= 2x^{2+1} \cdot y$ Theorem 6.1

    c. $= 2x^3 y$ Basic numeral for $2 + 1$

1. $x^3 \cdot x^2$

    a. $= x^{3+2}$

    b. $= x^5$

2. $y^3 \cdot y^5$

    a. $= y^{3+5}$

    b. $= y^8$

3. $\dfrac{x^7}{x^2}$

    a. $= x^{7-2}$

    b. $= x^5$

4. $\dfrac{y^9}{y^3}$

    a. $= y^{9-3}$

    b. $= y^6$

5. $(4r)(r^2 s^3)$

    a. $= (4r \cdot r^2) \cdot s^3$

    b. $= 4r^{1+2} \cdot s^3$

    c. $= 4r^3 s^3$

6. $(5p^2)(p^3 q^2)$

    a. $= (5p^2 \cdot p^3) \cdot q^2$

    b. $= 5p^{2+3} \cdot q^2$

    c. $= 5p^5 q^2$

Example. $\dfrac{x^2 y^4}{xy^2}$

    a. $= \dfrac{x^2}{x} \cdot \dfrac{y^4}{y^2}$ Theorem 4.7

    b. $= x^{2-1} \cdot y^{4-2}$ Theorem 6.2

    c. $= xy^2$ Basic numerals for $2 - 1$ and $4 - 2$

7. $\dfrac{15x^4}{3x^2}$

    a. $= \dfrac{15}{3} \cdot \dfrac{x^4}{x^2}$

    b. $= \dfrac{15}{3} \cdot x^{4-2}$

    c. $= \dfrac{15}{3} \cdot x^2$

    d. $= 5 \cdot x^2$

8. $\dfrac{-9x^5}{3x}$

    a. $= \dfrac{-9}{3} \cdot \dfrac{x^5}{x}$

    b. $= -\left(\dfrac{9}{3}\right) \cdot \dfrac{x^5}{x}$

    c. $= -3 \cdot \dfrac{x^5}{x}$

    d. $= -3 \cdot x^{5-1}$

    e. $= -3x^4$

In the following problems, do as much of the work as possible mentally. Write each product in simpler form.

Examples. a. $x^3 \cdot x^4$ b. $(2xy^2)(3x^2y)$ c. $(4x^2y)(2xy^2)(3x^3y^2)$

Solutions. a. x^7 b. $6x^3y^3$ c. $24x^6y^5$

9. $y^2 \cdot y^4$ 10. $z^3 \cdot z$ 11. $x^4 \cdot x^5$

12. $p \cdot p^4$ 13. $(-x^2)(x^3)$ 14. $(-y^3)(-y^4)$

15. $(2xy)(3xy^2)$ 16. $(3r^2s)(2rs^2)$ 17. $(5pq^2)(-7pq^2)$

18. $(-2s^2t)(3st^2)$ 19. $(-4x^2y^2)(-2xy^3)$ 20. $(-5wz^3)(-4w^3z)$

21. $(4x)(2x)(3x)$ 22. $(-y)(5y)(-2y)$ 23. $(2x^2)(-3x)(4x^3)$

24. $(3x^3)(-2x^2)(-6x)$ 25. $(xy^2)(-x^2y)(-3xy)$ 26. $(-4xy^3)(-xy^2)(-3x^3)$

Write each quotient in simpler form.

Examples. a. $\dfrac{x^5}{x^3}$ b. $\dfrac{2x^4}{x}$ c. $\dfrac{6x^2y^5}{3xy^2}$ d. $\dfrac{-14x^3y^2z^4}{2xyz^2}$

Solutions. a. x^2 b. $2x^3$ c. $2xy^3$ d. $-7x^2yz^2$

27. $\dfrac{r^2}{r}$ 28. $\dfrac{4x^2}{2x}$ 29. $\dfrac{-10x^4}{5x^2}$ 30. $\dfrac{-8y^3}{2y^2}$

31. $\dfrac{8r^5s^4}{-4r^3s^2}$ 32. $\dfrac{4h^5j^3}{-4hj}$ 33. $\dfrac{4x^4yz^3}{2xz^2}$ 34. $\dfrac{18x^2y^3z^2}{9xyz}$

35. $\dfrac{-25x^4yz^4}{5xz}$ 36. $\dfrac{-27r^2s^3t}{3rs^2}$ 37. $\dfrac{16x^4yz^3}{-8x^3z^2}$ 38. $\dfrac{33r^4s^4t^3}{-11rs^2t}$

39. $\dfrac{x^{14}y^{10}}{x^6y^9}$ 40. $\dfrac{y^{12}z^{14}}{y^5z}$ 41. $\dfrac{-14x^4y^3z^2}{-28x^3y^2z}$ 42. $\dfrac{-56x^4y^7z^2}{-28x^3y^2z}$

Simplify each expression.

Examples. a. $4x^3 - x(3x^2)$ b. $(3x)(xy^3) + 4x^2y^3$

Solutions. a. $4x^3 - x(3x^2) = 4x^3 - 3x^3$ b. $3x(xy^3) + 4x^2y^3 = 3x^2y^3 + 4x^2y^3$
$$= x^3 \qquad\qquad\qquad\qquad = 7x^2y^3$$

43. $5t^4 + 2t(3t^3)$

44. $3y^2(2y) - 6y^3$

45. $4p^2(-2q^2) + (3pq)(2pq)$

46. $(-2x^2y)(-2xy^2) - (-3x^3y)(y^2)$

47. $(3x)(y^2) + (3x^2)(y) - x^2y$

48. $(x)(x^2) + (y^2)(2y) - x^3$

49. $x + x(xy) + x^2y$

50. $y - (x^2)(xy) + 2x^3y$

51. $(4x)(2x^2) - (x^2)(x^3) - (7x^2)(x)$

52. $(2y)(y^3) - (2y^4)(3y) + (3y^2)(-y^2)$

53. $(3x)(-3x^2)(5x^3) + (4x^2)(5x^2)(4x^2)$

54. $(6y^2)(3y^3)(-2y) - (-7y^2)(5y)(2y^3)$

B

55. Write the law of exponents for multiplication in symbolic form.

56. Write the law of exponents for division in symbolic form.

Rewrite each expression with a single exponent for each variable. Assume no denominator equals zero.

57. $(x^3)^2$

58. $(x^2)^4$

59. $(2y)^2$

60. $(xy)^3$

61. $\left(\dfrac{2}{y}\right)^2$

62. $\left(\dfrac{x}{y}\right)^3$

63. $\left(\dfrac{xy}{rs}\right)^3$

64. $\left(\dfrac{x^2y}{rs^2}\right)^2$

65. Using Exercises 57 and 58 as guides, suggest another symbol for $(x^m)^n$, $(m, n \in N)$ which does not contain parentheses.

66. Using Exercises 59 and 60 as guides, suggest another symbol for $(xy)^m$, $(m \in N)$ which does not contain parentheses.

67. Using Exercises 61 and 62 as guides, suggest another symbol for $(x/y)^m$ $(m \in N)$ which does not contain parentheses.

6.5   Integer Exponents; Scientific Notation

Now let us turn to the quotient a^m/a^n for those cases in which it is not true that $m > n$ as required in Theorem 6.2.

First consider the quotient, $a^m/a^n = a^{m-n}$, $(m, n \in N)$ where $m = n$. In this case, if Theorem 6.2 were to hold,

$$\frac{a^m}{a^n} = a^{m-n} = a^{n-n} = a^0.$$

Furthermore, if $m = n$, then $a^m = a^n$ and using Theorem 4.1, we have

$$\frac{a^m}{a^n} = \frac{a^n}{a^n} = a^n \cdot \frac{1}{a^n} = 1.$$

Therefore, if we wish Theorem 6.2 to hold for the case where $m = n$, we must use a^0 as a symbol for the multiplicative identity element 1. This suggests the following definition.

Definition 6.2. For all $a \in R$ $(a \neq 0)$,
$$a^0 = 1.$$

Examples. a. $7^0 = 1$ b. $(1{,}798{,}241)^0 = 1$ c. $(0.00012)^0 = 1$

Now, if we assume Theorem 6.2 to hold for the case where $m < n$, we have, for example,

$$\frac{x^4}{x^7} = x^{4-7} = x^{-3}.$$

However, the symbol, x^{-3}, has not been defined. What meaning shall we give to it? We know that

$$\frac{x^4}{x^7} = x^4 \cdot \frac{1}{x^7}$$

$$= x^4 \cdot \left(\frac{1}{x^4} \cdot \frac{1}{x^3}\right)$$

$$= \left(x^4 \cdot \frac{1}{x^4}\right) \cdot \frac{1}{x^3} = \frac{1}{x^3}.$$

Therefore, if we want Theorem 6.2 to hold for the quotient, x^4/x^7, then x^{-3} must be defined to be equal to $1/x^3$. This example suggests the following definition.

Definition 6.3. For all $a \in R$ $(a \neq 0)$ and $m \in J$,
$$a^{-m} = \frac{1}{a^m}.$$

Examples

a. $5^{-2} = \dfrac{1}{5^2}$ b. $\dfrac{1}{x^{-3}} = \dfrac{1}{\frac{1}{x^3}} = 1 \cdot \dfrac{x^3}{1} = x^3$ c. $\left(\dfrac{a}{b}\right)^{-1} = \dfrac{1}{\frac{a}{b}} = 1 \cdot \dfrac{b}{a} = \dfrac{b}{a}$

Observe that a^{-1} and $1/a$ are not polynomials, but that we have now given meaning to a^m/a^n for *all* $m, n \in J$.

Consider the following powers of 10 as special cases of the definitions we have made:

$$10^0 = 1,$$

$$10^1 = 10, \qquad 10^{-1} = \frac{1}{10} = 0.1,$$

$$10^2 = 100, \qquad 10^{-2} = \frac{1}{10^2} = \frac{1}{100} = 0.01,$$

$$10^3 = 1000, \qquad 10^{-3} = \frac{1}{10^3} = \frac{1}{1000} = 0.001.$$

Any rational number can be written as the product of a number between 1 and 10 and a power of 10. Such exponential form is called **scientific notation.**

Examples.   a. $3476 = 3.476 \times 1000 = 3.476 \times 10^3$

b. $27.02 = 2.702 \times 10 = 2.702 \times 10^1$

c. $0.0215 = 2.15 \times \frac{1}{100} = 2.15 \times 10^{-2}$

If you consider the numbers used by scientists, you will see the advantage of scientific notation. For example, the computations for the path of a space vehicle involve the mass of the Earth which is approximately

$5,977,000,000,000,000,000,000,000,000$ grams $= 5.977 \times 10^{27}$ g.,

the speed of light is approximately

$29,979,300,000$ centimeters per second $= 2.99793 \times 10^{10}$ cm./sec.,

the mass of a hydrogen atom is approximately

$0.0000000000000000000000001673$ gram $= 1.673 \times 10^{-24}$ g.,

and the Ångstrom unit, a number used in measuring wavelengths of light, is approximately

0.00000001 centimeter $= 1 \times 10^{-8}$ cm.

Recall that you can multiply a number by a power of 10 by moving the decimal point to the *right* a number of places equal to the exponent of the power, and that you can divide by moving the decimal point to the *left* a number of places equal to the exponent of the power. Hence, numbers can be readily written in a variety of ways as powers of 10. Some of these are often more useful than scientific notation form.

Examples.   a. $2340 = 234 \times 10$

$$= 23.4 \times 10^2$$
$$= 2.34 \times 10^3$$

b. $0.081 = 0.81 \times 10^{-1}$

$$= 8.1 \times 10^{-2}$$
$$= 81 \times 10^{-3}$$
$$= 810 \times 10^{-4}$$

Numbers expressed in scientific notation can easily be written as basic numerals.

Examples.   a. $4.72 \times 10^3 = 4.72 \times 1000 = 4720$

b. $2.012 \times 10^{-2} = 2.012 \times \dfrac{1}{100} = 0.02012$

c. $3.064 \times 10^0 = 3.064 \times 1 = 3.064$

Exercise 6.5

A

State the axiom, definition, or theorem that justifies each statement.

Examples   a. 3^{-2} b. $\dfrac{2^{-1}}{3^{-2}}$

$= \dfrac{1}{3^2}$	Definition 6.3
$= \dfrac{1}{3 \cdot 3}$	Definition 6.1
$= \dfrac{1}{9}$	Basic numeral

$= \dfrac{\frac{1}{2}}{\frac{1}{3^2}}$	Definition 6.3
$= \dfrac{1}{2} \cdot \dfrac{1}{\frac{1}{3^2}}$	Theorem 4.1
$= \dfrac{1}{2} \cdot \dfrac{3^2}{1}$	Theorem 4.8
$= \dfrac{3^2}{2}$	Theorem 4.7
$= \dfrac{3 \cdot 3}{2}$	Definition 6.1
$= \dfrac{9}{2}$	Basic numeral

1. $4^{-1} = \dfrac{1}{4}$ 2. $7^{-1} = \dfrac{1}{7}$ 3. $5^0 = 1$ 4. $(-2)^0 = 1$

5. 2^{-3} 6. 4^{-2} 7. $\left(\dfrac{1}{2}\right)^{-1}$

   a. $= \dfrac{1}{2^3}$ a. $= \dfrac{1}{4^2}$ a. $= \dfrac{1}{\dfrac{1}{2}}$

   b. $= \dfrac{1}{2 \cdot 2 \cdot 2}$ b. $= \dfrac{1}{4 \cdot 4}$ b. $= 1 \cdot \dfrac{2}{1}$

   c. $= \dfrac{1}{8}$ c. $= \dfrac{1}{16}$ c. $= 1 \cdot 2$

                                                   d. $= 2$

8. $\left(\dfrac{3}{5}\right)^{-1}$ 9. $\dfrac{2^0}{3^{-1}}$ 10. $\dfrac{3^{-2}}{4^0}$

   a. $= \dfrac{1}{\dfrac{3}{5}}$ a. $= \dfrac{1}{3^{-1}}$ a. $= \dfrac{3^{-2}}{1}$

   b. $= 1 \cdot \dfrac{5}{3}$ b. $= \dfrac{1}{\dfrac{1}{3}}$ b. $= 3^{-2}$

   c. $= \dfrac{5}{3}$ c. $= 1 \cdot \dfrac{3}{1}$ c. $= \dfrac{1}{3^2}$

                              d. $= 3$ d. $= \dfrac{1}{3 \cdot 3}$

                                                   e. $= \dfrac{1}{9}$

In the following problems, do as much of the work as possible mentally. Write each expression equivalently without using negative or zero exponents.

Examples.   a. $3 \cdot 2^{-1} = 3 \cdot \dfrac{1}{2}$ b. $\dfrac{3^{-2} \cdot 2^0}{3^2} = \dfrac{2^0}{3^2 \cdot 3^2}$

                     $= \dfrac{3}{2}$ $= \dfrac{1}{3^4} = \dfrac{1}{81}$

11. $2 \cdot 3^{-2}$

12. $5 \cdot 2^{-3}$

13. $\dfrac{2^{-1}}{2}$

14. $\dfrac{5}{5^{-2}}$

15. $3^2 \cdot 4^{-3}$

16. $5^{-1} \cdot 2^2$

17. $\left(\dfrac{1}{3}\right)^{-2}$

18. $\left(\dfrac{1}{2}\right)^{-3}$

19. $\dfrac{4^{-1}}{2^{-2}}$

20. $\dfrac{3^{-2}}{9^{-3}}$

21. $\dfrac{5 \cdot 3^{-1}}{4^0}$

22. $\dfrac{2^{-3} \cdot 4^{-2}}{5^0}$

Express each power equivalently with a positive exponent. All variables represent nonzero real numbers.

Examples. a. $x^{-3} \cdot x^5 = \dfrac{1}{x^3} \cdot x^5$ b. $\dfrac{x^{-1}}{y^{-4}} = \dfrac{\dfrac{1}{x}}{\dfrac{1}{y^4}}$

$\qquad\qquad\qquad = \dfrac{x^5}{x^3}$ $\qquad\qquad\qquad = \dfrac{1}{x} \cdot \dfrac{y^4}{1}$

$\qquad\qquad\qquad = x^{5-3}$ $\qquad\qquad\qquad = \dfrac{y^4}{x}$

$\qquad\qquad\qquad = x^2$

23. $x^{-3} \cdot x^4$

24. $x^2 \cdot x^{-3}$

25. $\dfrac{x^{-2}}{x^5}$

26. $\dfrac{x^5}{x^{-2}}$

27. $\dfrac{x^{-1}}{y^{-1}}$

28. $\dfrac{x^{-2}}{y^{-3}}$

29. $x^{-2} \cdot y^4$

30. $x^{-2} \cdot y^{-3}$

Express each of the following in scientific notation.

Examples

a. $752{,}000 = 7.52 \times 100{,}000$ b. $0.0000321 = \dfrac{3.21}{100{,}000} = \dfrac{3.21}{10^5}$

$\qquad\qquad = 7.52 \times 10^5$ $\qquad\qquad\qquad\qquad = 3.21 \times 10^{-5}$

31. 14

32. 247

33. 7,145

34. 21,607

35. 62,000

36. 17,632,000

37. 0.5

38. 0.039

39. 0.0026

40. 0.000702

41. 0.00000516

42. 0.0000001428

Write each of the following in decimal notation.

Examples

a. $1.25 \times 10^4 = 1.25 \times 10{,}000$ b. $3.72 \times 10^{-7} = \dfrac{3.72}{10^7} = \dfrac{3.72}{10{,}000{,}000}$

$= 12{,}500$ $= 0.000000372$

43. 1.02×10^3 44. 1.72×10^4 45. -2.47×10^2

46. -5×10^6 47. 2.07×10^8 48. 1.23×10^9

49. 2.02×10^{-2} 50. 4.7×10^{-3} 51. 5.45×10^{-5}

52. 6.643×10^{-6} 53. -2.2×10^{-4} 54. -3.73×10^{-7}

B

Express each quotient as a power of 10.

55. $\dfrac{10^3 \times 10^{-7} \times 10^2}{10^{-2} \times 10^4}$

56. $\dfrac{10^2 \times 10^5 \times 10^{-3}}{10^2 \times 10^2}$

Express each quotient in scientific notation.

57. $\dfrac{0.00084 \times 0.093}{0.00021 \times 0.0031}$

58. $\dfrac{0.65 \times 50 \times 3.3}{1.3 \times 0.011 \times 0.5}$

6.6 Products of Polynomials

In Chapter 2 we observed that the distributive law is the basis for writing expressions for products in which one factor is a sum as the sum of two products; that is, if a, b, $c \in R$,

$$a \cdot (b + c) = a \cdot b + a \cdot c.$$

Since polynomials represent real numbers for all real number replacements of the variables, we can apply the distributive law to polynomials. For example,

$$2x^2y(xy + 3y^2) = 2x^2y \cdot xy + 2x^2y \cdot 3y^2$$
$$= 2x^3y^2 + 6x^2y^3.$$

The distributive law can be extended to cases in which the sum consists of more than two terms.

Examples.   a. $3x^2y(x^2 + xy + y^2)$ b. $-2xy^3(x^3 - 2xy + 3y)$

Solutions. The diagrams show a suggested order in which to consider the products.

a.

$$= \overset{1}{3x^2y \cdot x^2} + \overset{2}{3x^2y \cdot xy} + \overset{3}{3x^2y \cdot y^2}$$

$3x^2y \ (x^2 \quad + \quad xy \quad + \quad y^2)$

$$= 3x^4y + 3x^3y^2 + 3x^2y^3.$$

b.

$$= \overset{1}{-2xy^3 \cdot x^3} - \overset{2}{2xy^3(-2xy)} - \overset{3}{2xy^3 \cdot 3y}$$

$-2xy^3 \ (x^3 \quad - \quad 2xy \quad + \quad 3y)$

$$= -2x^4y^3 + 4x^2y^4 - 6xy^4.$$

Now consider the product of two binomials such as $(x + 3)(x + 7)$. Since a polynomial represents a real number for each real number replacement of its variable, we can view the first binomial as a single number in applying the left-hand distributive law to obtain

$$(x + 3)(x + 7) = (x + 3) \cdot x + (x + 3) \cdot 7.$$

The right-hand distributive law can then be applied to obtain

$$(x + 3) \cdot x + (x + 3) \cdot 7 = x \cdot x + 3 \cdot x + x \cdot 7 + 3 \cdot 7 \qquad (1)$$

$$= x^2 + 3x + 7x + 21$$

$$= x^2 + 10x + 21.$$

Observe from the right-hand member of Equation (1) that each term of one binomial factor in the original product has been multiplied by each term of the other factor.

Example. Express the product $(x + 5)(x + 3)$ as a polynomial.

Solution. The diagram shows a suggested order of multiplication.

$(x \ + \ 5) \ (x \ + \ 3)$

$$= \overset{1}{x \cdot x} + \overset{2}{5 \cdot x} + \overset{3}{3 \cdot x} + \overset{4}{5 \cdot 3}$$

$$= x^2 + 5x + 3x + 15$$

$$= x^2 + 8x + 15$$

Exercise 6.6

A

State the axiom, definition, or theorem that justifies each statement

Example

$$3x(x + 1) = 3x \cdot x + 3x \cdot 1 \qquad \text{Distributive law}$$
$$= 3x^2 + 3x \cdot 1 \qquad \text{Definition 6.1}$$
$$= 3x^2 + 3x \qquad \text{Multiplicative identity law}$$

1. $3(x + 1) = 3x + 3$

2. $2(y + 3) = 2y + 6$

3. $2p(x + p)$

   a. $= 2px + 2pp$

   b. $= 2px + 2p^2$

4. $5q(q + x)$

   a. $= 5qq + 5qx$

   b. $= 5q^2 + 5qx$

5. $-6(-x + y)$

   a. $= -6(-x) + (-6)y$

   b. $= 6x + (-6)y$

   c. $= 6x - 6y$

6. $3y(x + y)$

   a. $= 3y(x) + 3y(y)$

   b. $= 3 \cdot yx + 3 \cdot yy$

   c. $= 3 \cdot xy + 3 \cdot yy$

   d. $= 3xy + 3y^2$

Write each product equivalently as a polynomial in simplest form. Do as much of the work as possible mentally.

Examples. a. $x^2(x^2 + xy + y^2)$ b. $st(2s - 3t)$

Solutions. a.

$$x^2 \quad (x^2 \quad + \quad xy \quad + \quad y^2) = x^4 + x^3y + x^2y^2$$

    b.

$$st \quad (2s \quad - \quad 3t) = 2s^2t - 3st^2$$

7. $3x(4x - y^2)$

8. $x^2(6x - 4y)$

9. $xy(x - y)$

10. $x^2y(x + 2y)$

11. $-x(-x^2 - x + 2)$

12. $-y(y^2 - x + 3)$

13. $-3y^2(-2y^2 - 3y + 1)$ 14. $-p(-4p^2 - 2p + 3)$

15. $(x^2 - 3x - 2)x$ 16. $(2y^3 - y + 1)2y^2$

State the axiom, definition, or theorem that justifies each statement.

Example. $(x + 5)(x - 7)$

$= (x + 5)x + (x + 5)(-7)$	Distributive law
$= x \cdot x + 5 \cdot x + x(-7) + 5(-7)$	Distributive law
$= x^2 + 5x + x(-7) + 5(-7)$	Definition 6.1
$= x^2 + 5x + (-7)x + 5(-7)$	Commutative law of multiplication
$= x^2 + 5x + (-7)x - 35$	Theorem 3.5
$= x^2 + [5 + (-7)]x - 35$	Distributive law
$= x^2 + (-2)x - 35$	Theorem 3.3
$= x^2 - 2x - 35$	Theorem 3.5

17. $(x + 2)(x + 3)$

  a. $= (x + 2)x + (x + 2)3$

  b. $= x \cdot x + 2 \cdot x + x \cdot 3 + 2 \cdot 3$

  c. $= x^2 + 2 \cdot x + x \cdot 3 + 6$

  d. $= x^2 + 2 \cdot x + 3 \cdot x + 6$

  e. $= x^2 + (2 + 3)x + 6$

  f. $= x^2 + 5x + 6$

18. $(z - 3)(z + 1)$

  a. $= (z - 3)z + (z - 3)1$

  b. $= (z - 3)z + z - 3$

  c. $= z \cdot z + (-3) \cdot z + z - 3$

  d. $= z^2 + (-3) \cdot z + z - 3$

  e. $= z^2 - 3z + z - 3$

  f. $= z^2 + (-3 + 1)z - 3$

  g. $= z^2 + (-2)z - 3$

  h. $= z^2 - 2z - 3$

Write each product equivalently as a polynomial in simplest form. Do as much of the work as possible mentally.

Examples.   a. $(x + 4)(x - 6)$ b. $(2x + 3)(5x - 2)$

Solutions on page 172.

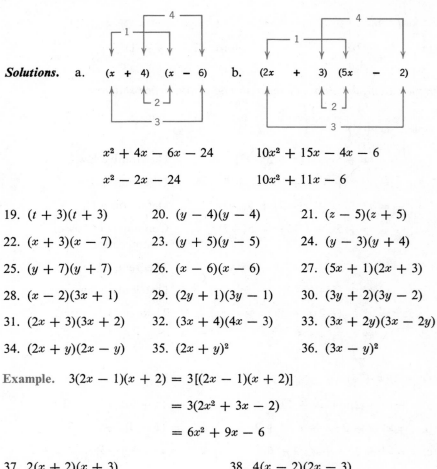

Solutions.   a. $(x + 4)$ $(x - 6)$ b. $(2x + 3)$ $(5x - 2)$

$$x^2 + 4x - 6x - 24 \qquad\quad 10x^2 + 15x - 4x - 6$$

$$x^2 - 2x - 24 \qquad\qquad 10x^2 + 11x - 6$$

19. $(t + 3)(t + 3)$ 20. $(y - 4)(y - 4)$ 21. $(z - 5)(z + 5)$

22. $(x + 3)(x - 7)$ 23. $(y + 5)(y - 5)$ 24. $(y - 3)(y + 4)$

25. $(y + 7)(y + 7)$ 26. $(x - 6)(x - 6)$ 27. $(5x + 1)(2x + 3)$

28. $(x - 2)(3x + 1)$ 29. $(2y + 1)(3y - 1)$ 30. $(3y + 2)(3y - 2)$

31. $(2x + 3)(3x + 2)$ 32. $(3x + 4)(4x - 3)$ 33. $(3x + 2y)(3x - 2y)$

34. $(2x + y)(2x - y)$ 35. $(2x + y)^2$ 36. $(3x - y)^2$

Example. $3(2x - 1)(x + 2) = 3[(2x - 1)(x + 2)]$

$$= 3(2x^2 + 3x - 2)$$

$$= 6x^2 + 9x - 6$$

37. $2(x + 2)(x + 3)$ 38. $4(x - 2)(2x - 3)$

39. $2(3x + 1)(x - 3)$ 40. $6(3x + 2)(3x - 2)$

41. $7(x + 7)(x - 7)$ 42. $x(x - 2)(2x + 5)$

43. $y(y + 2)(y - 1)$ 44. $3t(t + 1)^2$

45. $6s(2s - 1)^2$ 46. $xy(y - 3)(2y + 1)$

47. $2rs(2r + 3s)(3r + 2s)$ 48. $x^2y(x + y)(x - y)$

B

Write each product equivalently as a polynomial.

Example. $(x + y)(x^2 + 2xy + y^2)$
$$= x \cdot x^2 + x \cdot 2xy + x \cdot y^2 + y \cdot x^2 + y \cdot 2xy + y \cdot y^2$$
$$= x^3 + 2x^2y + xy^2 + x^2y + 2xy^2 + y^3$$
$$= x^3 + 3x^2y + 3xy^2 + y^3$$

49. $(x + 2)(x^2 + x + 1)$ 50. $(2x^2 - x + 1)(x - 2)$

51. $(x^2 - 5x + 6)(x^2 + 5x + 6)$ 52. $(2x^2 + 3x + 4)(4x^2 + 3x + 2)$

53. $(x^2 + xy + y^2)(x - y)$ 54. $(x - 1)(x^4 + x^3 + x^2 + x + 1)$

Products like the square of a binomial, $(a + b)^2$, or $(a - b)^2$, or that of the sum and difference of the same two numbers, $(a + b)(a - b)$, occur so frequently that rules for writing these products directly are useful. Use the distributive law to rewrite each product equivalently as a polynomial, and then formulate such a rule for each of the following three types.

55. $(x + y)^2$; $(ax + by)^2$ 56. $(x - y)^2$; $(ax - by)^2$

57. $(x + y)(x - y)$; $(ax + by)(ax - by)$

6.7 Factored Forms

Recall that we expressed composite numbers such as 6 in a factored form as $2 \cdot 3$. We also expressed the factored form of a monomial such as $3x^2$ as $3xx$. Rewriting algebraic expressions in equivalent forms is frequently much easier if any polynomials in the expression can first be expressed in a factored form. The distributive law enables us to write

$$ab + ac = a(b + c).$$

The right-hand member $a(b + c)$ is said to be the factored form of $ab + ac$. Thus, by the distributive law,

$$4x^2 + 8x = 4x \cdot x + 4x \cdot 2 = 4x(x + 2).$$

Of course, we can also write

$$4x^2 + 8x = 4(x^2 + 2x), \quad \text{or} \quad 4x^2 + 8x = 4x^2\left(1 + \frac{2}{x}\right),$$

or any other of an infinite variety of such expressions, each of which is the same product. We are, however, primarily interested in writing the factored form of a polynomial in a unique manner (except for signs and order of factors), which is called the **completely factored form.** A polynomial with integral numerical coefficients is in completely factored form if:

1. It is written as a product of polynomials with integral coefficients, and
2. No polynomial, other than a monomial, in the factored form contains another polynomial factor with integral coefficients.

Since only factors which are polynomials are to be used, the exponents of all variables must belong to the set of whole numbers. In addition, since the numerical coefficients are to be restricted to integers, we cannot have such factored forms as

$$x + 2 = 2\left(\frac{x}{2} + 1\right), \quad \text{or} \quad x^2 + 4 = x\left(x + \frac{4}{x}\right).$$

In the factoring of polynomials, it is not customary for a monomial factor to be written in completely factored form. Thus,

$$12x^3y^2 - 6x^2y = 6x^2y \cdot 2xy - 6x^2y \cdot 1 = 6x^2y(2xy - 1)$$

and the right-hand member is considered to be the completely factored form. It is not necessary to write the right-hand member as

$$2 \cdot 3 \cdot x \cdot x \cdot y \cdot (2xy - 1).$$

The commutative law of multiplication, $a \cdot b = b \cdot a$, and the theorems on the multiplication of positive and negative numbers,

$$-a \cdot b = (-a) \cdot (b) = (a) \cdot (-b),$$

and

$$a \cdot b = (-a) \cdot (-b) = (a) \cdot (b),$$

make the choice of signs and order of the factors in a factored form of a product arbitrary; the form that seems most "logical" is the one that should be used. However, this is not always easy to determine. For example,

$$x(2 - 3x - x^2) \quad \text{and} \quad -x(x^2 + 3x - 2)$$

are the same product, but it would be difficult to establish that one form is more logical than the other.

If each term of a polynomial of two or more terms contains the same monomial as a factor, the common monomial factor can be factored from such a polynomial by an application of the distributive law in the form

$ab + ac = a(b + c)$. For example, by inspection we note that each term of the polynomial

$$4x^3 + 8x^2 + 12x$$

contains $4x$ as a factor. Therefore, we write

$$4x^3 + 8x^2 + 12x = 4x(\qquad)$$

and insert the appropriate polynomial within the parentheses. This polynomial should be determined by inspection. We simply ask ourselves which monomials, when multiplied by $4x$, will give the products $4x^3$, $8x^2$, and $12x$. In this case we obtain

$$4x^3 + 8x^2 + 12x = 4x(x^2 + 2x + 3).$$

Exercise 6.7

A

Write each polynomial in completely factored form.

Example. $9x^2y + 12x^2y^2 - 3xy$

Solution.   By inspection we note that each term contains 3, x, and y as factors. Hence, $3xy$ is the common monomial factor and we first write

$$3xy(\quad + \quad - \quad).$$

By inspection we determine the terms in the parentheses such that the sum of products of $3xy$ and each term equals the polynomial $9x^2y + 12x^2y^2 - 3xy$. The completely factored form is

$$3xy(3x + 4xy - 1).$$

1. $3x + 6 = 3(?)$

2. $4s - 2 = 2(?)$

3. $6x^2 - 2x = 2x(?)$

4. $5y^2 + 10y = 5y(?)$

5. $3xy^2 + 2x^2y = xy(?)$

6. $7s^2t - 2st^2 = st(?)$

7. $9xy^2 + 6y$

8. $6p^2 - 15pq$

9. $3r + 6s - 9t$

10. $xy + xz - xw$

11. $px + py - pz$

12. $x^2 - x + xy$

13. $2x + xy + x^2$

14. $3x^2 + 6xy - 9x$

15. $18xy - 27x + 45y$

16. $x^3y + x^2y^2 - xy^3$

17. $2rs - 4r^2s^2 - 8r^3s^3$

18. $3p^2q + 9p^2q^2 - 12pq$

In Exercises 19 to 34, complete the missing factor.

Examples.   a. $-2x + 2xy = -2x(?)$ b. $x - 1 = -1(?)$

Solutions.   a. $-2x(1 - y)$ b. $-1(-x + 1)$, or $-1(1 - x)$

19. $-x - x^2 = -x(?)$ 20. $-4s - st = -s(?)$

21. $-x^3y + xy^2 = -xy(?)$ 22. $-8x^3 - 4x^2 - 4x = -4x(?)$

23. $-3x^2 - 6x + 3 = -3(?)$ 24. $-xy^3 - xy^2 - xy = -xy(?)$

25. $-x + x^2 - x^3 = -x(?)$ 26. $-9r^2 + 12r^2s - 3rs = -3r(?)$

27. $-p^2q + 2pq - pq^2 = -pq(?)$ 28. $-xy - 2x^2y + 3x^3y^2 = -xy(?)$

29. $x - 2 = -1(?)$ 30. $x^2 - 1 = -1(?)$

31. $x^2 - 2x = -1(?)$ 32. $x^2 + 3x - 4 = -1(?)$

33. $-m^2 - m - 1 = -1(?)$ 34. $-x^2 - xy - y^2 = -1(?)$

B

Factor each polynomial and then show that both the original polynomial and the factored form are equivalent for the specified replacement(s) of the variable(s).

35. $x^2 + 2x$; $x = 2$ 36. $x^3 - 2x^2 + 2x$; $x = 1$

37. $3x^2y - 5xy$; $x = 3, y = 2$ 38. $6x^3y + 4x^2y^2$; $x = 2, y = 1$

6.8 Factoring Trinomials—I

When the terms of a polynomial which contains a single variable are arranged in descending order of their degrees, we shall refer to the polynomial as being in **standard form.** Thus a polynomial written in the form $ax^2 + bx + c$ ($a \neq 0$) is a second-degree polynomial in standard form. Such a polynomial is called a **quadratic trinomial.**

Recall from Section 6.6 that several applications of the distributive law enable us to write

$$(x + 3)(x + 7) = x^2 + 3x + 7x + 21$$
$$= x^2 + 10x + 21,$$

where the coefficient 10 of the first-degree term, $10x$, in the right-hand member is the sum of the constants 3 and 7 in the binomial factors of the left-hand member. The term 21 is the product of 3 and 7.

More generally, if p and q are integers,

$$(x + p)(x + q) = x^2 + px + qx + pq$$
$$= x^2 + (p + q)x + pq,$$

where again the coefficient of the first-degree term $p + q$ in the right-hand member is the sum of the constants p and q in the left-hand member and the term pq is the product of p and q. Using the symmetric property of equality, we have

$$x^2 + (p + q)x + pq = (x + p)(x + q),$$

which suggests that a trinomial of the form $x^2 + bx + c$ is factorable into two binomials with integers as constants if there are two integers whose sum equals b and whose product equals c.

Examples. Factor the trinomials over the set of integers if possible.

a. $x^2 + 2x - 15$ b. $x^2 + 2x + 3$

Solutions

a. By inspection we determine that there are two integers, 5 and -3, whose sum is 2 and whose product is -15. Hence,

$$x^2 + 2x - 15 = (x + 5)(x - 3).$$

b. By inspection we observe that the possible factors of 3 are 3 and 1, or -3 and -1. Since neither $3 + 1$ nor $-3 + (-1)$ is 2, the trinomial is not factorable over the integers.

Multiplication is commutative. Hence, the order of the factors is immaterial and the factored form of $x^2 + 2x - 15$ in Example a above could also be expressed as $(x - 3)(x + 5)$.

Here, as in Section 6.7, we are concerned with writing polynomials in *completely factored form*. If the terms of a trinomial contain a common monomial factor, the factorization is accomplished more easily if the monomial is first factored from the polynomial as discussed in Section 6.7.

Example. Factor $4x^3 + 20x^2 + 24x$.

Solution. By inspection, we note that each term contains $4x$ as a factor. Thus,

$$4x^3 + 20x^2 + 24x = 4x(x^2 + 5x + 6).$$

We next observe that the integers 2 and 3 have a sum 5 and a product 6. Thus,

$$4x(x^2 + 5x + 6) = 4x(x + 2)(x + 3),$$

which is the completely factored form.

Now consider the polynomial $x^2 - 4$. This is a binomial, but we can view it as a trinomial with 0 as the coefficient of the first-degree term. Thus, to factor $x^2 + 0x - 4$, we can see by inspection that -2 and 2 are two numbers whose sum is 0 and whose product is -4. Hence,

$$x^2 - 4 = (x - 2)(x + 2).$$

Writing a polynomial in factored form usually involves several applications of the distributive law. In future sections we shall simply cite this law as the justification for rewriting any expression in which a polynomial of more than one term is expressed in factored form.

Exercise 6.8

A

Factor each trinomial over the set of integers if possible.

Examples.   a. $x^2 + 9x + 18$ b. $x^2 + 4x + 5$

Solutions

a. Two integers whose sum is 9 and whose product is 18 are 6 and 3. Hence,

$$x^2 + 9x + 18 = (x + 6)(x + 3).$$

b. The polynomial is not factorable over the integers because there are not two integers whose sum is 4 and whose product is 5.

1. $x^2 + 5x + 6$	2. $x^2 + 6x + 8$	3. $x^2 + 7x + 12$
4. $x^2 + 7x + 10$	5. $x^2 - 3x - 10$	6. $x^2 - x - 6$
7. $x^2 - 2x - 8$	8. $x^2 - x - 12$	9. $x^2 + x - 6$
10. $x^2 + 3x - 10$	11. $x^2 + x - 12$	12. $x^2 + 2x - 8$
13. $x^2 + 3x + 7$	14. $x^2 + 6x + 4$	15. $x^2 - 2x + 3$

16. $x^2 - 7x + 5$

17. $x^2 - 6x + 8$

18. $x^2 - 7x + 12$

19. $x^2 - 7x + 10$

20. $x^2 - 5x + 6$

21. $x^2 + 4x + 4$

22. $x^2 + 2x + 1$

23. $x^2 + 8x + 16$

24. $x^2 + 10x + 25$

25. $x^2 + 7x + 6$

26. $x^2 + 9x + 8$

27. $x^2 - 8x + 9$

28. $x^2 - 4x + 16$

29. $x^2 + x - 20$

30. $x^2 + x - 56$

31. $x^2 - 7x - 78$

32. $x^2 - 12x - 45$

Example. $x^2 - 25$

Solution. $x^2 - 25 = x^2 + 0x - 25$. Since two numbers whose product is -25 and whose sum is 0 are -5 and 5,

$$x^2 - 25 = (x - 5)(x + 5).$$

33. $x^2 - 16$

34. $x^2 - 9$

35. $x^2 - 36$

36. $x^2 - 49$

Example. $3x^2 + 12x + 9$

Solution. Note that each term of the polynomial contains 3 as a factor. Thus,

$$3x^2 + 12x + 9 = 3(x^2 + 4x + 3) = 3(x + 3)(x + 1).$$

37. $5x^2 + 20x + 15$

38. $2x^2 - 2x - 24$

39. $5x^2 + 10x - 40$

40. $3x^2 + 9x - 30$

41. $6x^2 + 36x + 24$

42. $5x^2 - 10x + 15$

43. $7x^2 + 14x + 7$

44. $5x^2 - 30x + 45$

45. $7x^2 - 70x + 175$

46. $9x^2 - 72x + 144$

47. $12x^2 + 84x + 72$

48. $8x^2 - 80x + 72$

6.9 Factoring Trinomials—II

In Section 6.8 we factored trinomials of the form $ax^2 + bx + c$ where $a = 1$. In this section we factor some polynomials in which $a \neq 1$. First let us consider a product such as $(2x + 3)(3x + 5)$. Figure 6.1a (page 180) shows how we name the terms of the binomials and Figure 6.1b shows the various products of these terms. Observe that by several applications of the distributive law, we have

$$(2x + 3)(3x + 5) = 6x^2 + 9x + 10x + 15$$
$$= 6x^2 + 19x + 15.$$

Note that the coefficient of the second-degree term of the trinomial ($6x^2$) is the product of the coefficients of the first terms of the binomial factors, the

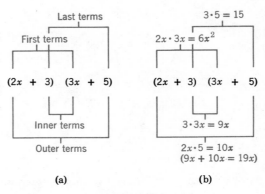

Figure 6.1

constant term (15) is the product of the last terms, and that the coefficient of the first-degree term (19x) is the sum of the product of the constants in the inner terms (3 · 3) and the product of the constants in the outer terms (2 · 5). Note also that the product of the two constants in this sum (9 + 10) is equal to the product of the coefficient of the second-degree term (6) and the constant term (15). That is, 9 · 10 = 6 · 15 = 90.

The preceding discussion suggests that a quadratic trinomial with integral coefficients can be factored into two binomials with integral coefficients if two integers exist whose sum equals the coefficient of the first-degree term and whose product equals the product of the coefficient of the second-degree term and the constant term. If no such integers exist, the trinomial is not factorable over the set of integers. In this book, we consider only factorization over this set.

In many cases the binomial factors of a quadratic trinomial can be determined by inspection. However, if this cannot be done easily, a careful consideration of the coefficients of the terms containing the variable and the constant term, as in the discussion above, offers a means by which the multiplication operation can be reversed to obtain the binomial factors, if they exist in the set of integers. In the following examples, it is assumed that factoring by inspection would be difficult.

Example. Factor $12x^2 + 17x + 6$.

Solution. We note that $12 \cdot 6 = 72$ and that 9 and 8 are two factors of 72 whose sum is 17. We can then write

$$12x^2 + 17x + 6 = 12x^2 + 9x + 8x + 6.$$

We now observe that the first two terms of the right-hand member contain the common factor $3x$ and that the last two terms contain the common factor 2. Hence

$$12x^2 + 9x + 8x + 6 = 3x(4x + 3) + 2(4x + 3).$$

The right-hand member now consists of two terms each containing the factor $(4x + 3)$; thus

$$3x(4x + 3) + 2(4x + 3) = (3x + 2)(4x + 3).$$

We can verify by multiplication that this last expression is a factored form of $12x^2 + 17x + 6$.

Example.   Factor $2x^2 + 5x + 7$.

Solution.   We observe that $2 \cdot 7 = 14$. The possible pairs of factors of 14 are: $-1, -14$; $1, 14$; $-2, -7$; and $2, 7$. Since each sum of these pairs of factors is not equal to 5, the trinomial is not factorable.

If the product of the coefficient of the second-degree term and the constant term is negative, then for any pair of factors of this product, one factor must be positive and the other must be negative. Recall that the sum of two numbers having different signs is the difference of the absolute values and has the same sign as the number with the greater absolute value.

Example. Factor $6x^2 + 5x - 6$.

Solution.   We note that $6 \cdot (-6) = -36$, and that 9 and -4 are two numbers whose product is -36 and whose sum is 5. We can then write

$$6x^2 + 5x - 6 = 6x^2 + 9x - 4x - 6$$

$$= 3x(2x + 3) - 2(2x + 3)$$

$$= (3x - 2)(2x + 3).$$

We are generally concerned with the completely factored form of a polynomial. Hence, we must always consider the possibility of a common monomial factor.

Example.   Factor $4x^3 + 14x^2 + 12x$ completely.

Solution on page 182.

Solution. Inspection of the polynomial reveals that $2x$ is a common factor. Then

$$4x^3 + 14x^2 + 12x = 2x(2x^2 + 7x + 6)$$

$$= 2x(2x + 3)(x + 2).$$

If the factored form of $2x^2 + 7x + 6$ is not evident by inspection, we can use the technique above. Thus, $2 \cdot 6 = 12$ and 4 and 3 are two factors of 12 whose sum is 7. Hence,

$$2x^2 + 7x + 6 = 2x^2 + 4x + 3x + 6$$

$$= 2x(x + 2) + 3(x + 2)$$

$$= (2x + 3)(x + 2),$$

from which we have

$$4x^3 + 14x^2 + 12x = 2x(2x + 3)(x + 2).$$

The above technique of factoring can also be applied to the factorization of certain quadratic trinomials in two variables.

Example. Factor $6x^2 + 13xy + 5y^2$.

Solution. The product $6 \cdot 5$ or 30 has two factors, 10 and 3, whose sum is 13, and the trinomial is factorable. If the binomial factors are not evident by inspection, we can write

$$6x^2 + 13xy + 5y^2 = 6x^2 + 10xy + 3xy + 5y^2$$

$$= 2x(3x + 5y) + y(3x + 5y)$$

$$= (2x + y)(3x + 5y).$$

Exercise 6.9

A

Factor each quadratic trinomial. If a trinomial is not factorable, state this as your answer.

Examples.   a. $2x^2 + 5x + 2$ b. $2x^2 + 3x + 2$

Solutions.   a. Since $2 \cdot 2 = 4 \cdot 1$ and $4 + 1 = 5$,

$$2x^2 + 5x + 2 = 2x^2 + 4x + x + 2$$
$$= 2x(x + 2) + 1(x + 2)$$
$$= (2x + 1)(x + 2).$$

b. Not factorable; $2 \cdot 2$ has no two integral factors whose sum is 3.

1. $3x^2 + 7x + 4$
2. $5x^2 + 9x + 4$
3. $4x^2 + 12x + 5$

4. $6x^2 + 13x + 6$
5. $3x^2 + 4x + 12$
6. $5x^2 + 8x + 2$

7. $5x^2 - x - 4$
8. $3x^2 - x - 4$
9. $6x^2 + 5x - 6$

10. $5x^2 + 3x - 2$
11. $3x^2 - 8x + 4$
12. $4x^2 - 9x + 5$

13. $4x^2 + 4x + 1$
14. $3x^2 + 7x + 2$
15. $4x^2 + 12x + 9$

16. $5x^2 + 12x + 7$
17. $9x^2 + 3x + 2$
18. $4x^2 + 2x + 3$

19. $5x^2 - 2x - 7$
20. $8x^2 - 3x - 5$
21. $6x^2 + x - 7$

22. $6x^2 + 7x - 5$
23. $4x^2 - 9$ (*Hint:* $4x^2 - 9 = 4x^2 + 0x - 9$.)

24. $9x^2 - 16$
25. $2x^2 - 7$
26. $9x^2 + 4$

Examples.   a. $2x^2 + 5xy - 3y^2$
   $= (2x - y)(x + 3y)$
b. $4m^2 + 16mn + 15n^2$
   $= (2m + 3n)(2m + 5n)$

27. $4s^2 + 5st + t^2$
28. $9y^2 - 2yz - 2z^2$

29. $3x^2 - 7xy + 2y^2$
30. $16r^2 + 40rs + 25s^2$

31. $16y^2 - 25z^2$
32. $49x^2y^2 - 16m^2n^2$

Examples.   a. $2x^2 + 10x + 12$
   $= 2(x^2 + 5x + 6)$
   $= 2(x + 2)(x + 3)$
b. $9m^2 - 15m - 6$
   $= 3(3m^2 - 5m - 2)$
   $= 3(3m + 1)(m - 2)$

33. $8xy^2 - 6xy - 2x$
34. $4m^2n + 12mn - 72n$

35. $18x^2 - 9x - 27$
36. $4x^3 - 10x^2y - 6xy^2$

37. $50xy^3 - 2x^3y$
38. $3x^3 - 27x$

39. $27xy^3 - 3x^3y$
40. $x^3 - x^5$

41. $4x^3 - 10x^2y - 6xy^2$
42. $6x^3y^2 + 21x^2y + 9x$

B

Factor each polynomial if possible.

43. $21x^2 + 38x + 16$

44. $12x^2 - 13x - 35$

45. $21x^2 - 2x - 55$

46. $56x^2 + 113xy + 56y^2$

47. $289x^2 - 361y^2$

48. $576x^2 - 729y^2$

49. $39x^2 - 65x - 26$

50. $55x^2y + 77xy - 66y$

6.10 Quotients

The quotient of two polynomials can be written as a fraction. If the denominator is an exact factor of the numerator, the quotient can be expressed as a polynomial. Recall that division by zero is undefined. Hence, the replacement sets for variables must be restricted so that no denominator can equal zero. For example, the quotient

$$\frac{5x^3 - 10x^2 + 15x}{5x}$$

is defined if and only if $x \neq 0$. We shall use this fraction to illustrate two ways in which some quotients can be expressed as polynomials. If the numerator is factored, the fraction can be expressed as

$$\frac{5x(x^2 - 2x + 3)}{5x},$$

which upon application of the fundamental principle of fractions reduces to

$$x^2 - 2x + 3 \qquad (x \neq 0).$$

Also, by using Theorem 4.1, the fraction can be rewritten equivalently as

$$\frac{5x^3 - 10x^2 + 15x}{5x} = (5x^3 - 10x^2 + 15x) \cdot \frac{1}{5x}$$

$$= \frac{5x^3}{5x} - \frac{10x^2}{5x} + \frac{15x}{5x}$$

$$= x^2 - 2x + 3 \qquad (x \neq 0).$$

Now consider

$$\frac{x^2 + 8x + 12}{x + 2} \qquad (x \neq -2)$$

which has a binomial as the denominator. Factoring the numerator, we obtain

$$\frac{(x + 6)(x + 2)}{x + 2},$$

which, by the fundamental principle of fractions, reduces to

$$x + 6 \qquad (x \neq - 2).$$

The polynomial form for the above quotient $(x^2 + 8x + 12)/(x + 2)$ can also be found by a process which is similar to that used in long division in arithmetic. Consider the following:

1. $\begin{array}{r} x \\ x + 2 \overline{\smash{)}\, x^2 + 8x + 12} \end{array}$ 1. Divide x, of the divisor, into x^2.

2. $\begin{array}{r} x \\ x + 2 \overline{\smash{)}\, x^2 + 8x + 12} \\ \underline{x^2 + 2x} \\ 6x + 12 \end{array}$ 2. Multiply $x + 2$, the complete divisor, by x and subtract the result from the dividend.

3. $\begin{array}{r} x + 6 \\ x + 2 \overline{\smash{)}\, x^2 + 8x + 12} \\ \underline{x^2 + 2x} \\ 6x + 12 \end{array}$ 3. Divide x, of the divisor, into $6x$, the first term of the new dividend, $6x + 12$.

4. $\begin{array}{r} x + 6 \\ x + 2 \overline{\smash{)}\, x^2 + 8x + 12} \\ \underline{x^2 + 2x} \\ 6x + 12 \\ \underline{6x + 12} \\ 0 \end{array}$ 4. Multiply $x + 2$, the complete divisor, by 6 and subtract the result from the second dividend.

Hence,

$$\frac{x^2 + 8x + 12}{x + 2} = x + 6 \qquad (x \neq -2).$$

Exercise 6.10

A

Express each quotient as a polynomial. In Problems 1 to 6, follow Example a. In Problems 7 to 12, follow Example b. State the restrictions, if any, on the variables.

Examples.   a. $\dfrac{x^3y^2 - x^2y + x}{x}$ b. $\dfrac{6x^4 - 6x^3 + 12x^2}{6x}$

$$= \frac{x(x^2y^2 - xy + 1)}{x} \qquad\qquad = \frac{6x^4}{6x} - \frac{6x^3}{6x} + \frac{12x^2}{6x}$$

$$= x^2y^2 - xy + 1 \quad (x \neq 0) \qquad = x^3 - x^2 + 2x \quad (x \neq 0)$$

1. $\dfrac{4y^2 - y}{y}$ 2. $\dfrac{2x^3 + x^2 + 3x}{x}$ 3. $\dfrac{6xy^2 - 2xy + x^2y}{xy}$

4. $\dfrac{4x^2y^2 - 2xy^2 + 2x^2y}{2xy}$ 5. $\dfrac{9x^2y^3 - 3xy^2 + 3xy}{-3xy}$ 6. $\dfrac{16xy - 4x - 4}{-4}$

7. $\dfrac{6x^2 + 3x - 9}{3}$ 8. $\dfrac{8t^2 + 2t + 2}{2}$ 9. $\dfrac{4x^3 + 2x^2 + 2x}{2x}$

10. $\dfrac{2y^4 - 4y^3 + 2y^2 - 8y}{2y}$ 11. $\dfrac{x^3y^2 - x^2y + xy}{-xy}$ 12. $\dfrac{y^4 + 2y^3 - y^2 - y}{-y}$

In Problems 13 *to* 18 *express each quotient as a polynomial by using the fundamental principle of fractions as illustrated in Example* a. *In Problems* 19 *to* 30 *use the long division method as illustrated in Example* b. *Assume all divisors to be non-zero.*

Examples.   a. $\dfrac{x^2 + 5x + 6}{x + 3}$ b. $\dfrac{x^2 + 5x + 4}{x + 1}$

Solutions.   a. $\dfrac{x^2 + 5x + 6}{x + 3}$

b.
$$\begin{array}{r} x + 4 \\ x + 1 \overline{\smash{\big)}\ x^2 + 5x + 4} \\ \underline{x^2 + x} \\ 4x + 4 \\ \underline{4x + 4} \\ 0 \end{array}$$

$$= \frac{(x + 2)(x + 3)}{x + 3}$$

$$= x + 2$$

13. $\dfrac{x^2 + 5x - 6}{x - 1}$ 14. $\dfrac{y^2 + y - 6}{y - 2}$ 15. $\dfrac{r^2 + 3r - 4}{r + 4}$

16. $\dfrac{x^2 - 4}{x - 2}$ 17. $\dfrac{p^2 + 4p + 3}{p + 1}$ 18. $\dfrac{2x^2 - x - 1}{x - 1}$

19. $\dfrac{2x^2 - 9x + 4}{2x - 1}$ 20. $\dfrac{4x^2 + 4x + 1}{2x + 1}$ 21. $\dfrac{4x^2 + 4x - 3}{2x - 1}$

22. $\dfrac{4x^2 - 8x - 5}{2x + 1}$ 23. $\dfrac{4x^2 - 12x + 5}{2x - 1}$ 24. $\dfrac{6x^2 - 5x - 6}{3x + 2}$

25. $\dfrac{6x^2 + 10x - 4}{3x - 1}$
26. $\dfrac{9x^2 + 9x - 4}{3x + 4}$
27. $\dfrac{8x^2 + 6x - 2}{4x - 1}$

28. $\dfrac{12x^2 + 2x - 2}{4x + 2}$
29. $\dfrac{10x^2 - 8x - 2}{5x + 1}$
30. $\dfrac{15x^2 + 4x - 4}{5x - 2}$

B

Express each quotient as a polynomial by using the long division method as illustrated in Example b above. State the restrictions on the variables so that the quotients represent real numbers.

31. $\dfrac{x^3 + 3x^2 + 7x + 5}{x + 1}$
32. $\dfrac{x^4 + 3x^3 + 3x^2 + 3x + 2}{x + 2}$

33. $\dfrac{x^3 - 1}{x - 1}$
34. $\dfrac{x^3 + 1}{x + 1}$
35. $\dfrac{x^3 + x^2 + 4}{x + 2}$

36. $\dfrac{x^3 - x - 6}{x - 2}$
37. $\dfrac{x^4 - 1}{x - 1}$
38. $\dfrac{x^4 - 1}{x + 1}$

CHAPTER SUMMARY

6.1 If $a \in R$ and $m \in N$, then $a^m = a \cdot a \cdot a \cdots a$ (m factors) where m is an **exponent** and indicates how many times the number a appears as a factor in the product called the **power.** The number, a, is called the **base** of the power.

6.2 Some polynomials are named according to the number of terms they contain. The degree of a polynomial in one variable is the exponent of the highest power of the variable in any of its terms. A constant is of degree zero.

By the closure laws for addition and multiplication, polynomials represent real numbers for real number replacements of the variable.

By definition, operations are performed in the following specified order:

a. Operations within parentheses. Operations above and/or below a fraction bar.

b. Evaluation of powers, then multiplications or divisions in the order in which they occur.

c. Additions or subtractions in the order in which they occur.

The symbol $P(x)$ represents a polynomial in x and $P(a)$ is the value of the polynomial when x is replaced by a. If $P(c) = 0$, then c is a **zero** of $P(x)$.

6.3 Expressions that represent the same number for the same replacement(s) of the variable(s) are **equivalent expressions.** Sums and differences of polynomials can be rewritten as equivalent expressions by using the axioms and theorems applicable to real numbers.

6.4 The following theorems are applicable to certain algebraic expressions in exponential form:

If $a \in R$ and $m, n \in N$, then

$$a^m \cdot a^n = a^{m+n} \quad and \quad \frac{a^m}{a^n} = a^{m-n} \quad (a \neq 0, m > n).$$

6.5 At times the symbols a^0 and a^{-m} are convenient to use and are defined as follows:

If $a \in R$ and $m \in N$, then

$$a^0 = 1 \quad and \quad a^{-m} = \frac{1}{a^m} \quad (a \neq 0).$$

Any rational number can be expressed in **scientific notation** as the product of a number between 1 and 10 and a power of 10.

6.6–6.8 The distributive law justifies rewriting the product of polynomials equivalently as a polynomial.

6.9 A polynomial of the form $ax^2 + bx + c$ $(a \neq 0)$ is factorable if $a \cdot c$ is the product of two factors whose sum is b.

6.10 The quotient of two polynomials can be written equivalently as a polynomial if the divisor is a factor of the dividend.

CHAPTER REVIEW

A

6.1 1. Write $72x^2y^3$ in completely factored form.
     2. Write $4rrssttt$ in exponential form.

6.2 3. Give the name and degree of the polynomial $x^4 + 7x - 2$.

4. State the exponent and numerical coefficient of each term of the polynomial $4x^7 + 13x^2$.

5. Write $\dfrac{2^2 \cdot (3 + 1)}{4} - \dfrac{12}{3} + 5$ as a basic numeral.

6. Write $\dfrac{3^3}{9} + \dfrac{6 - 4^2}{2}$ as a basic numeral.

7. If $P(x) = 2x^2 - 5x + 7$, what is $P(-1)$?
8. Find $P(0)$ if $P(x) = 5x^3 - 17x^2 + 21x - 5$.
9. Find the zeros of $(x - 1)(x + 2)$ by inspection.

6.3 10. Express $(x^2 + 2xy + 5y^2) + (3x^2 - 7xy - 2y^2) - (2x^2 - 5xy + y^2)$ as a polynomial in simple form.

6.4 11. Express the product $3x^2y \cdot 4xy^3$ in simple form.

12. Express the quotient $\dfrac{-27r^4s}{-3r^2s^3}$ $(r, s \neq 0)$ in simple form.

6.5 13. Express the quotient $\dfrac{x^{-2}y^3}{x^0}$ $(x \neq 0)$ without negative or zero exponents.

Express each rational number in scientific notation.

14. 37,200   15. 0.000234

Express each rational number in decimal notation.

16. 4.21×10^{-3} 17. 5.76×10^6

6.6 18. Write the polynomial form of the product $(2x + 3)(3x - 7)$.

6.7–6.9 *Factor each polynomial completely when possible. If the polynomial is not factorable, so state.*

19. $6x^2y + 3xy - 18xy^2$ 20. $6x^2 + 13x + 6$

21. $49x^2y^2 - 121z^2$ 22. $x^2 + 3x + 1$

6.10 23. Express the quotient $4x^3 - 2x^2 + 8x$ divided by $2x$ as a polynomial.
24. Express the quotient of $3x^2 + 2x - 5$ divided by $x - 1$ as a polynomial.

B

25. If $P(x) = 2x^2 - 7x + 3$, find $P(2) \cdot P(3)$.
26. If $P(x) = x + 2$ and $Q(x) = x - 3$, find $P(2) - Q(1)$.

27. Express the quotient $\dfrac{3x^{-3}y^2}{xy^{-3}}$ $(x, y \neq 0)$ without negative exponents.

Compute by using scientific notation.

28. $\dfrac{0.0054 \times 0.05 \times 300}{0.0015 \times 0.027 \times 20}$

29. $\dfrac{810 \times 50 \times 17}{15{,}300 \times 90 \times 125}$

30. Express $60x^2 + 68x - 168$ in completely factored form.
31. Express the quotient $(x^5 - 1) \div (x - 1)$ as a polynomial.
32. Express the quotient $(x^6 - 1) \div (x + 1)$ as a polynomial.

7
chapter

Rational Expressions

In Chapter 4 we discussed *quotients of integers* and called

$$\left\{ \frac{a}{b} \,\middle|\, a, b \in J, b \neq 0 \right\}$$

the set of rational numbers; the symbol a/b was called a fraction. At that time we studied several theorems which enabled us to rewrite a fraction in equivalent forms.

Expressions for *quotients of polynomials* are called **rational expressions.** These include polynomials such as

$$x^2 + 3x + 1, \qquad 2 - x, \qquad \text{and} \qquad x^2 + xy + y^2,$$

which we can view as representing quotients with divisor 1, and also include expressions such as

$$\frac{x}{x + 2}, \qquad \frac{x^2 - x - 1}{x}, \qquad \text{and} \qquad \frac{1}{x^2 + xy + y^2}$$

which are fractions.

7.1 Reducing to Lower Terms

In this section we establish several theorems that enable us to rewrite rational expressions in equivalent forms which can be more useful than the original form. First, however, we establish the conditions for equality.

From the laws of closure for addition and multiplication, rational expressions of the form $N(x)/D(x)$, where $N(x)$ and $D(x)$ represent polynomials in one or more variables, represent real numbers for real number replacements of the variable(s) for which $D(x) \neq 0$. Therefore, *many axioms and theorems about real numbers apply to rational expressions.* The proofs of the following theorems follow directly from the laws of closure and the related Theorems 4.1, 4.2, and 4.3 applicable to real numbers.

Theorem 7.1. If $P(x)$ and $Q(x)$ are polynomials, $[Q(x) \neq 0]$, then

$$\frac{P(x)}{Q(x)} = P(x) \cdot \frac{1}{Q(x)} .$$

Examples

a. $\dfrac{x + 5}{x - 3} = (x + 5)\dfrac{1}{x - 3}$ $(x \neq 3)$

b. $\dfrac{x^2 + 4x + 7}{x^2 - x - 12} = (x^2 + 4x + 7)\dfrac{1}{x^2 - x - 12}$ $(x \neq -3 \text{ or } 4)$

Theorem 7.2. If $P(x)$, $Q(x)$, $R(x)$, $S(x)$ are polynomials $[Q(x), S(x) \neq 0]$, then

$$\frac{P(x)}{Q(x)} = \frac{R(x)}{S(x)} \quad \text{if} \quad P(x) \cdot S(x) = Q(x) \cdot R(x)$$

and

$$P(x) \cdot S(x) = Q(x) \cdot R(x), \quad \text{if} \quad \frac{P(x)}{Q(x)} = \frac{R(x)}{S(x)} .$$

A rational expression is said to be in *simplest form,* or in *lowest terms,* if the numerator and denominator have no factors in common. If the polynomials do have common factors, the expression can be rewritten in lower terms by a direct application of the following theorem.

Theorem 7.3. If $P(x)$, $Q(x)$, $R(x)$ are polynomials $[Q(x), R(x) \neq 0]$, then

$$\frac{P(x) \cdot R(x)}{Q(x) \cdot R(x)} = \frac{P(x)}{Q(x)} \quad \text{and} \quad \frac{P(x)}{Q(x)} = \frac{P(x) \cdot R(x)}{Q(x) \cdot R(x)} .$$

We shall refer to this theorem as the **fundamental principle of fractions.**

Examples

a. $\dfrac{5a - 15}{15} = \dfrac{5(a - 3)}{5 \cdot 3}$

$= \dfrac{a - 3}{3}$

b. $\dfrac{x^2 + 2x + 1}{x^2 - 1} = \dfrac{(x + 1)(x + 1)}{(x - 1)(x + 1)}$

$= \dfrac{x + 1}{x - 1} \qquad (x \neq 1, -1)$

Sometimes a rational expression may be in lowest terms but we wish to rewrite the expression equivalently as the sum of a polynomial and a rational expression. If the divisor is a polynomial of two or more terms, as in Example b below, the process is similar to that described in Section 6.10 except that there is a remainder.

Examples. a. $\dfrac{-xy + 1}{y}$

b. $\dfrac{x^2 + x + 1}{x + 2} \qquad (x \neq -2)$

Solutions. a. $\dfrac{-xy + 1}{y}$

$= (-xy + 1) \cdot \dfrac{1}{y}$

$= \dfrac{-xy}{y} + \dfrac{1}{y}$

$= -x + \dfrac{1}{y}$

b.
$$
\begin{array}{r}
x - 1 \\
x + 2 \overline{\smash{\big)}\ x^2 + x + 1} \\
\underline{x^2 + 2x} \\
-x + 1 \\
\underline{-x - 2} \\
3
\end{array}
$$

$\dfrac{x^2 + x + 1}{x + 2} = x - 1 + \dfrac{3}{x + 2}$

Exercise 7.1

A

State the axiom, definition, or theorem that justifies each statement. Assume no denominator equals zero.

1. $\dfrac{3}{5x} = 3 \cdot \dfrac{1}{5x}$

2. $\dfrac{2 \cdot x}{5 \cdot x} = \dfrac{2}{5}$

3. $\dfrac{5x + 10}{5}$

    a. $= \dfrac{5(x + 2)}{5}$

    b. $= x + 2$

4. $\dfrac{2x^2 - 4x + 5}{2x}$

    a. $= (2x^2 - 4x + 5) \cdot \dfrac{1}{2x}$

    b. $= \dfrac{2x^2}{2x} - \dfrac{4x}{2x} + \dfrac{5}{2x}$

    c. $= x - 2 + \dfrac{5}{2x}$

5. $\dfrac{2x^2 - x - 10}{x + 2}$

  a. $= \dfrac{(2x - 5)(x + 2)}{x + 2}$

  b. $= 2x - 5$

6. $\dfrac{6x^2 + 23x + 21}{3x + 7}$

  a. $= \dfrac{(3x + 7)(2x + 3)}{3x + 7}$

  b. $= 2x + 3$

Reduce each fraction to lowest terms and state any restrictions on the variables.

Examples.   a. $\dfrac{2x^3 + 4x^2 + 2x}{2x}$ b. $\dfrac{2x + 2y}{2x}$ c. $\dfrac{y^2 - y}{y^2 + y - 2}$

Solutions.   Factor the numerator and/or the denominator.

  a. $\dfrac{2x(x^2 + 2x + 1)}{2x}$ b. $\dfrac{2(x + y)}{2x}$ c. $\dfrac{y(y - 1)}{(y + 2)(y - 1)}$

Apply the fundamental principle of fractions.

$$x^2 + 2x + 1 \quad (x \neq 0) \qquad \dfrac{x + y}{x} \quad (x \neq 0) \qquad \dfrac{y}{y + 2} \quad (y \neq -2, 1)$$

7. $\dfrac{9x^3 - 6x^2 + 3x}{3x}$

8. $\dfrac{s^3 - 3s^2 + 2s}{s}$

9. $\dfrac{3x^2 - 3x}{3x}$

10. $\dfrac{3x^3 - 6x^2 + 3x}{-3x}$

11. $\dfrac{12x^3 + 8x^2 - 4x}{4x}$

12. $\dfrac{8x^2y - 4xy + 6x}{2x}$

13. $\dfrac{x^2 + 5x - 14}{x - 2}$

14. $\dfrac{t^2 + 5t + 6}{t + 3}$

15. $\dfrac{x^2 - 4xy + 4y^2}{x^2 - 4y^2}$

16. $\dfrac{t^2 - 3t}{t^2 - 2t - 3}$

17. $\dfrac{x^2 + x - 6}{x^2 - 9}$

18. $\dfrac{x^2 + 5x + 6}{x^2 - 4}$

19. $\dfrac{p^2 + 6p + 9}{p^2 + 2p - 3}$

20. $\dfrac{z^2 + 5z + 6}{z^2 + 6z + 9}$

Write each of the following as the sum of a polynomial and a rational expression. Assume all denominators to be non-zero.

Examples.   a. $\dfrac{6x^3 + 3x^2 + 2}{3x}$ b. $\dfrac{x^2 + 5x + 6}{x - 2}$

Solutions. a. $\dfrac{6x^3 + 3x^2 + 2}{3x}$

b. $x - 2\,\overline{\smash{\big)}\,x^2 + 5x + 6}$

$$= \dfrac{6x^3}{3x} + \dfrac{3x^2}{3x} + \dfrac{2}{3x}$$

$$\dfrac{x^2 - 2x}{7x + 6}$$

$$= 2x^2 + x + \dfrac{2}{3x}$$

$$\dfrac{7x - 14}{20}$$

$$\dfrac{x^2 + 5x + 6}{x - 2} = x + 7 + \dfrac{20}{x - 2}$$

21. $\dfrac{3x^2 + 2x + 1}{x}$

22. $\dfrac{6y^2 + 3y - 2}{3}$

23. $\dfrac{8x^2 + 4x - 1}{4x}$

24. $\dfrac{6t^4 - 3t^2 + 2}{3t^2}$

25. $\dfrac{x^2 + 3x - 9}{x + 5}$

26. $\dfrac{x^2 + 5x - 7}{x + 6}$

27. $\dfrac{x^2 - 6x - 10}{x - 7}$

28. $\dfrac{3x^2 - 8x - 1}{x - 3}$

29. $\dfrac{2x^2 - x - 2}{x + 1}$

30. $\dfrac{4x^2 - 4x - 5}{x - 2}$

31. $\dfrac{6x^2 + 5x + 4}{2x + 1}$

32. $\dfrac{4x^2 + 2x + 9}{2x - 3}$

B

33. $\dfrac{6x^3 + x^2 + x + 1}{3x + 2}$

34. $\dfrac{4x^3 - 3x^2 + 2x - 1}{x^2 + x + 2}$

35. $\dfrac{6x^3 + 8x^2 + 5x - 3}{3x^2 + x + 1}$

36. $\dfrac{8y^3 + 4y^2 + 4y + 1}{4y^2 + 2y + 2}$

Assume $\dfrac{P(x)}{Q(x)} = \dfrac{R(x)}{S(x)}$. *State whether each statement is true or false for all polynomials for which denominators do not equal zero.*

37. $\dfrac{P(x) \cdot Q(x)}{P(x)} = Q(x)$

38. $\dfrac{P(x) + Q(x)}{P(x)} = Q(x)$

39. $\dfrac{Q(x)}{P(x)} = \dfrac{S(x)}{R(x)}$

40. $\dfrac{P(x)}{R(x)} = \dfrac{Q(x)}{S(x)}$

41. $\dfrac{P(x) + Q(x)}{Q(x)} = \dfrac{R(x) + S(x)}{S(x)}$

42. $\dfrac{P(x)}{S(x)} = \dfrac{Q(x)}{R(x)}$

43. Can a rational expression always be written as a polynomial? Explain.

7.2 Building to Higher Terms

Sometimes it is necessary to write a rational expression as an equivalent expression with a specified numerator or denominator. The expression can be rewritten in higher terms by a direct application of Theorem 7.3, the fundamental principle of fractions, by multiplying both numerator and denominator by the same non-zero polynomial.

Example.   Rewrite the rational expression in the left member as an equivalent expression with the specified denominator.

$$\frac{x + 2}{x + 1} = \frac{?}{(x + 1)(x + 4)} \qquad (x \neq -1, -4)$$

Solution.   Apply the fundamental principle of fractions and multiply both numerator and denominator by $x + 4$. Thus,

$$\frac{x + 2}{x + 1} = \frac{(x + 2)(x + 4)}{(x + 1)(x + 4)}.$$

Exercise 7.2

A

State the axiom, definition, or theorem that justifies each lettered statement. Assume no denominator equals zero.

1. $\dfrac{2}{3} = \dfrac{?}{15}$

   $(15 = 3 \cdot 5)$

   a. $\dfrac{2}{3} = \dfrac{2 \cdot 5}{3 \cdot 5}$

   b. $\quad = \dfrac{10}{15}$

2. $\dfrac{5}{x + 1} = \dfrac{?}{x^2 - 1}$

   $[x^2 - 1 = (x + 1)(x - 1)]$

   a. $\dfrac{5}{x + 1} = \dfrac{5(x - 1)}{(x + 1)(x - 1)}$

   b. $\quad = \dfrac{5x - 5}{x^2 - 1}$

3. $\dfrac{x+1}{x-1} = \dfrac{x^2 + 2x + 1}{?}$

$\qquad [x^2 + 2x + 1 = (x+1)(x+1)]$

a. $\dfrac{x+1}{x-1} = \dfrac{(x+1)(x+1)}{(x-1)(x+1)}$

b. $\qquad = \dfrac{x^2 + 2x + 1}{x^2 - 1}$

4. $\dfrac{2x+3}{5x-1} = \dfrac{6x^2 + 5x - 6}{?}$

$\qquad [6x^2 + 5x - 6 = (2x+3)(3x-2)]$

a. $\dfrac{2x+3}{5x-1} = \dfrac{(2x+3)(3x-2)}{(5x-1)(3x-2)}$

b. $\qquad = \dfrac{6x^2 + 5x - 6}{15x^2 - 13x + 2}$

Write each rational expression as an equivalent expression with the given denominator or numerator. List any restrictions on the variables.

Examples

a. $\dfrac{3}{x-y} \,;\, \dfrac{?}{x^2 - y^2}$

b. $\dfrac{2x-3}{x} \,;\, \dfrac{2x^2 - x - 3}{?}$

Solutions. Factor $x^2 - y^2$.

a. $\dfrac{3}{x-y} \,;\, \dfrac{?}{(x+y)(x-y)}$

$\dfrac{3}{x-y} = \dfrac{3(x+y)}{(x-y)(x+y)}$

$\qquad = \dfrac{3x + 3y}{x^2 - y^2} \quad (x \ne y, -y)$

Factor $2x^2 - x - 3$.

b. $\dfrac{2x-3}{x} \,;\, \dfrac{(2x-3)(x+1)}{?}$

$\dfrac{2x-3}{x} = \dfrac{(2x-3)(x+1)}{x(x+1)}$

$\qquad = \dfrac{2x^2 - x - 3}{x^2 + x} \quad (x \ne 0, -1)$

5. $\dfrac{1}{3} \,;\, \dfrac{?}{3(x+y)}$

6. $\dfrac{1}{5} \,;\, \dfrac{?}{10(p+q)}$

7. $\dfrac{x-1}{3} \,;\, \dfrac{?}{9(x+1)}$

8. $\dfrac{y-2}{2} \,;\, \dfrac{?}{6(y+1)}$

9. $\dfrac{5}{x+y} \,;\, \dfrac{?}{x^2 - y^2}$

10. $\dfrac{x+1}{-2} \,;\, \dfrac{x^2 + 3x + 2}{?}$

11. $\dfrac{2x+y}{5} \,;\, \dfrac{4x^2 - y^2}{?}$

12. $\dfrac{3x}{y+2} \,;\, \dfrac{?}{y^2 - y - 6}$

13. $\dfrac{s+3}{5r} \,;\, \dfrac{s^2 + s - 6}{?}$

14. $\dfrac{-3}{y+2} \,;\, \dfrac{?}{y^2 + 6y + 8}$

15. $\dfrac{7x}{y-x}$; $\dfrac{?}{y^2-x^2}$

16. $\dfrac{3y}{y-2}$; $\dfrac{?}{y^2-3y+2}$

17. $\dfrac{x-2}{x+1}$; $\dfrac{?}{x^2+5x+4}$

18. $\dfrac{y-4}{y+1}$; $\dfrac{?}{3y^2-y-4}$

19. $\dfrac{2x-1}{x-2}$; $\dfrac{4x^2+4x-3}{?}$

20. $\dfrac{3y+4}{y-5}$; $\dfrac{9y^2-16}{?}$

21. $\dfrac{x+2y}{x-2y}$; $\dfrac{?}{x^2-4xy+4y^2}$

22. $\dfrac{3x-2y}{2x-3y}$; $\dfrac{?}{2x^2-xy-3y^2}$

23. $\dfrac{1}{x-1}$; $\dfrac{?}{1-x}$

24. $\dfrac{3x}{x-2}$; $\dfrac{?}{4-2x}$

25. $\dfrac{2x-1}{x^2}$; $\dfrac{3-6x}{?}$

26. $\dfrac{4x-3}{x+1}$; $\dfrac{15-20x}{?}$

B

27. $\dfrac{3x+5}{x^2}$; $\dfrac{36x^2-3x-105}{?}$

28. $\dfrac{x^2+x+1}{3x+3}$; $\dfrac{?}{3x^2-3}$

29. $\dfrac{5x-7}{5x+7}$; $\dfrac{?}{25x^2+70x+49}$

30. $\dfrac{x^2+2x+1}{x-2}$; $\dfrac{?}{x^2-x-2}$

31. $\dfrac{x+1}{x-1}$; $\dfrac{x^3+1}{?}$

32. $\dfrac{x-1}{x+1}$; $\dfrac{x^3-1}{?}$

33. $\dfrac{2x+1}{x+2}$; $\dfrac{2x^3+x^2-4x-2}{?}$

34. $\dfrac{3x-1}{x-4}$; $\dfrac{3x^3+2x^2-4x+1}{?}$

35. $\dfrac{x^2+1}{2x-3}$; $\dfrac{1-x^4}{?}$

36. $\dfrac{3x+2}{x^3+1}$; $\dfrac{?}{1-x^6}$

7.3 Sums

Since rational expressions represent real numbers for all real number replacements for the variables for which denominators do not equal zero, we have the following theorem directly from Theorem 4.4.

Theorem 7.4. If $P(x)$, $Q(x)$, and $R(x)$ are polynomials $[R(x) \neq 0]$, then

$$\frac{P(x)}{R(x)} + \frac{Q(x)}{R(x)} = \frac{P(x)+Q(x)}{R(x)}.$$

Thus the sum of rational expressions can be rewritten equivalently as a single fraction if the rational expressions have the same denominator. In case the denominators are not the same, each expression can be "built up" so that all do have the same denominator, preferably the least common multiple of the original denominators.

Recall from Section 4.3 that the least common multiple of a set of numbers is the product of the individual prime factors of the numbers, each factor being used the *greatest* number of times it appears in any number.

The least common multiple of a set of polynomials can be found in a similar manner. Although a prime polynomial was not defined, the conditions for a polynomial to be in completely factored form were stated in Section 6.7. Thus the least common multiple of a set of polynomials is the product of all of the distinct factors, each factor being used the greatest number of times it appears in any polynomial.

Example. Find the least common multiple of

$$x^2 + 2x + 1, \qquad x^2 - 1, \qquad \text{and} \qquad x^2 + x - 2.$$

Solution. Factor each polynomial.

$$x^2 + 2x + 1 = (x + 1)(x + 1)$$
$$x^2 - 1 = (x + 1)(x - 1)$$
$$x^2 + x - 2 = (x + 2)(x - 1)$$

Hence, the least common multiple is

$$(x + 1)(x + 1)(x - 1)(x + 2).$$

The factor $x + 1$ is used twice since it appears two times as a factor of the first polynomial.

Now we are prepared to write a sum of two rational expressions with unlike denominators equivalently as a single fraction. For example, to write the sum of two fractions with unlike denominators, such as

$$\frac{x}{x + 3} + \frac{2}{x - 2},$$

equivalently as a single fraction, we must first apply the fundamental principle of fractions to each fraction to obtain an equivalent fraction so that both fractions have the same denominator. We note by inspection that $(x + 3)(x - 2)$ is the least common multiple of the denominators. Thus,

$$\frac{x(x - 2)}{(x + 3)(x - 2)} + \frac{2(x + 3)}{(x - 2)(x + 3)} = \frac{x(x - 2) + 2(x + 3)}{(x + 3)(x - 2)}$$

$$= \frac{x^2 - 2x + 2x + 6}{(x + 3)(x - 2)}$$

$$= \frac{x^2 + 6}{(x + 3)(x - 2)}, \qquad (x \neq -3, 2).$$

Often the factored form of a polynomial is more useful. Therefore, as in the denominator of the example above, results are frequently left in this form.

The following example illustrates a case where the denominators of two fractions must be factored in order to find their least common multiple.

Example. $\dfrac{x + 4}{x^2 - x - 6} + \dfrac{x - 1}{x^2 + 5x + 6}$

$$= \frac{x + 4}{(x + 2)(x - 3)} + \frac{x - 1}{(x + 2)(x + 3)}$$

$$= \frac{(x + 4)(x + 3)}{(x + 2)(x - 3)(x + 3)} + \frac{(x - 1)(x - 3)}{(x + 2)(x + 3)(x - 3)}$$

$$= \frac{(x + 4)(x + 3) + (x - 1)(x - 3)}{(x + 2)(x - 3)(x + 3)}$$

$$= \frac{x^2 + 7x + 12 + x^2 - 4x + 3}{(x + 2)(x - 3)(x + 3)}$$

$$= \frac{2x^2 + 3x + 15}{(x + 2)(x - 3)(x + 3)} \qquad (x \neq -2, 3, -3).$$

In general, we prefer a rational expression to be in lowest terms. Since the numerator $(2x^2 + 3x + 15)$ cannot be factored, the numerator and denominator have no factors in common. Thus the expression is in its lowest terms.

Exercise 7.3

A

State the axiom or theorem that justifies each statement. Assume that no denominator equals zero.

1. $\dfrac{2x}{x+1} + \dfrac{3x}{x+1}$

  a. $= \dfrac{2x+3x}{x+1}$

  b. $= \dfrac{(2+3)x}{x+1}$

  c. $= \dfrac{5x}{x+1}$

2. $\dfrac{6(xy)}{1-y} + \dfrac{5(xy)}{1-y}$

  a. $= \dfrac{6(xy)+5(xy)}{1-y}$

  b. $= \dfrac{(6+5)(xy)}{1-y}$

  c. $= \dfrac{11xy}{1-y}$

3. $\dfrac{3x}{x-y} + \dfrac{4x}{x-y}$

  a. $= \dfrac{3x+4x}{x-y}$

  b. $= \dfrac{(3+4)x}{x-y}$

  c. $= \dfrac{7x}{x-y}$

4. $\dfrac{5x}{xy-3} + \dfrac{6x}{xy-3}$

  a. $= \dfrac{5x+6x}{xy-3}$

  b. $= \dfrac{(5+6)x}{xy-3}$

  c. $= \dfrac{11x}{xy-3}$

Write each sum equivalently as a single fraction in lowest terms.

Examples. a. $\dfrac{x+2}{2x-3} + \dfrac{2x-1}{2x-3}$

$= \dfrac{x+2+2x-1}{2x-3}$

$= \dfrac{3x+1}{2x-3}$

  b. $\dfrac{2x+3y}{x+y} + \dfrac{x-5y}{x+y}$

$= \dfrac{2x+3y+x-5y}{x+y}$

$= \dfrac{3x-2y}{x+y}$

5. $\dfrac{x+1}{2y} + \dfrac{x}{2y}$

6. $\dfrac{y+1}{x} + \dfrac{y-1}{x}$

7. $\dfrac{2x-y}{x} + \dfrac{x-y}{x}$

8. $\dfrac{3x+1}{3y} + \dfrac{2x-5}{3y}$

9. $\dfrac{3x-1}{4y} + \dfrac{2x-1}{4y}$

10. $\dfrac{x-y}{y} + \dfrac{x+y}{y}$

11. $\dfrac{3}{x+2y} + \dfrac{2}{x+2y}$

12. $\dfrac{x+3}{x+3y} + \dfrac{x-1}{x+3y}$

13. $\dfrac{x+1}{x^2-2x+1} + \dfrac{5-3x}{x^2-2x+1}$

14. $\dfrac{y+4}{y^2+y-2} + \dfrac{2y-3}{y^2+y-2}$

Find the least common multiple, in factored form, of each set of polynomials.

Example. $(x+1)^2,\ x^2-1,\ x^2-3x+2$

Solution.   Write each polynomial in factored form.

$$(x+1)(x+1),\quad (x+1)(x-1),\quad (x-1)(x-2)$$

Use each factor the greatest number of times it appears in any polynomial.

$$(x+1)(x+1)(x-1)(x-2)\qquad\text{or}\qquad (x+1)^2(x-1)(x-2).$$

15. $2xy,\ 6y^2$

16. $12xy,\ 24x^3y^2$

17. $6xy,\ 8x^2,\ 3xy^2$

18. $7x,\ 8y,\ 6z$

19. $x-y,\ x+y,\ x^2-y^2$

20. $6(x+y)^2,\ 3x^2-3y^2$

21. $x+2,\ x^2-4,\ x^2-5x+6$

22. $x^2-x-2,\ x^2-4x+4$

23. $4y^2-4,\ (y-1)^2,\ y+1$

24. $3x^2-3,\ x^2+2x+1,\ 4x+4$

Write each sum equivalently as a single fraction. Assume that no denominator equals zero.

Examples.   a. $\dfrac{3}{2x} + \dfrac{2}{3x}$ b. $\dfrac{2}{x+3} + \dfrac{1}{x-3}$

Solutions.   Find the least common multiple of the denominators.

a. $2 \cdot 3 \cdot x$ or $6x$ b. $(x+3)(x-3)$

Build each fraction to an equivalent fraction with the new denominator.

a. $\dfrac{3}{2x} + \dfrac{2}{3x} = \dfrac{3 \cdot 3}{2x \cdot 3} + \dfrac{2 \cdot 2}{3x \cdot 2} = \dfrac{9+4}{6x} = \dfrac{13}{6x}$

b. $\dfrac{2}{x+3} + \dfrac{1}{x-3} = \dfrac{2(x-3)}{(x+3)(x-3)} + \dfrac{1(x+3)}{(x-3)(x+3)}$

$$= \dfrac{2x-6+x+3}{(x+3)(x-3)} = \dfrac{3x-3}{(x+3)(x-3)}$$

25. $\dfrac{4}{x} + \dfrac{3}{y}$

26. $\dfrac{2}{xy} + \dfrac{2}{x}$

27. $\dfrac{2}{x^2} + \dfrac{3}{xy}$

28. $\dfrac{x+2}{3} + \dfrac{x-3}{9}$

29. $\dfrac{y-2}{6} + \dfrac{y+1}{3}$

30. $\dfrac{x-3}{4} + \dfrac{5-x}{10}$

31. $\dfrac{2x-y}{2} + \dfrac{x+y}{3}$

32. $\dfrac{3p+q}{3} + \dfrac{p-2q}{6}$

33. $\dfrac{7}{x+2} + \dfrac{4}{x+5}$

34. $\dfrac{2}{t+2} + \dfrac{3}{t+3}$

35. $\dfrac{2}{x+3} + \dfrac{1}{x-3}$

36. $\dfrac{7}{y-5} + \dfrac{3}{y+5}$

37. $\dfrac{2}{x+y} + \dfrac{1}{x-y}$

38. $\dfrac{1}{r-s} + \dfrac{3}{r+s}$

39. $\dfrac{1}{x^2-1} + \dfrac{1}{x^2+2x+1}$

40. $\dfrac{3}{y^2-1} + \dfrac{1}{y^2-2y+1}$

41. $\dfrac{1}{y^2-16} + \dfrac{2}{y+4}$

42. $\dfrac{3}{y^2-5y+4} + \dfrac{2}{y^2-1}$

43. $\dfrac{x}{x^2-4} + \dfrac{2x}{x^2-5x+6}$

44. $\dfrac{y}{y^2-16} + \dfrac{y+1}{y^2-5y+4}$

45. $\dfrac{x+y}{x^2+xy-2y^2} + \dfrac{x-y}{2x+4y}$

46. $\dfrac{x+3y}{x^2+2xy+y^2} + \dfrac{x-y}{x^2+4xy+3y^2}$

7.4 Differences

Theorem 4.5 implies that if any two of the three signs associated with a fraction are changed, the resulting fraction represents the same rational number as the original fraction. From the laws of closure, this theorem can be restated to apply to rational expressions.

Theorem 7.5. If $P(x)$ and $Q(x)$ are polynomials $[Q(x) \neq 0]$, then

$$\frac{P(x)}{Q(x)} = -\frac{-P(x)}{Q(x)} = -\frac{P(x)}{-Q(x)} = \frac{-P(x)}{-Q(x)},$$

and

$$-\frac{P(x)}{Q(x)} = \frac{-P(x)}{Q(x)} = \frac{P(x)}{-Q(x)} = -\frac{-P(x)}{-Q(x)}.$$

Examples

a. $\dfrac{y-x}{4} = -\dfrac{-(y-x)}{4} = -\dfrac{y-x}{-4} = \dfrac{-(y-x)}{-4} = \dfrac{-y+x}{-4} = \dfrac{x-y}{-4}$

b. $-\dfrac{1}{x-y} = \dfrac{-1}{x-y} = \dfrac{1}{-(x-y)} = -\dfrac{-1}{-(x-y)} = \dfrac{1}{-x+y} = \dfrac{1}{y-x}$

Note that if the sign of a polynomial is changed, then the sign of every term of the polynomial is changed.

Theorem 4.6 can now be restated to apply to a difference of two rational expressions.

Theorem 7.6. If $P(x)$, $Q(x)$, and $R(x)$ are polynomials $[R(x) \neq 0]$, then

$$\frac{P(x)}{R(x)} - \frac{Q(x)}{R(x)} = \frac{P(x) - Q(x)}{R(x)}.$$

The following example illustrates an application of this theorem.

Example. $\dfrac{x+y}{x-y} - \dfrac{x-y}{x+y} = \dfrac{(x+y)(x+y)}{(x-y)(x+y)} - \dfrac{(x-y)(x-y)}{(x+y)(x-y)}$

$$= \frac{(x+y)(x+y) - (x-y)(x-y)}{(x+y)(x-y)}$$

$$= \frac{x^2 + 2xy + y^2 - (x^2 - 2xy + y^2)}{(x+y)(x-y)}$$

$$= \frac{x^2 + 2xy + y^2 - x^2 + 2xy - y^2}{(x+y)(x-y)}$$

$$= \frac{4xy}{(x+y)(x-y)} \qquad (x \neq y, -y).$$

Exercise 7.4

A

State the axiom or theorem that justifies each statement. Assume that no denominator equals zero.

1. $\dfrac{3x}{x+2} - \dfrac{2x}{x+2}$

2. $\dfrac{5xy}{1-x} - \dfrac{3xy}{1-x}$

a. $= \dfrac{3x - 2x}{x + 2}$

a. $= \dfrac{5xy - 3xy}{1 - x}$

b. $= \dfrac{(3 - 2)x}{x + 2}$

b. $= \dfrac{5(xy) - 3(xy)}{1 - x}$

c. $= \dfrac{x}{x + 2}$

c. $= \dfrac{(5 - 3)(xy)}{1 - x}$

d. $= \dfrac{2xy}{1 - x}$

3. $\dfrac{4s}{r - s} - \dfrac{2s}{r - s}$

4. $\dfrac{6x}{2x - 3} - \dfrac{7x}{2x - 3}$

a. $= \dfrac{4s - 2s}{r - s}$

a. $= \dfrac{6x - 7x}{2x - 3}$

b. $= \dfrac{4s + (-2s)}{r - s}$

b. $= \dfrac{6x + (-7x)}{2x - 3}$

c. $= \dfrac{[4 + (-2)]s}{r - s}$

c. $= \dfrac{[6 + (-7)]x}{2x - 3}$

d. $= \dfrac{2s}{r - s}$

d. $= \dfrac{-x}{2x - 3}$

Write each difference as a single fraction.

Examples. a. $\dfrac{3x + 1}{2x - 5} - \dfrac{x - 3}{2x - 5}$

b. $\dfrac{3x + 4y}{x + y} - \dfrac{2x - y}{x + y}$

$= \dfrac{3x + 1 - (x - 3)}{2x - 5}$

$= \dfrac{3x + 4y - (2x - y)}{x + y}$

$= \dfrac{3x + 1 - x + 3}{2x - 5}$

$= \dfrac{3x + 4y - 2x + y}{x + y}$

$= \dfrac{2x + 4}{2x - 5}$

$= \dfrac{x + 5y}{x + y}$

5. $\dfrac{y + 2}{3x} - \dfrac{y}{3x}$

6. $\dfrac{x + 2}{y} - \dfrac{x - 3}{y}$

7. $\dfrac{3x - 2}{x} - \dfrac{x + 1}{x}$

8. $\dfrac{2x + 3}{y} - \dfrac{3x + 4}{y}$

9. $\dfrac{2y - 1}{4x} - \dfrac{y + 2}{4x}$

10. $\dfrac{3x - 1}{4y} - \dfrac{2x - 1}{4y}$

11. $\dfrac{x-y}{y} - \dfrac{x+y}{y}$ 12. $\dfrac{y+x}{x} - \dfrac{y-x}{x}$ 13. $\dfrac{x-3}{x+2y} - \dfrac{x+2}{x+2y}$

14. $\dfrac{x+2}{x+3y} - \dfrac{x-1}{x+3y}$ 15. $\dfrac{2x-1}{x^2-3x+2} - \dfrac{x+3}{x^2-3x+2}$

16. $\dfrac{3y+5}{y^2-2y+1} - \dfrac{2y-1}{y^2-2y+1}$

Write each difference as a single fraction. Assume that no denominator equals zero

Examples. a. $\dfrac{4}{3y} - \dfrac{1}{2y}$ b. $\dfrac{2}{x+3} - \dfrac{1}{x-3}$

Solutions. Find the least common multiple of the denominators.

  a. $3 \cdot 2 \cdot y$, or $6y$ b. $(x+3)(x-3)$

Build each fraction to an equivalent fraction with the new denominator.

a. $\dfrac{4 \cdot 2}{3y \cdot 2} - \dfrac{1 \cdot 3}{2y \cdot 3} = \dfrac{8-3}{6y}$ b. $\dfrac{2(x-3)}{(x+3)(x-3)} - \dfrac{1(x+3)}{(x-3)(x+3)}$

$= \dfrac{5}{6y}$ $= \dfrac{2x-6-(x+3)}{(x+3)(x-3)}$

$= \dfrac{2x-6-x-3}{(x+3)(x-3)} = \dfrac{x-9}{(x+3)(x-3)}$

17. $\dfrac{3}{x} - \dfrac{2}{y}$ 18. $\dfrac{4}{xy} - \dfrac{5}{y}$

19. $\dfrac{3}{y^2} - \dfrac{1}{xy}$ 20. $\dfrac{x+3}{3} - \dfrac{x-2}{9}$

21. $\dfrac{x-3}{6} - \dfrac{x+2}{3}$ 22. $\dfrac{y-4}{5} - \dfrac{5-y}{10}$

23. $\dfrac{3x-y}{2} - \dfrac{x+y}{3}$ 24. $\dfrac{3r+s}{3} - \dfrac{r+2s}{6}$

25. $\dfrac{x+3y}{2} - \dfrac{x-y}{3}$ 26. $\dfrac{3}{x-y} - \dfrac{2}{x+y}$

27. $\dfrac{3}{w+3} - \dfrac{2}{w+2}$ 28. $\dfrac{6}{y-1} - \dfrac{5}{y-2}$

Write five equivalent fractions for each given fraction. Assume that no denominator equals zero.

Examples

a. Using Theorem 7.5,

$$\frac{x - 2y}{3} = -\frac{-(x - 2y)}{3} = -\frac{x - 2y}{-3}$$

$$= \frac{-(x - 2y)}{-3} = \frac{-x + 2y}{-3} = \frac{2y - x}{-3}.$$

b. Using Theorem 7.5,

$$-\frac{2}{y - x} = \frac{-2}{y - x} = \frac{2}{-(y - x)} = -\frac{-2}{-(y - x)} = \frac{2}{-y + x} = \frac{2}{x - y}.$$

29. $\dfrac{2x - y}{3}$

30. $\dfrac{x - 3y}{4}$

31. $-\dfrac{3y - x}{5}$

32. $-\dfrac{y - 5x}{2}$

33. $\dfrac{3}{2y - x}$

34. $\dfrac{5}{x - 4y}$

Write each difference as a single fraction. Assume no denominator equals zero.

35. $\dfrac{3}{3 - x} - \dfrac{1}{x - 3}$

36. $\dfrac{8}{y - 2} - \dfrac{2}{2 - y}$

37. $\dfrac{1}{x - 4} - \dfrac{2}{4 - x}$

38. $\dfrac{3}{y - 5} - \dfrac{1}{5 - y}$

39. $\dfrac{5}{x - y} - \dfrac{1}{y - x}$

40. $\dfrac{2}{y - x} - \dfrac{3}{x - y}$

Example. $\dfrac{1}{x + 3} - \dfrac{2}{x^2 - 9} = \dfrac{1}{x + 3} - \dfrac{2}{(x + 3)(x - 3)}$

$$= \dfrac{1(x - 3)}{(x + 3)(x - 3)} - \dfrac{2}{(x + 3)(x - 3)}$$

$$= \dfrac{x - 3 - 2}{(x + 3)(x - 3)} = \dfrac{x - 5}{(x + 3)(x - 3)}$$

41. $\dfrac{1}{x^2 + 2x + 1} - \dfrac{1}{x^2 - 1}$

42. $\dfrac{2}{y^2 - 1} - \dfrac{1}{y^2 - 2y + 1}$

43. $\dfrac{2}{x^2 - 1} - \dfrac{4}{x^2 + 5x + 4}$

44. $\dfrac{3}{y^2 - 5y + 4} - \dfrac{3}{y^2 - 1}$

45. $\dfrac{2x}{x^2 - 4} - \dfrac{x}{x^2 - x - 6}$

46. $\dfrac{2y}{y^2 - 16} - \dfrac{y + 1}{y^2 + 3y - 4}$

47. $\dfrac{x}{x^2 + x - 2} - \dfrac{2x}{x^2 + 2x - 3}$

48. $\dfrac{3x}{x^2 - 2x - 3} - \dfrac{x}{x^2 + 3x + 2}$

7.5 Products and Quotients

Since rational expressions represent real numbers for all real number replacements for the variables, excluding any numbers which produce zero denominators, we have the following from Theorems 4.7, 4.8, and 4.9.

Theorem 7.7.   If $P(x)$, $Q(x)$, $R(x)$, $S(x)$ are polynomials $[Q(x), S(x) \neq 0]$, then

$$\frac{P(x)}{Q(x)} \cdot \frac{R(x)}{S(x)} = \frac{P(x) \cdot R(x)}{Q(x) \cdot S(x)}.$$

Theorem 7.8.   If $P(x)$ and $Q(x)$ are polynomials $[P(x), Q(x) \neq 0]$, then

$$\frac{1}{P(x)/Q(x)} = \frac{Q(x)}{P(x)}.$$

Theorem 7.9.   If $P(x)$, $Q(x)$, $R(x)$, $S(x)$ are polynomials $[Q(x), R(x), S(x) \neq 0]$, then

$$\frac{P(x)}{Q(x)} \div \frac{R(x)}{S(x)} = \frac{P(x)}{Q(x)} \cdot \frac{S(x)}{R(x)}.$$

These theorems justify rewriting products and quotients of polynomials in equivalent forms which often are more useful.

Examples

a. $\dfrac{4}{5} \cdot \dfrac{x+3}{x-2} = \dfrac{4(x+3)}{5(x-2)}$

$\qquad = \dfrac{4x+12}{5x-10}$

b. $\dfrac{\dfrac{1}{x+2}}{x-3} = \dfrac{x-3}{x+2}$

c. $\dfrac{2x-1}{3x+2} \div \dfrac{x-2}{2x+3} = \dfrac{(2x-1)(2x+3)}{(3x+2)(x-2)} = \dfrac{4x^2+4x-3}{3x^2-4x-4}$

Exercise 7.5

A

Justify each statement with an appropriate axiom, definition, or theorem. Assume no denominator equals zero.

1. $\dfrac{x+2}{x+3} \cdot \dfrac{x+4}{x+5} = \dfrac{(x+2)(x+4)}{(x+3)(x+5)}$

2. $\dfrac{3-y}{x-4} \div \dfrac{2-x}{y+2} = \dfrac{3-y}{x-4} \cdot \dfrac{y+2}{2-x}$

3. $\dfrac{p+5}{q-2} \cdot \dfrac{q-2}{p+3}$

   a. $\quad = \dfrac{(p+5)(q-2)}{(q-2)(p+3)}$

   b. $\quad = \dfrac{(p+5)(q-2)}{(p+3)(q-2)}$

   c. $\quad = \dfrac{p+5}{p+3}$

4. $\dfrac{(s+2)(t-1)}{s+3} \div \dfrac{t-1}{s+2}$

   a. $\quad = \dfrac{(s+2)(t-1)}{s+3} \cdot \dfrac{s+2}{t-1}$

   b. $\quad = \dfrac{(s+2)(t-1)(s+2)}{(s+3)(t-1)}$

   c. $\quad = \dfrac{(s+2)(s+2)}{s+3}$

   d. $\quad = \dfrac{(s+2)^2}{s+3}$

Write each product or quotient equivalently as a single fraction. Assume that no denominator equals zero.

Examples

a. $\dfrac{16}{x^2} \cdot \dfrac{3x^3 \cdot y}{4} = \dfrac{48x^3 \cdot y}{4x^2}$

$\qquad = 12xy$

b. $\dfrac{x+1}{x-1} \div \dfrac{x+2}{x+1} = \dfrac{x+1}{x-1} \cdot \dfrac{x+1}{x+2}$

$\qquad = \dfrac{x^2+2x+1}{x^2+x-2}$

5. $\dfrac{5(x+5)}{2x} \cdot \dfrac{4x}{2(x+5)}$

6. $\dfrac{15x^2y}{2} \cdot \dfrac{3}{45xy^2}$

7. $\dfrac{3y}{2(2x-3y)} \cdot \dfrac{2x-3y}{12x}$

8. $\dfrac{-21r^2s}{8t} \cdot \dfrac{-14t^2}{3rs}$

9. $\dfrac{16y^2}{6} \div \dfrac{4y}{3}$

10. $\dfrac{-x^2y^2}{u^2v^2} \div \dfrac{xy^2}{u^2v}$

11. $\dfrac{2(x-y)}{xy} \div \dfrac{4(x-y)}{xy}$

12. $\dfrac{6(y+2)}{3(y+3)} \div \dfrac{4(y+2)}{5(y+3)}$

Examples

a. $\dfrac{4y^2-1}{y^2-16} \cdot \dfrac{y^2-4y}{2y+1}$

b. $\dfrac{x^2-x-6}{x^2+2x-15} \div \dfrac{x-2}{x+5}$

Solutions.   Factor all factorable polynomials in numerators and denominators.

a. $\dfrac{(2y+1)(2y-1)}{(y+4)(y-4)} \cdot \dfrac{y(y-4)}{(2y+1)}$

b. $\dfrac{(x+2)(x-3)}{(x+5)(x-3)} \div \dfrac{(x-2)}{(x+5)}$

$= \dfrac{(2y+1)(2y-1)y(y-4)}{(y+4)(y-4)(2y+1)}$

$= \dfrac{(x+2)(x-3)(x+5)}{(x+5)(x-3)(x-2)}$

$= \dfrac{y(2y-1)}{y+4}$

$= \dfrac{x+2}{x-2}$

13. $\dfrac{5x+25}{2x} \cdot \dfrac{4x}{2x+10}$

14. $\dfrac{2s-4t}{8s+24t} \cdot \dfrac{2s+6t}{4s-8t}$

15. $\dfrac{x^2-3x-10}{x^2+2x-35} \cdot \dfrac{x^2+4x-21}{x^2+9x+14}$

16. $\dfrac{6r^2-r-2}{12r^2+5r-2} \cdot \dfrac{8r^2-6r+1}{4r^2-1}$

17. $\dfrac{x+2y}{x-2y} \cdot \dfrac{x^2-xy-2y^2}{x^2+5xy+6y^2}$

18. $\dfrac{p^2+p-6}{2p^2+6p} \cdot \dfrac{8p^2}{p^2-5p+6}$

19. $\dfrac{x^2-x}{x+2} \cdot \dfrac{4x+8}{5x-5}$

20. $\dfrac{y^2+y}{2y+1} \cdot \dfrac{10y+5}{3y+3}$

21. $\dfrac{u^2-4}{u^2-5u+6} \cdot \dfrac{u^2-2u-3}{u^2+3u+2}$

22. $\dfrac{x^2+3x}{x^2-3x-4} \cdot \dfrac{x^2-5x+4}{x^2+2x-3}$

23. $\dfrac{x^2-9y^2}{16x^2-y^2} \div \dfrac{x+3y}{4x-y}$

24. $\dfrac{4x^2-y^2}{x^2-4y^2} \div \dfrac{2x-y}{x-2y}$

25. $\dfrac{s^2 - 6s + 5}{s^2 + 8s + 7} \div \dfrac{s^2 - 3s - 10}{s^2 + 3s + 2}$

26. $\dfrac{t^2 - 8t + 15}{t^2 + 9t + 14} \div \dfrac{t^2 + 4t - 21}{t^2 - 6t - 16}$

27. $\dfrac{2x^2 - x - 28}{3x^2 - x - 2} \div \dfrac{4x^2 + 16x + 7}{3x^2 + 11x + 6}$

28. $\dfrac{r^2 + 7r + 10}{r^2 + 7r + 12} \div \dfrac{r^2 + 6r + 5}{r^2 + 8r + 16}$

29. $\dfrac{x^2 + 3x}{x^2 - 3x - 4} \div \dfrac{x^2 + 2x - 3}{x^2 - 5x + 4}$

30. $\dfrac{y^2 - y - 20}{y^2 - 6y + 5} \div \dfrac{y^2 - 7y + 10}{y^2 + y - 2}$

31. $\dfrac{x^2 + x - 2}{x^2 + 5x - 6} \div \dfrac{x^2 - 3x - 4}{x^2 - x - 12}$

32. $\dfrac{x^2 - 4}{x^2 - 5x + 6} \div \dfrac{x^2 + 3x + 2}{x^2 - 2x - 3}$

33. $\dfrac{t^2 - 4}{t^2 - 1} \cdot \dfrac{1 - t}{2t^2 + 4t}$

34. $\dfrac{9 - s^2}{4 - s^2} \cdot \dfrac{s - 2}{9s - 3s^2}$

35. $\dfrac{x^2 + xy}{x^2 - xy} \div \dfrac{x + y}{4y - 4x}$

36. $\dfrac{2x^2 - xy}{3x^2 - xy} \div \dfrac{2x - y}{4y - 12x}$

B

37. $\left(\dfrac{x + 3}{x} \cdot \dfrac{-2}{x + 3}\right) \div \dfrac{4x^2}{x^2 + 3x}$

38. $\left(\dfrac{x - 2}{3x} \cdot \dfrac{1}{2x - 4}\right) \div \dfrac{2x + 1}{5x^2}$

39. $\dfrac{x^2 - 1}{x + 3} \div \left(\dfrac{x + 4}{x^2} \cdot \dfrac{2x + 2}{x + 4}\right)$

40. $\dfrac{x - 4}{9 - x^2} \div \left(\dfrac{2x^2}{x + 1} \cdot \dfrac{2x - 8}{2x + 6}\right)$

7.6 Complex Fractions

A complex fraction is a fraction that contains fractions in either its numerator, denominator, or both. Thus a complex fraction can denote a quotient involving rational expressions. Some examples are

$$\dfrac{\dfrac{2}{x + 1}}{\dfrac{5}{x - 1}}, \qquad \dfrac{1 + \dfrac{1}{y}}{2 - \dfrac{1}{y}}, \qquad \text{and} \qquad 1 + \dfrac{1}{1 + \dfrac{1}{2}}.$$

When a complex fraction is written equivalently as a quotient of two polynomials, it is said to be simplified. There are several ways to simplify complex fractions. In some cases, they can be simplified by a direct application of the fundamental principle of fractions.

Example. Simplify $\dfrac{1 + \dfrac{1}{y}}{2 - \dfrac{1}{y}}$ $(y \neq 0)$.

Solution. Multiplying both numerator and denominator of the complex fraction by the least common multiple of the denominators of the fractions in the numerator and the denominator of the complex fraction yields

$$\frac{1 + \dfrac{1}{y}}{2 - \dfrac{1}{y}} = \frac{\left(1 + \dfrac{1}{y}\right)y}{\left(2 - \dfrac{1}{y}\right)y}.$$

Applying the distributive law to the right-hand member we have

$$\frac{\left(1 + \dfrac{1}{y}\right)y}{\left(2 - \dfrac{1}{y}\right)y} = \frac{1 \cdot y + \dfrac{1}{y} \cdot y}{2 \cdot y - \dfrac{1}{y} \cdot y} = \frac{y + 1}{2y - 1}.$$

An alternative method of simplifying complex fractions is to change the expressions in both numerator and denominator to single fractions and then treat the entire expression as a quotient. Using the same example as above, we have

$$\frac{1 + \dfrac{1}{y}}{2 - \dfrac{1}{y}} = \frac{\dfrac{y}{y} + \dfrac{1}{y}}{\dfrac{2y}{y} - \dfrac{1}{y}} = \frac{\dfrac{y + 1}{y}}{\dfrac{2y - 1}{y}} = \frac{y + 1}{y} \cdot \frac{y}{2y - 1} = \frac{y + 1}{2y - 1}.$$

Exercise 7.6

A

Simplify each complex fraction. Assume no denominator is equal to zero.

Examples

a. $\dfrac{\dfrac{3}{5}}{\dfrac{3}{2}}$

b. $\dfrac{1 - \dfrac{1}{3}}{2 + \dfrac{2}{3}}$

c. $\dfrac{4 - \dfrac{1}{x^2}}{2 - \dfrac{1}{x}}$

Solutions. Multiply each term of both numerator and denominator of the complex fraction by the least common multiple of the denominators of all fractions in the numerator and denominator. Apply the distributive law where applicable and then reduce to lowest terms.

a.
$$\dfrac{10 \left(\dfrac{3}{5}\right)}{10 \left(\dfrac{3}{2}\right)}$$

$$= \dfrac{6}{15}$$

$$= \dfrac{2}{5}$$

b.
$$\dfrac{3 \cdot 1 \quad 3 \cdot \dfrac{1}{3}}{3 \cdot 2 + 3 \cdot \dfrac{2}{3}}$$

$$= \dfrac{3 - 1}{6 + 2}$$

$$= \dfrac{2}{8}$$

$$= \dfrac{1}{4}$$

c.
$$\dfrac{x^2 \cdot 4 - x^2 \cdot \dfrac{1}{x^2}}{x^2 \cdot 2 - x^2 \cdot \dfrac{1}{x}}$$

$$= \dfrac{4x^2 - 1}{2x^2 - x}$$

$$= \dfrac{(2x + 1)(2x - 1)}{x(2x - 1)}$$

$$= \dfrac{2x + 1}{x}$$

1. $\dfrac{\dfrac{4}{9}}{\dfrac{3}{2}}$

2. $\dfrac{\dfrac{5}{6}}{\dfrac{3}{2}}$

3. $\dfrac{\dfrac{5}{8}}{\dfrac{15}{16}}$

4. $\dfrac{\dfrac{3}{5}}{\dfrac{4}{3}}$

5. $\dfrac{\dfrac{2}{3}}{2 - \dfrac{2}{3}}$

6. $\dfrac{1 + \dfrac{1}{6}}{1 - \dfrac{1}{3}}$

7. $\dfrac{3 + \dfrac{1}{5}}{2 - \dfrac{3}{5}}$

8. $\dfrac{3 + \dfrac{3}{4}}{5 - \dfrac{1}{2}}$

9. $\dfrac{x - \dfrac{1}{x}}{x + \dfrac{1}{x}}$

10. $\dfrac{1 + \dfrac{1}{2y}}{1 - \dfrac{1}{4y^2}}$

11. $\dfrac{x + \dfrac{x}{y}}{1 - \dfrac{1}{y}}$

12. $\dfrac{\dfrac{2}{y} + \dfrac{1}{3y}}{\dfrac{y}{2} + \dfrac{3y}{4}}$

13. $\dfrac{\dfrac{x}{3y} - \dfrac{1}{2}}{\dfrac{4}{3y} - \dfrac{1}{2}}$

14. $\dfrac{2 - \dfrac{x}{y}}{2 - \dfrac{y}{x}}$

15. $\dfrac{\dfrac{2}{x} + \dfrac{3}{2x}}{5 + \dfrac{1}{x}}$

16. $\dfrac{x + 2}{1 - \dfrac{4}{x^2}}$

17. $\dfrac{1 + \dfrac{1}{x}}{x - \dfrac{1}{x}}$

18. $\dfrac{2x - \dfrac{1}{3}}{3x + \dfrac{1}{6}}$

19. $\dfrac{3 - \dfrac{3}{y}}{y - \dfrac{1}{y}}$

20. $\dfrac{x - \dfrac{x}{y}}{y - \dfrac{y}{x}}$

B

Example. Simplify $\dfrac{x}{x + \dfrac{2}{2 + \dfrac{1}{2}}}$

Solution. Write $2 + \dfrac{1}{2}$ as a basic fraction $\left(\dfrac{4}{2} + \dfrac{1}{2} = \dfrac{5}{2}\right)$. Then,

$$\frac{x}{x + \dfrac{2}{2 + \dfrac{1}{2}}} = \frac{x}{x + \dfrac{2}{\dfrac{5}{2}}} = \frac{x}{x + \dfrac{4}{5}}.$$

Write $x + \dfrac{4}{5}$ as a basic fraction $\left(\dfrac{5x}{5} + \dfrac{4}{5} = \dfrac{5x + 4}{5}\right)$. Then,

$$\frac{x}{x + \dfrac{2}{2 + \dfrac{1}{2}}} = \frac{x}{x + \dfrac{4}{5}} = \frac{x}{\dfrac{5x + 4}{5}} = x \cdot \frac{5}{5x + 4} = \frac{5x}{5x + 4}.$$

21. $\dfrac{1 - \dfrac{1}{2}}{1 + \dfrac{2}{2}}$

22. $1 - \dfrac{1}{1 - \dfrac{1}{2}}$

23. $3 - \dfrac{2}{1 - \dfrac{1}{x + 1}}$

24. $1 + \dfrac{1}{1 - \dfrac{1}{y + 1}}$

25. $2 + \dfrac{x}{1 - \dfrac{2}{x + 3}}$

26. $4 - \dfrac{2x}{1 - \dfrac{4}{2x - 1}}$

CHAPTER SUMMARY

7.1–7.5 The following laws are applicable to rational expressions. In each statement $P(x)$, $Q(x)$, $R(x)$, $S(x)$ are polynomials for which no denominator equals zero.

$$\frac{P(x)}{Q(x)} = P(x) \cdot \frac{1}{Q(x)} \qquad\qquad \text{Theorem 7.1}$$

$$\frac{P(x)}{Q(x)} = \frac{R(x)}{S(x)}, \quad \text{if } P(x) \cdot S(x) = Q(x) \cdot R(x);$$

$$P(x) \cdot S(x) = Q(x) \cdot R(x), \quad \text{if } \frac{P(x)}{Q(x)} = \frac{R(x)}{S(x)} \qquad \text{Theorem 7.2}$$

$$\frac{P(x)}{Q(x)} = \frac{P(x) \cdot R(x)}{Q(x) \cdot R(x)} \quad \text{or} \quad \frac{P(x) \cdot R(x)}{Q(x) \cdot R(x)} = \frac{P(x)}{Q(x)} \qquad \text{Theorem 7.3}$$

$$\frac{P(x)}{R(x)} + \frac{Q(x)}{R(x)} = \frac{P(x) + Q(x)}{R(x)} \qquad\qquad \text{Theorem 7.4}$$

$$\frac{P(x)}{Q(x)} = -\frac{-P(x)}{Q(x)} = -\frac{P(x)}{-Q(x)} = \frac{-P(x)}{-Q(x)}, \quad \text{and} \qquad \text{Theorem 7.5}$$

$$-\frac{P(x)}{Q(x)} = \frac{-P(x)}{Q(x)} = \frac{P(x)}{-Q(x)} = -\frac{-P(x)}{-Q(x)}$$

$$\frac{P(x)}{Q(x)} - \frac{R(x)}{Q(x)} = \frac{P(x) - R(x)}{Q(x)} \qquad\qquad \text{Theorem 7.6}$$

$$\frac{P(x)}{Q(x)} \cdot \frac{R(x)}{S(x)} = \frac{P(x) \cdot R(x)}{Q(x) \cdot S(x)} \qquad\qquad \text{Theorem 7.7}$$

$$\frac{1}{P(x)/Q(x)} = \frac{Q(x)}{P(x)} \qquad\qquad \text{Theorem 7.8}$$

$$\frac{P(x)}{Q(x)} \div \frac{R(x)}{S(x)} = \frac{P(x)}{Q(x)} \cdot \frac{S(x)}{R(x)} \qquad\qquad \text{Theorem 7.9}$$

7.6 Sometimes a complex fraction can be rewritten as a fraction which contains no other fractions in its numerator or its denominator by a direct application of the fundamental principle of fractions.

CHAPTER REVIEW

A

7.1 *Reduce each fraction to lowest terms; state any restrictions on the variables.*

1. $\dfrac{x^2 - 25}{x^2 - 4x - 5}$
2. $\dfrac{x^2 - 6x - 7}{x^2 + 2x + 1}$

7.2 *In the following exercises, assume no denominator equals zero. Build each fraction to an equivalent fraction with the specified denominator.*

3. $\dfrac{x + 1}{x + 2}; \quad \dfrac{?}{x^2 - 4}$
4. $\dfrac{3}{x - 3}; \quad \dfrac{?}{3x^2 - 27}$

7.3 *Find the least common multiple in factored form of each set of polynomials.*

5. $x^2 + 7x + 6, \quad x^2 - x, \quad 3x^2 - 3$

6. $x^2 + 3x + 2, \quad x^2 + 4x + 4, \quad x^2 + 2x + 1$

Write each expression as a single fraction in lowest terms.

7. $\dfrac{5x - 2}{3} + \dfrac{6x - 5}{5}$

8. $\dfrac{3x + 2y}{2} + \dfrac{2x - 3y}{3}$

9. $\dfrac{x + 3}{x^2 - 4} + \dfrac{4}{x + 2}$

10. $\dfrac{1 - x}{x^2 - x - 6} + \dfrac{5}{x - 3}$

7.4 11. $\dfrac{x}{x^2 - 9} - \dfrac{x - 3}{x^2 + 4x + 3}$

12. $\dfrac{1}{y^2 - 16} - \dfrac{1}{y^2 - 5y + 4}$

7.5 13. $\dfrac{6x - 12}{3x + 9} \cdot \dfrac{5x + 15}{4x - 8}$

14. $\dfrac{3y}{4xy - 6y} \cdot \dfrac{2x - 3}{12x}$

15. $\dfrac{5x + 25}{2x} \div \dfrac{2x + 10}{4x}$

16. $\dfrac{4y - 18}{3y + 2} \div \dfrac{6y - 27}{6y + 4}$

17. $\dfrac{x - 1}{3x + 2} \cdot \dfrac{5}{1 - x}$

18. $\dfrac{r^2 - 2rs}{r} \div \dfrac{2s - r}{s}$

7.6 19. $\dfrac{\dfrac{x^2 y}{2}}{\dfrac{xy}{4}}$

20. $\dfrac{\dfrac{2}{x} - y}{\dfrac{1}{x} + y}$

21. $\dfrac{x - \dfrac{1}{x}}{1 + \dfrac{1}{x}}$

22. $\dfrac{\dfrac{x}{y} + 1}{\dfrac{x}{y} - 1}$

B

23. $\dfrac{x + 1}{x^2 - 2x + 1} + \dfrac{5 - 3x}{x^2 - 1}$

24. $\dfrac{5x}{x^2 + 3x + 2} - \dfrac{3x - 6}{x^2 + 4x + 4}$

25. $\dfrac{x^2 - x - 20}{x^2 + 7x + 12} \cdot \dfrac{(x + 3)^2}{(x - 5)^2}$

26. $\dfrac{4s^2 - 1}{s^2 - 16} \cdot \dfrac{s^2 - 4s}{2s + 1}$

27. $\dfrac{y^2 - y - 6}{y^2 + 2y - 15} \div \dfrac{y^2 - 4}{y^2 - 25}$

28. $\dfrac{t^2 + 7t + 10}{t^2 + 7t + 12} \div \dfrac{t^2 + 6t + 5}{t^2 + 8t + 16}$

29. $x - \dfrac{x}{1 - \dfrac{1}{x}}$

30. $2y + \dfrac{3}{3 - \dfrac{2y}{y - 1}}$

8
chapter

First-Degree Equations and Inequalities—One Variable

Many problems are concerned with relationships between numbers or between numerical measures of physical quantities. There are three basic steps in solving such problems if the solution is not immediately evident. First, the quantitative ideas are expressed symbolically; second, the mathematical model, which generally takes the form of an equation or an inequality is solved; and third, the solution is interpreted in terms of the physical problem. Our initial concern is to study the mathematics associated with simple equations. Later, we will study inequalities and then give our attention to the important task of formulating these models for certain problems.

8.1 Equations

Recall from Definition 2.5 that a **solution** of an equation or an inequality is a replacement for the variable which forms a true statement, and that the set of all solutions is its **solution set.** A solution of an equation is also called a **root** of the equation.

Also, recall that an equation is a **conditional equation** if it is true for one or more members of the replacement set of its variable, but not for all members. It is an **identity** if it is true for all permissible replacements of the variable. In this context, a permissible replacement is any replacement for which both members of the equation are real numbers. Thus,

$$x + 1 = x + 1 \qquad \text{and} \qquad \frac{x^3}{x} = x^2$$

are identities. Note that 0 is not a permissible replacement for x in $\frac{x^3}{x} = x^2$ because the left-hand member is not defined for $x = 0$.

In this section, we are primarily concerned with *first-degree conditional equations in one variable*, such as

$$x + 3 = 7, \qquad 2x - 5 = 6, \qquad \text{and} \qquad 4 - 3x = \frac{1}{2} x.$$

Although we shall not prove it here, such an equation has *one and only one* member in its solution set. Thus, when you find one solution you have found the entire solution set.

As we observed in previous chapters, in many simple cases, the solution set of an equation is evident by inspection. Thus, $x + 3 = 7$ suggests that we simply ask: "What number added to 3 equals 7?" The solution set is obviously {4}.

Examples.   Find the solution set of each equation by inspection.

a. $5 = x - 7$ b. $3 \cdot x = 24$ c. $x + 4 = x + 4$

Solutions

a. {12}   b. {8}   c. $\{x \mid x \in R\};$ identity

Exercise 8.1

A

State which of the following are conditional equations and which are identities. Find the solution set of conditional equations by inspection.

Examples.   a. $3(x - 1) = 3x - 3$ b. $9 = 2 + x$

Solutions.   a. Identity   b. Conditional; {7}

1. $x - 1 = 6$

2. $x - 4 = 6$

3. $x + 2 = 5$

4. $7 + y = 15$

5. $y - 2 = 11$

6. $13 - x = 4$

7. $3 = x - 5$

8. $18 = y - 6$

9. $11 = 6 - x$

10. $9 = 12 - y$

11. $x + 3 = 3 + x$

12. $y - 7 = -7 + y$

13. $5x = 25$

14. $21 = 7x$

15. $8x = x(5 + 3)$

16. $x(9 - 2) = 7x$

17. $\dfrac{y}{2} = 10$

18. $\dfrac{x}{3} = -1$

19. $\dfrac{x}{4} = -6$

20. $\dfrac{x}{5} = -12$

21. $2(x + 1) = 2x + 2$

22. $3(y - 2) = 3y - 6$

23. $x + 0 = x$

24. $0 + y = y$

25. $1 \cdot x = x$

26. $x + (-x) = 0$

Specify the permissible replacement set for which each equation is an identity.

Example. $\dfrac{5}{x} - \dfrac{2}{x} = \dfrac{3}{x}$.

Solution. Using Theorem 7.6, we obtain $\dfrac{3}{x} = \dfrac{3}{x}$, which we note by inspection is an identity for all $x \in R$ $(x \neq 0)$.

27. $\dfrac{1}{y} + \dfrac{2}{y} = \dfrac{3}{y}$

28. $\dfrac{4}{x} - \dfrac{3}{x} = \dfrac{1}{x}$

29. $\dfrac{x(x + 4)}{2(x + 4)} = \dfrac{x}{2}$

30. $\dfrac{y}{5} = \dfrac{y(y - 3)}{5(y - 3)}$

31. $\dfrac{x + 2}{4} \cdot \dfrac{4}{x + 2} = 1$

32. $\dfrac{y - 4}{2} \cdot \dfrac{2}{y - 4} = 1$

33. $\dfrac{2}{y - 4} = \dfrac{2(y + 5)}{(y - 4)(y + 5)}$

34. $\dfrac{x}{x + 2} = \dfrac{x(x - 3)}{(x + 2)(x - 3)}$

8.2 Equivalent Equations—I

In Section 8.1 and previous chapters we found solution sets of some simple first-degree equations by reading the equations carefully and thinking about what they said. However, not all equations, for example,

$$7 + x = 3 + 4(x + 1),$$

have solutions that are evident by inspection. To find solutions of this equation and others, we need some mathematical tools.

First, consider equations such as

$$2x + 3 = 11, \qquad 2x = 8, \qquad \text{and} \qquad x = 4,$$

all of which have the solution set {4}. Equations such as these, which have the same solution set, are said to be **equivalent equations.**

Example. $x = -2$ and $3x + 5 = x + 1$ are equivalent, because if we replace x with -2 in the first equation we have

$$-2 = -2 \qquad \text{(a true statement)},$$

and if we replace x with -2 in the second equation we have

$$3(-2) + 5 = (-2) + 1,$$

or

$$-1 = -1 \qquad \text{(also a true statement)},$$

and {−2} is therefore the solution set of both equations.

The previous example suggests a method to solve first-degree equations whose solutions are not immediately evident. If we can find the solution of one of two equivalent first-degree equations, then we have also found the solution of the other equation. The problem then is how to find equivalent equations. There are several ways. One method is by simplifying a given equation. Perhaps the solution of the equation $8x - 7x = 15$ is not evident to you. However, since $8x - 7x = x$, the substitution of x for $8x - 7x$ leads to the equivalent equation $x = 15$ whose solution set, {15}, is evident by inspection.

Example. Solve $13x - 12x = -2$.

Solution. Since $13x - 12x$ equals x, we can substitute x for $13x - 12x$ in the left-hand member to obtain $x = -2$, whose solution set is obviously {−2}. Since the two equations are equivalent, the solution set of the original equation is also {−2}.

There is a second method of generating equivalent equations. In an equation such as $x - 3 = 5$, for any value of x for which the equation is true, $x - 3$ and 5 both represent the number 5. Therefore, the addition of

3 to each member of the equation produces a true statement for the same replacement of x for which the original statement is true. Thus,

$$x - 3 = 5,$$
$$x - 3 + (3) = 5 + (3),$$
$$x + 0 = 5 + 3,$$
$$x = 8.$$

Thus, $x - 3 = 5$ and $x = 8$ are equivalent equations. Do you see that we added 3 to both members in order to leave the variable x as the only term in the left member?

The preceding example suggests the following theorem.

Theorem 8.1. If $P(x)$, $Q(x)$, and $R(x)$ are expressions that represent real numbers, then the equations
$$P(x) = Q(x) \tag{1}$$
and
$$P(x) + R(x) = Q(x) + R(x) \tag{2}$$
are equivalent.

Examples. Transform each equation to an equivalent equation in which the variable is the only term in one of the members of the equation and the variable does not appear in the other member.

a. $x + 5 = 3$ b. $4 = y - 2$ c. $2x + 2 = x + 5$

Solutions
a. Add (-5), the additive inverse of 5, to each member to obtain
$$x + 5 + (-5) = 3 + (-5),$$
$$x = -2.$$
b. Add $(+2)$, the additive inverse of -2, to each member to obtain
$$4 + (+2) = y - 2 + (+2),$$
$$6 = y.$$
c. Add $(-x - 2)$ to each member to obtain
$$2x + 2 + (-x - 2) = x + 5 + (-x - 2)$$
$$x = 3.$$

There is a third method of generating equivalent equations. In an equation such as $x/2 = 3$, for any value of x for which the equation is true, $x/2$ and 3 both represent the number 3. Therefore, the multiplication of each member by 2, the multiplicative inverse of $1/2$, produces a true statement for the same replacement of x for which the statement is true. Thus,

$$\frac{x}{2} = 3,$$

$$(2) \cdot \frac{x}{2} = (2) \cdot 3,$$

and

$$x = 6,$$

are equivalent equations, and since 6 is a solution of $x = 6$, it is also a solution of $x/2 = 3$.

The preceding example suggests the following theorem.

Theorem 8.2. If $P(x)$, $Q(x)$, and $R(x)$ are expressions and $R(x)$ is not zero, then the equations

$$P(x) = Q(x) \tag{3}$$

and

$$P(x) \cdot R(x) = Q(x) \cdot R(x) \tag{4}$$

are equivalent.

Note that (4) is *not* equivalent to (3) for $R(x) = 0$.

Examples. Transform each equation into an equivalent equation in which the variable is the only term in one of the members.

a. $\dfrac{x}{7} = 5$ b. $18 = 3x$ c. $\dfrac{2}{3}x = 8$

Solutions
a. Multiply each member by 7, the multiplicative inverse of $1/7$, to obtain

$$7 \cdot \frac{x}{7} = 7 \cdot 5,$$

$$x = 35.$$

b. Multiply each member by $1/3$, the multiplicative inverse of 3, to obtain

$$\frac{1}{3} \cdot 18 = \frac{1}{3} \cdot 3x,$$

$$6 = x.$$

c. Multiply each member by $3/2$, the multiplicative inverse of $2/3$, to obtain

$$\frac{3}{2} \cdot \frac{2}{3} x = \frac{3}{2} \cdot 8,$$

$$x = 12.$$

As in the examples above, to solve a first-degree equation in one variable in which the solution set cannot be obtained simply by inspection, we form an equivalent equation whose solution is obvious. We generate equivalent equations by:

1. Simplifying one or both members of the equation.
2. Adding the same expression, representing a real number, to each member.
3. Multiplying each member by the same expression representing a non-zero real number.

Of course, we can add any expression or multiply by any nonzero expression, but in general we wish to obtain an equation in which the variable appears as the *only term* in one of the members and does not appear in the other member since the solution set is more easily determined in this form. As you may have noted, this can be done by adding inverses of the other *terms* in a member, or by multiplying by the multiplicative inverse of the *coefficient* of the variable, or both.

Example.   Solve $25 = 7x - 3$.

Solution.   If the solution is not evident by inspection, you can generate the equivalent equations

$$25 + 3 = 7x - 3 + 3, \qquad \text{Theorem 8.1}$$
$$28 = 7x + 0, \qquad \text{Additive inverse law}$$
$$28 = 7x. \qquad \text{Identity law for addition}$$

Solution continued on page 224.

If the solution set is still not evident, you can generate the equivalent equations

$$\frac{1}{7} \cdot 28 = \frac{1}{7} \cdot 7x, \qquad \text{Theorem 8.2}$$

$$4 = 1 \cdot x, \qquad \text{Multiplicative inverse law}$$

$$4 = x. \qquad \text{Identity law for multiplication}$$

The solution set is now obvious. Since the last equation is equivalent to $25 = 7x - 3$, the solution set of this equation is also $\{4\}$.

Exercise 8.2

A

In Problems 1 to 10, state which axiom or theorem implies that the second given equation is equivalent to the first equation. Use the symbolic form of the axiom or the theorem.

Example. $4x - 7 = 9$,
$\qquad\qquad 4x - 7 + 7 = 9 + 7$

Solution.   Theorem 8.1: If $P(x) = Q(x)$, then $P(x) + R(x) = Q(x) + R(x)$ is an equivalent equation.

1. $2x + 5 = 6$,
   $\qquad 2x + 5 - 5 = 6 - 5$

2. $3y - 4 = 12$,
   $\qquad 3y - 4 + 4 = 12 + 4$

3. $x + 3x = 16$,
   $\qquad (1 + 3)x = 16$

4. $3y + 2y = 25$,
   $\qquad (3 + 2)y = 25$

5. $x - 4 = 13$,
   $\qquad x - 4 + 4 = 13 + 4$

6. $3y + 5 = 11$,
   $\qquad 3y + 5 - 5 = 11 - 5$

7. $8x + 4 = -3 + x$,
   $\qquad 8x + 4 - x - 4 = -3 + x - x - 4$

8. $3x - 7 = 6 - x$,
   $\qquad 3x - 7 + 7 + x = 6 - x + 7 + x$

9. $-27 = 12x - 3x$,
   $\qquad -27 = (12 - 3)x$

10. $7 = 9y - 2y$,
    $\qquad 7 = (9 - 2)y$

Solve. If the solution set is evident by inspection, write it directly. If not, generate equivalent equations until the solution set is obvious.

Examples

a. $x + 2 = 6$ b. $7x - 4 = 4x + 11$

Solutions

a. By inspection, the solution set is {4}.

b. $7x - 4 + (-4x + 4) = 4x + 11 + (-4x + 4)$
$$3x = 15$$
The solution set is {5}.

11. $4 = x - 5$ 12. $5 - x = x + 3$ 13. $3 - y = 5 + 4y$

14. $3x = 4 + 2x$ 15. $4x + 3 = 7x - 7$ 16. $4x - 2 = 3x - 2$

17. $4x + 4 = 2x + 4$ 18. $4x - 6 = 0$ 19. $0 = 7x + 5$

20. $4x - 3 = 2x - 6$ 21. $\dfrac{1}{2}x = 5$ 22. $\dfrac{4}{3}y = -8$

23. $\dfrac{-2x}{3} = -12$ 24. $15 = \dfrac{-5y}{3}$ 25. $-8 = \dfrac{2y}{5}$

26. $\dfrac{4x}{5} = -16$ 27. $\dfrac{y}{3} = 0$ 28. $\dfrac{-2x}{5} = 0$

B

29. $2x - 3 + 2x = 4 - x + 8$ 30. $6y + 5 - 7y = 10 - 2y + 3$

31. $5z + 3 - z = 10 + z + 2$ 32. $5t + 7 - 2t - 16 = 0$

33. $\dfrac{x + 1}{2} = x + 2$ 34. $\dfrac{y - 4}{2} = y - 3$

35. $\dfrac{3x + 2x}{2} = \dfrac{10 + 4}{2} - 2$ 36. $\dfrac{4x + x}{3} = 10$

37. $\dfrac{3x + 5x}{2} = 3$ 38. $\dfrac{8x + 4x}{2} = \dfrac{8 + 10}{3}$

39. $\dfrac{5x - 3x}{2} = 2x - 4$ 40. $\dfrac{6x - x}{4} = x + 7$

8.3 Equivalent Equations—II

Quite frequently, as in the example below, an application of the distributive law facilitates the procedure of generating equivalent equations before using Theorems 8.1 and 8.2.

Example.   Solve $7(x - 3) - (2x + 2) = 7$.

Solution.   By first applying the distributive law to each term in the left member, we generate the following sequence of equivalent equations.

$$7x - 21 - 2x - 2 = 7,$$
$$5x - 23 = 7,$$
$$5x = 30,$$
$$x = 6.$$

Hence, the solution set is $\{6\}$.

Any application of Theorems 8.1 and 8.2 results in an equivalent equation. However, you must be careful in the application of Theorem 8.2 because multiplication by zero is excluded. In cases where you multiply by an expression that contains a variable, you should notice any replacements for the variable for which the expression has a value of zero. The new equation may not be equivalent to the original. The following example illustrates the point.

Example.   Solve $\dfrac{x}{x - 3} = \dfrac{3}{x - 3} + 2$.   (1)

Solution.   Each member can be multiplied by $(x - 3)$ if $x - 3 \neq 0$, or $x \neq 3$, to obtain

$$(x - 3) \cdot \frac{x}{x - 3} = (x - 3) \cdot \frac{3}{x - 3} + (x - 3) \cdot 2,$$ (2)
$$x = 3 + 2x - 6,$$
$$6 - 3 = 2x - x,$$
$$3 = x.$$

However, for $x = 3$, Equations (1) and (2) are not equivalent. We multiplied each member of (1) by a polynomial having a value of zero. Substituting 3 for x in (1), we obtain

$$\frac{3}{0} = \frac{3}{0} + 2,$$

and neither member is defined. Hence, the solution set of (1) is $\varnothing$.

The lesson here is clear. In any transformation which involves multiplying each member by an expression $R(x)$ which involves a variable, you must check any number obtained as a solution against the restricted values of the variable.

An equation of the form

$$\frac{P(x)}{Q(x)} = \frac{R(x)}{S(x)} \qquad [Q(x),\ S(x) \neq 0]$$

is sometimes called a **proportion.** In this context, $P(x)$ and $S(x)$ are called the **extremes** and $Q(x)$ and $R(x)$ are called the **means** of the proportion. Observe by Theorem 8.2 that the equation above is equivalent to

$$Q(x) \cdot R(x) = P(x) \cdot S(x),$$

that is, *the product of the means equals the product of the extremes.*

Examples. Solve.

a. $\dfrac{x}{9} = \dfrac{5}{3}$

b. $\dfrac{5}{x} = \dfrac{4}{x + 5}$

Solutions. Apply Theorem 8.2.

a. $3x = 45$

$\qquad \{15\}$

b. $5x + 25 = 4x$

$\qquad 5x - 4x = -25$

$\qquad\qquad \{-25\}$

Exercise 8.3

A

Solve each equation. If the solution set is evident by inspection, write it directly. If not, generate equivalent equations until the solution set is obvious.

Example. $3(6 - 2x) = 10 - 7(x + 1)$

Solution. Apply the distributive

$$18 - 6x = 10 - 7x - 7$$

$$18 - 6x = 3 - 7x$$

Solution continued on page 228.

Add $7x - 18$ to each member.

$$x = -15$$

The solution set is $\{-15\}$.

1. $3(x + 6) = 21$ 2. $4(x + 2) = 4$

3. $6 = 3(3x - 1)$ 4. $5 = 5(3x + 2)$

5. $x - (9 - x) = 3$ 6. $3x - (4 - x) = 0$

7. $4(8 - 3x) = 32 - 8(x + 2)$ 8. $7x - 3(3x + 6) = 7(6 + x)$

Example. $5 - (x - 2)(x + 3) = 9 - x^2$

Solution. Apply the distributive law.

$$5 - (x^2 + x - 6) = 9 - x^2$$

$$5 - x^2 - x + 6 = 9 - x^2$$

Add $x^2 - 11$ to each member.

$$-x = -2$$

Multiply each member by (-1).

$$x = 2$$

The solution set is $\{2\}$.

9. $(x - 2)(x - 2) = x^2 - 12$ 10. $(3x + 2)(x - 4) = 3x^2 - 6x + 8$

11. $(x - 3)^2 = x^2 - 9$ 12. $(x - 2)^2 = x^2 - 12$

13. $5 - (x - 4)(x + 3) = 12 - x^2$ 14. $3 - (x + 1)(3x - 2) = -3x^2$

15. $2 + 4x - x^2 = - (x - 3)(x + 4)$ 16. $0 = 2x^2 + (3 - 2x)(x + 1)$

Example. $\dfrac{3x}{4} - x = \dfrac{5}{3}$

Solution. Multiply each member by the least common denominator 12..

$$12 \left(\frac{3x}{4} - x \right) = 12 \left(\frac{5}{3} \right)$$

Simplify each member.

$$9x - 12x = 20$$

$$-3x = 20$$

Multiply each member by $(-1/3)$.

$$-\frac{1}{3}(-3x) = -\frac{1}{3}(20)$$

$$x = -\frac{20}{3}$$

The solution set is $\{-20/3\}$.

17. $\dfrac{7x}{2} - 1 = -\dfrac{1}{2} + x$

18. $2 + \dfrac{x}{10} = \dfrac{5}{3}$

19. $\dfrac{x}{5} - \dfrac{x}{6} = 2$

20. $\dfrac{x}{2} + \dfrac{x}{3} - \dfrac{x}{4} = 14$

21. $\dfrac{4x - 3}{3} + 3 = \dfrac{2}{3}$

22. $\dfrac{3x - 2}{6} = \dfrac{x + 2}{3}$

23. $\dfrac{3x}{4} - \dfrac{3x + 5}{6} = 1$

24. $\dfrac{1}{5}(x + 4) + \dfrac{1}{3}(x - 5) = \dfrac{1}{3}(x - 2)$

Example. $\dfrac{3}{4} = 8 - \dfrac{2x + 11}{x - 5}$

Solution. Determine the least common denominator which is $4(x - 5)$. Multiply each member by $4(x - 5)$.

$$4(x - 5) \cdot \frac{3}{4} = 4(x - 5)\left(8 - \frac{2x + 11}{x - 5}\right)$$

$$3(x - 5) = 4 \cdot 8(x - 5) - 4(2x + 11)$$

Apply the distributive law.

$$3x - 15 = 32x - 160 - 8x - 44$$

Simplify the right-hand member.

$$3x - 15 = 24x - 204$$

Add $-3x + 204$ to each member.

$$189 = 21x$$

Multiply each member by $1/21$.

$$9 = x$$

Since we have multiplied each member of the equation (step 2) by a polynomial, $4(x - 5)$, in which the variable appears, we check, and observe that this polynomial $4(x - 5)$ is *not equal* to zero for the solution, 9. Therefore, $\{9\}$ is the solution set of the original equation.

25. $\dfrac{5}{x} + \dfrac{4}{x} = \dfrac{9}{2}$

26. $\dfrac{2}{x} + \dfrac{3}{x} = \dfrac{1}{5}$

27. $\dfrac{x}{x + 2} = 2 - \dfrac{9}{x + 2}$

28. $\dfrac{x}{x - 2} = \dfrac{2}{x - 2} + 7$

29. $\dfrac{5}{x - 3} = \dfrac{x + 2}{x - 3} + 3$

30. $\dfrac{1}{3x + 3} + \dfrac{1}{x + 1} = \dfrac{1}{6}$

Examples.   a. $\dfrac{x}{x + 2} = \dfrac{3}{5}$

b. $\dfrac{x + 1}{3} = \dfrac{x^2 + x}{3x + 1}$

Solutions.   a. $5x = 3(x + 2)$
$5x = 3x + 6$
$2x = 6$
$\{3\}$

b. $(x + 1)(3x + 1) = 3(x^2 + x)$
$3x^2 + 4x + 1 = 3x^2 + 3x$
$x + 1 = 0$
$\{-1\}$

31. $\dfrac{x}{x + 3} = \dfrac{4}{7}$

32. $\dfrac{5}{x - 3} = \dfrac{2}{x}$

33. $\dfrac{x + 3}{x + 2} = \dfrac{10}{9}$

34. $\dfrac{x - 7}{2} = \dfrac{x + 2}{11}$

35. $\dfrac{x}{2} = \dfrac{x^2 + 3}{2x - 3}$

36. $\dfrac{x + 2}{4} = \dfrac{x^2 + 2}{4x - 2}$

B

37. $\dfrac{x^2 + 9}{x^2 - 4} = \dfrac{x}{x + 2} + \dfrac{3}{x - 2}$

38. $\dfrac{4x}{4x^2 - 9} + \dfrac{4}{2x - 3} = \dfrac{1}{2x + 3}$

39. $\dfrac{x^2 - 3x + 4}{x + 2} = \dfrac{4}{x + 2} + x - 3$

40. $\dfrac{x + 2}{x - 1} + x + 1 = \dfrac{(x + 1)^2}{x - 1}$

41. $\dfrac{2x + 3}{x + 4} - \dfrac{x - 2}{x + 1} = 1$

42. $\dfrac{3x - 2}{2x - 3} + \dfrac{x + 1}{x - 1} = \dfrac{5x + 2}{2x - 3}$

43. Why can you tell by inspection that $\dfrac{x + 3}{x + 4} = 1$ has no solution?

44. For what value of k will the equation $2x - 3 = \dfrac{x + 5}{k}$ have as its solution set $\{-2\}$?

8.4 Solving Equations for Specified Symbols

You can now solve first-degree equations when the solution set is not evident by inspection. To reinforce this knowledge, we shall now rewrite equations with more than one variable as equivalent equations. In this case, however, we shall *not* be looking for number replacements for any of the variables.

For example, the equation, $d = rt$, can be transformed to the equivalent equation, $d/r = t$, by multiplying each member of the first equation by $1/r$ $(r \neq 0)$. This process is referred to as "solving for the variable t in terms of the variables d and r," or simply, "solving for t." In solving such equations for one variable in terms of the others, it is customary to write the specified variable as the left-hand member of the solved equation. Thus, in the example above, the symmetric law of equality is invoked to write $t = d/r$ $(r \neq 0)$. Although we can view d/r as a solution of the original equation $d = rt$ and express the solution set as $\{d/r\}$, we do not do so. We simply view the equations $d = rt$ and $t = d/r$ as equivalent equations.

Exercise 8.4

A

Solve each equation for the specified variable in terms of the other variables. In Problems 1 to 6, justify the equivalency of each equation wherever Theorem 8.1 or 8.2 is invoked. Indicate any restrictions on the variables.

Example. $\dfrac{2x}{3} + 3a = 0$ for x.

Solution. Add $-3a$, the additive inverse of $3a$, to each member.

$$\frac{2x}{3} = -3a \qquad \text{Theorem 8.1}$$

Multiply each member by $3/2$, the reciprocal of $2/3$.

$$\frac{3}{2} \cdot \frac{2x}{3} = \frac{3}{2} \cdot (-3a) \qquad \text{Theorem 8.2}$$

$$\left(\frac{3}{2} \cdot \frac{2}{3}\right) x = \frac{-9a}{2}$$

$$x = \frac{-9a}{2}$$

1. $b = y + a - c$ for y.

2. $4x - 4a = x - a$ for x.

3. $4y = a^2b - 4y$ for y.

4. $3x + b^2 = 0$ for x.

5. $\dfrac{3x}{4} - c = b$ for x.

6. $\dfrac{2}{3}x + a = 0$ for x.

7. $x + y = 7$ for y.

8. $2y - x = 4$ for y.

9. $p = a + b + c$ for b.

10. $x + by + c = 0$ for x.

11. $I = prt$ for r.

12. $V = lwh$ for w.

13. $E = mc^2$ for m.

14. $V = \pi r^2h$ for h.

15. $p = 2l + 2w$ for w.

16. $x + by + c = 0$ for y.

17. $v = a + gt$ for t.

18. $t = a - bp$ for b.

19. $y = \dfrac{k}{x}$ for x.

20. $p = \dfrac{k}{v}$ for v.

Example. $xy + y = 1$ for y.

Solution.   Factor y from each term in the left-hand member.

$$y(x + 1) = 1$$

Multiply each member by $\dfrac{1}{x + 1}$, the reciprocal of $x + 1$.

$$y = \frac{1}{x + 1} \quad (x \neq -1)$$

21. $4y + xy = -2$ for y.

22. $5y - xy = 3$ for y.

23. $3x = \dfrac{-xy}{4} + 1$ for x.

24. $2x = \dfrac{xy}{5} - 3$ for x.

B

25. $x = \dfrac{3}{y - 1}$ for y.

26. $l = a + (n - 1)d$ for d.

27. $s = \dfrac{a}{1 - r}$ for r.

28. $s = \dfrac{a - ar^n}{1 - r}$ for a.

29. $A = \dfrac{h}{2}(b + c)$ for b.

30. $\dfrac{1}{r} = \dfrac{1}{a} + \dfrac{1}{b}$ for r.

8.5 Solution of Inequalities

Recall that the graphs of solution sets of inequalities in one variable can be displayed on the number line. For example, the graph of the solution set of the inequality $x > 5$ over the set of integers is shown in Figure 8.1a, and the graph of the solution set over the real numbers is shown in Figure 8.1b.

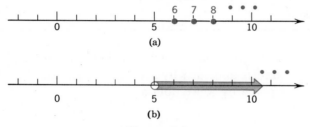

(a)

(b)

Figure 8.1

The solid line in Figure 8.1b implies that there are an infinite number of points between any two points. Note that an open dot has been placed at the point associated with 5 to indicate that 5 is not a solution.

As we observed with equations, sometimes we cannot determine the solution set of an inequality by inspection. Hence, we need to generate equivalent inequalities. In the following discussion we illustrate some properties of inequalities that suggest two theorems. These theorems help us to generate equivalent inequalities. Consider the inequality $2 < 4$. Adding 5 to each member, we obtain

$$2 + 5 < 4 + 5, \quad \text{or} \quad 7 < 9,$$

and adding -5 to each member, we obtain

$$2 + (-5) < 4 + (-5), \quad \text{or} \quad -3 < -1,$$

all of which are true statements. These examples are shown graphically in Figures 8.2a and 8.2b, respectively.

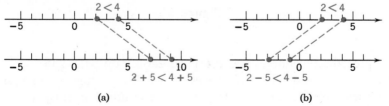

(a)

(b)

Figure 8.2

The preceding examples suggest the following theorem.

Theorem 8.3.   If $P(x)$, $Q(x)$, and $R(x)$ are expressions that represent real numbers, then the inequalities

$$P(x) < Q(x)$$

and

$$P(x) + R(x) < Q(x) + R(x)$$

are equivalent.

Now consider the *multiplication* of each member of an inequality, first by a positive real number and then by a negative real number. Multiplying each member of $2 < 4$ by 3, we obtain

$$3 \cdot 2 < 3 \cdot 4, \quad \text{or} \quad 6 < 12.$$

The *order* of the inequality remains the same for the transformed true statement; the left-hand member remains less than the right-hand member. However, if we multiply each member of $2 < 4$ by -3, we have -6 as the left-hand member and -12 as the right-hand member. Since

$$-6 > -12,$$

the order of the inequality is *reversed* for the transformed true statement.

These examples are shown graphically in Figures 8.3a and 8.3b, respectively.

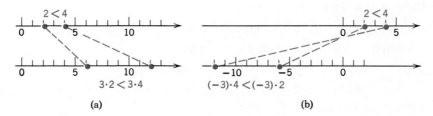

(a)   (b)

Figure 8.3

The preceding examples suggest Theorem 8.4 on page 235.

Theorems 8.3 and 8.4 are the bases by which we can generate equivalent inequalities whose solution sets can be determined by inspection. The proofs of these theorems require concepts which we have not discussed and will not be shown.

Theorem 8.4. If $P(x)$, $Q(x)$, and $R(x)$ are expressions that represent real numbers, then the inequalities

$$P(x) < Q(x)$$

and

$$P(x) \cdot R(x) < Q(x) \cdot R(x)$$

are equivalent for every real number replacement for x for which $R(x) > 0$; the inequalities

$$P(x) < Q(x)$$

and

$$P(x) \cdot R(x) > Q(x) \cdot R(x)$$

are equivalent for every real number replacement for x for which $R(x) < 0$.

Example.   Solve $2x + 1 \geq 9$, where $x \in J$, and graph the solution set.

Solution.   Add -1 to each member of $2x + 1 \geq 9$.

$$2x \geq 8$$

Multiply each member by $1/2$.

$$x \geq 4$$

The solution set is $\{x \mid x \geq 4, x \in J\}$.

Exercise 8.5

A

Solve and graph the solution set on a number line. In Problems 1 to 6, justify the equivalency of each inequality wherever Theorem 8.3 or 8.4 is invoked.

Example. $\dfrac{x + 5}{4} \leq 5 - x, \; x \in R$

Solution.   Multiply each member by 4, the multiplicative inverse of $1/4$.

$$4\left(\frac{x + 5}{4}\right) \leq 4(5 - x) \qquad \text{Theorem 8.4}$$

$$\left(4 \cdot \frac{1}{4}\right)(x + 5) \leq 20 - 4x$$

$$x + 5 \leq 20 - 4x$$

Solution continued on page 236.

Add $4x - 5$, the additive inverse of $-4x + 5$, to both members.

$$x + 5 + 4x - 5 \le 20 - 4x + 4x - 5 \qquad \text{Theorem 8.3}$$

$$5x \le 15$$

Multiply each member by $1/5$, the multiplicative inverse of 5.

$$\frac{1}{5} \cdot 5x \le \frac{1}{5} (15) \qquad \text{Theorem 8.4}$$

$$\left(\frac{1}{5} \cdot 5\right) x \le 3$$

$$x \le 3$$

The solution set is $\{x \mid x \le 3, x \in R\}$.

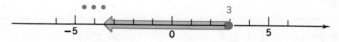

1. $x - 3 < 2, x \in J$

2. $x + 8 > 9, x \in J$

3. $4x < 8, x \in J$

4. $3x > -15, x \in J$

5. $2x - 6 \ge 8, x \in R$

6. $3x - 4 < 5, x \in R$

7. $\dfrac{2x}{3} > -4, x \in R$

8. $\dfrac{3x}{4} < 6, x \in R$

9. $4x - 3 > 4 + 3x, x \in J$

10. $3x + 4 \le 2x - 2, x \in J$

11. $3x + 2 < 4x + 1, x \in J$

12. $x + 7 \ge 2x - 1, x \in J$

13. $2 - 3x > -6 + 5x, x \in R$

14. $5 - 6x \le 3x + 4, x \in R$

15. $\dfrac{3x - 9}{3} > 0, x \in R$

16. $\dfrac{2x - 10}{7} < 0, x \in R$

17. $\dfrac{2x + 5}{3} \le 1, x \in R$

18. $\dfrac{4x - 3}{5} \ge -3, x \in R$

19. $\dfrac{6x - 8x}{6} > 2, x \in R$

20. $\dfrac{2x - 4x}{10} \le 1, x \in R$

21. $\dfrac{4x - 7x}{7} \le 3, x \in R$

22. $\dfrac{3x - 8x}{5} \le -8, x \in R$

23. $\dfrac{3x - 2x}{5} > x + 4, x \in R$

24. $\dfrac{5x - 4x}{3} < x - 2, x \in R$

8.6 Mathematical Models for Word Problems

Word problems generally state relationships between numbers. When we solve a word problem, we seek a number or numbers for which the stated relationship holds, whether the problem is explicitly concerned with numbers or with numerical measures of physical quantities. A mathematical model obtained by expressing the quantitative ideas symbolically is helpful. In Sections 8.7 and 8.8 we shall write models for word problems which are in the form of equations and inequalities. In this section, as we have done in previous sections, we shall simply write mathematical models for simple word phrases.

Examples.   Express each phrase symbolically. Use n as the variable.

a. The sum of a number and 25.
b. One more than 3 times a certain number.
c. The sum of two consecutive integers.
d. The sum of two consecutive even integers.

Solutions.   a. $n + 25$ b. $3n + 1$
c. $n + (n + 1)$. If n is an integer, then $n + 1$ represents the next consecutive integer.
d. $n + (n + 2)$. If n is an even integer, then $n + 2$ represents the next consecutive even integer.

It is necessary to understand that the replacement set for a variable is sometimes restricted by the conditions of the problem. Thus if n represents the *number* of boys in a class, the variable n can only be replaced by elements in the set of non-negative integers. Certainly you would not consider as a replacement set, either the set of integers, rational numbers, or real numbers.

Exercise 8.6

A

Write each phrase symbolically using n as the variable.

1. The sum of 14 and an integer.

2. The sum of an integer and three times itself.

3. An integer subtracted from three times itself.

4. An integer subtracted from the square of the next consecutive integer.

Examples. The sum of two real numbers is 12. If y represents the smaller number, represent in terms of y:

a. The larger number.   b. Three times the   c. Four times the larger
                              smaller number.   number.

Solutions   a. $12 - y$ b. $3y$ c. $4(12 - y)$

5. The sum of two real numbers is 23. If s represents the smaller number, represent in terms of s:

   a. The larger   b. Six times the   c. One-half of the larger
      number.   smaller number.   number.

6. The difference between two real numbers is 15. If x represents the larger number, represent in terms of x:

   a. The smaller   b. Three times the   c. Two-thirds of the
      number.   smaller number.   larger number.

Examples.   Represent (in cents) the value of each of the following collections of coins (i.e., the value of one coin times the number of that type of coin):

a. Two dimes   b. x dimes   c. Four nickels
d. $(n - 2)$ nickels   e. y quarters   f. $(y + 6)$ quarters

Solutions

a. $10(2)$ or 20 cents   b. $10x$ cents   c. $5(4)$ or 20 cents

d. $5(n - 2)$ cents   e. $25y$ cents   f. $25(y + 6)$ cents

7. Represent in cents the value of each collection of coins:

   a. 21 nickels.   b. $(y - 3)$ dimes.

   c. $(3t + 4)$ halves.   d. $\left(\dfrac{x}{2} - 4\right)$ quarters.

8. Represent in cents the value of each collection of bills:

   a. One five-dollar bill.   b. x ten-dollar bills.
   c. $(y + 5)$ twenty-dollar bills.   d. $(s - 4)$ fifty-dollar bills.

Examples. One truck has a capacity 4 tons greater than another truck. If x represents the capacity of the smaller truck, represent in terms of x:

a. The total tons hauled by the smaller truck in 5 trips.

b. The total tons hauled by the larger truck in six trips.

Solutions. a. $5x$ b. $6(x + 4)$

9. One moving van has a capacity 15 cubic feet greater than another moving van. If x represents the capacity of the smaller van in cubic feet, represent in terms of x:

a. The total number of cubic feet carried by the smaller van in 12 trips.

b. The total number of cubic feet carried by the larger van in three trips.

10. One rocket has a payload capacity 200 tons greater than another type of rocket. If x represents the capacity in tons of the larger rocket, represent in terms of x:

a. The total payload in tons carried by the larger rocket in six trips to Mars.

b. The total payload in tons carried by the smaller rocket in four trips to Venus.

In each sentence, select from the sets of numbers (a) *to* (d) *the most appropriate replacement set for the indicated variable.*

(a) {whole numbers}   (b) {integers}

(c) {positive real numbers}   (d) {real numbers}

11. There are a total of n people living in the United States.

12. There are y girls in the classroom.

13. There are n boys on the football squad.

14. The distance to school is m miles.

15. The length of a rectangle is x feet.

16. The altitude of a right triangle is x inches long.

17. The temperature (to the nearest degree) in the room is T degrees.

18. The atmospheric pressure is p inches of mercury.

19. The number n is associated with a point on a number line.

20. The distance to the moon is d miles (to the nearest mile).

8.7 Word Problems Concerning Numbers

In Section 8.6 you learned to write mathematical models for word phrases involving numerical quantities. In this section we write mathematical models in the form of equations for sentences involving numerical quantities. The equations can then be solved by techniques you have learned previously and the results interpreted in terms of the original word problem.

The following suggestions are offered as guides in the writing of mathematical models for solving word problems.

1. Write the quantities asked for in a problem in word phrases and represent these quantities by variables. Since we are concerned with equations having only one variable at this time, all relevant quantities should be represented in terms of one variable.

2. Write an equation or inequality which states a condition on the variable. The equation or inequality can be obtained from the problem itself and generally involves conditions on numbers. For example, "The sum of what number and 5 equals 9?" would produce the equation $5 + x = 9$.

3. Solve the resulting equation.

4. Interpret the solution in terms of the original word problem.

Example.   The sum of a certain integer and 14 is equal to three times the integer. What is the integer?

Solution. Let $x =$ an integer. Write a mathematical model relating known and unknown quantities.

$$x + 14 = 3x$$

Solve the equation.

$$14 = 2x$$
$$7 = x$$

The solution set of each equations is {7}.

Check: Is the sum of 7 and 14 equal to 3 times 7? Yes; the number we seek is therefore 7.

There may be more than one mathematical model that can be used for a particular word problem. In the answer section, we will give only one mathematical model for a problem even though there are others that can be used.

Example. One number exceeds a second by 3. Find the two numbers if their sum equals 47.

Solution. If x represents the larger number, then $x - 3$ represents the smaller number. From the conditions in the word problem, we have

$$x + (x - 3) = 47,$$

$$2x = 50,$$

$$x = 25.$$

The solution set of the equation is {25}. Since $x = 25$, it follows that $x - 3 = 22$ and the two numbers are 25 and 22.

Alternative Solution. If x represents the smaller number, then $x + 3$ represents the larger number. For these conditions on x we have

$$x + (x + 3) = 47,$$

$$2x = 44,$$

$$x = 22.$$

The solution set of the equation is {22}. Since $x = 22$, it follows that $x + 3 = 25$ and the two numbers are 22 and 25.

A solution of an equation that is the mathematical model of a word problem is not necessarily a solution of the stated problem. Consider the following.

Example. Find three consecutive *odd* integers whose sum is 18.

Solution. Let $n =$ an odd integer; then $n + 2 =$ the next consecutive odd integer, and $n + 4 =$ the next consecutive odd integer.
Write a mathematical model which describes the condition on n.

$$n + (n + 2) + (n + 4) = 18$$

Solve the equation.

$$3n + 6 = 18$$

$$3n = 12$$

$$n = 4$$

Solution continued on page 242.

Now, although 4 is a solution of each of these equivalent equations, 4 is not an *odd* integer and hence does not meet the condition stipulated in the problem. Thus, there are no such numbers.

Exercise 8.7

A

For each problem, use x as the variable.
a. Write a mathematical model. (There may be more than one model possible.)
b. Find the solution set of the resulting equation.
c. Interpret the solution in terms of the stated problem.

Example.   One integer is four more than another integer. The sum of the two integers is 246. Find the integers.

Solution.   a. Let $x =$ an integer; then $x + 4 =$ the second integer. A mathematical model is

$$x + (x + 4) = 246.$$

b. Solve for x.

$$2x = 242$$

$$x = 121$$

The solution set is $\{121\}$.

c. If $x = 121$, then $x + 4 = 125$.

Are 121 and 125 integers such that $121 + 125 = 246$? Yes.

Alternate Solution.   a. Let $x =$ an integer; then $246 - x =$ the second integer. If x is 4 more than the second integer, then a mathematical model is

$$x - (246 - x) = 4.$$

b. Solve for x.

$$x - 246 + x = 4$$

$$2x = 250$$

$$x = 125$$

The solution set is $\{125\}$.

c. If $x = 125$, then $246 - x = 121$. The integers are 121 and 125, as in the solution above.

1. The sum of two integers is 112. One of the integers is 72 less than the other integer. Find the integers.

2. The difference of two natural numbers is 138. One of the numbers is twice the other number. Find the numbers.
3. The sum of two integers is 135. One is four times the other. Find the integers.
4. The difference of two whole numbers is 84. One number is one-third of the other. Find the numbers.
5. The sum of two integers is 105. One integer is two-fifths of the other. Find the integers.
6. The difference of the squares of two whole numbers is 23. One number is one more than the other. Find the numbers.

Example. Find three consecutive integers whose sum is 360.

Solution.   a. Let $x = $ an integer; then

$$x + 1 = \text{the next consecutive integer};$$
$$x + 2 = \text{the next consecutive integer}.$$

The mathematical model is

$$x + (x + 1) + (x + 2) = 360.$$

b. Solve for x.

$$3x + 3 = 360$$
$$3x = 357$$
$$x = 119$$

The solution set is $\{119\}$.

c. If $x = 119$, then $x + 1 = 120$ and $x + 2 = 121$.
Does $119 + 120 + 121 = 360$? Yes.

7. Find four consecutive integers whose sum is 206.
8. Find three consecutive *even* integers whose sum is -132.
9. Find three consecutive *odd* integers whose sum is 237.
10. The difference of the squares of two consecutive integers is 17. Find the integers.
11. The difference of the squares of two consecutive positive *even* integers is 52. Find the integers.
12. Find three consecutive *odd* integers such that twice the first, plus three times the second, minus one-third of the third, equals 84.
13. One positive integer is three times another. If 5 is added to the smaller, the sum is equal to 9 subtracted from the larger. Find the integers.
14. One positive integer is four times another. A third integer is six more than the smaller of the other two. The sum of the three integers is equal to the smallest integer plus 46. What are the integers?

15. One positive integer is three less than a second integer. The larger plus four times the smaller equals 123. Find the integers.

16. One integer is four more than a second integer. Five times the second integer plus twice the first equals 113. Find the integers.

17. The sum of two integers is 25. Twice the larger integer is one more than 5 times the smaller. Find the integers.

18. If one-half of a certain integer is added to four times the integer itself, the result is 153. Find the integer.

B

19. The product of two consecutive even integers is 22 less than the square of the second one. Find the integers.

20. The quotient of two consecutive even integers is 12/13. Find the integers.

21. The sum of the squares of three consecutive integers is equal to three times the square of the first integer plus 59. Find the integers.

22. The sum of the squares of three consecutive even integers is equal to three times the square of the third integer minus 76. Find the integers.

23. Find two integers in the ratio of 9:7 whose sum exceeds their difference by 56. (*Hint:* Let $9x$ represent one integer; then $7x$ represents the other.)

24. Find three integers in the ratio 3:2:1 such that twice the first integer plus five times the difference between the second and third integers is equal to 33.

25. Find three consecutive even integers such that 3/2 of the sum of the first and the second equals the third decreased by 9.

26. The denominator of a certain fraction is 6 less than three times the numerator. If 10 is subtracted from each of the numerator and denominator, the resulting fraction has the value 1/10. Find the fraction.

8.8 Word Problems Concerning Physical Quantities

Many situations arise which involve numerical measures of physical quantities, such as lengths, areas, weights, speeds, etc. In this section we solve problems concerned with such quantities. These word problems may be solved by first expressing relationships between the measures of these quantities in terms of a mathematical model and then interpreting the solution(s) of the resulting equation or inequality in terms of the physical quantities expressed in the problem.

The suggestions given on page 240 are also helpful in the writing of models for word problems concerned with physical quantities. In addition, where applicable, a sketch of any figures described in a problem should be made and all parts labeled with the appropriate numeral or variable. While mathematical models for problems which involve conditions on numbers are obtained from the problems themselves, problems concerned with measures of physical quantities sometimes require that we use as mathematical models, relationships that are part of our general background, such as $A = \pi r^2$, $C = 2\pi r$, $d = rt$, etc.

Exercise 8.8

A

For each problem:
a. Write a mathematical model.
b. Find the solution set of the resulting equation.
c. Interpret the solution in terms of the stated problem.

General Problems

Example. A 20-foot chain is cut into two pieces so that one piece is 4 feet shorter than the other. How long is each piece?

Solution. a. Let $x =$ the first piece; then $x + 4 =$ the longer piece. A mathematical model is

$$x + x + 4 = 20.$$

b. Solve for x.

$$2x + 4 = 20$$
$$2x = 16$$
$$x = 8$$

The solution set is {8}.

c. The shorter piece, represented by x, is 8 feet long and the longer, $x + 4$, is 12 feet long. Does $8 + 12 = 20$? Yes.

1. A 36-foot rope is cut into two pieces so that one piece is 10 feet longer than the other. How long is each piece?

2. A 72-foot chain is cut into three pieces so that the length of the first piece is twice the length of the second piece and the length of the second piece is one-third as long as the third piece. Find the length of the three pieces.

3. The admission at a football game was $2.10 for adults and $0.75 for children. The receipts were $293.25 for 166 paid admissions. How many adults and children attended the game?

4. The admission at a concert was $5.00 for adults and $2.50 for children. There were 3,600 in attendance and the gate receipts were $12,750. How many adults and children came to the concert?

5. In an election, there was a total of 72,100 votes cast. The 18 to 21 age group cast 6,900 less votes than all others. How many votes did the 18 to 21 age group cast?

6. A rocket scientist earns $30 per hour while on the job, and his assistant earns $20 per hour. Following their work on a rocket launch, they were paid a total of $1,450. The assistant worked 12 hours less than the scientist. How much did each receive?

Geometry Problems

To aid you in the solution of problems involving geometric figures, we have listed several basic figures and formulas in the Appendix on page 394.

Example. The length of a rectangle is 12 feet greater than its width and its perimeter is 48 feet. Find the dimensions.

Solution. a. Let $x =$ the width; then $x + 12 =$ the length. Make a sketch.

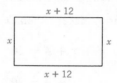

A mathematical model is

$$x + x + (x + 12) + (x + 12) = 48.$$

b. Solve for x.

$$4x + 24 = 48$$
$$4x = 24$$
$$x = 6$$

The solution set of the equation is $\{6\}$.

c. The width represented by x is therefore 6 feet and the length, $x + 12$, is 18 feet. Does $6 + 6 + 18 + 18 = 48$? Yes.

7. The width of a rectangle is 8 feet less than the length. Its perimeter is 76 feet. Find the dimensions.

8. The perimeter of an equilateral triangle is 12 inches more than the length of a side. Find the length of a side.

9. The perimeter of a triangle is 50 inches. The first side is 2 inches less than the second side and 3 inches less than the third side. What is the length of each side?

10. When each side of a square is decreased in length by 6 inches, the area is decreased by 84 square inches. Find the length of the side of the original square.

Coin Problems

In solving coin problems, keep in mind that the *value* of a number of coins is equal to the product of the value of a single coin and the number of coins. Furthermore, the value of the entire collection equals the sum of the values of the various kinds of coins in the collection. You will find it helpful to work in terms of cents, rather than dollars.

Example. A collection of coins consisting of nickels and dimes has a value of $2.90. How many nickels and dimes are there in the collection if there are 19 more nickels than dimes?

Solution. a. Let $x =$ the number of dimes; then $x + 19$ represents the number of nickels.

A mathematical model relating the *value* of the nickels and the *value* of the dimes to the *value* of the entire collection is:

$$\begin{bmatrix} \text{Value of nickels} \\ \text{in cents} \end{bmatrix} + \begin{bmatrix} \text{Value of dimes} \\ \text{in cents} \end{bmatrix} = \begin{bmatrix} \text{Value of collection} \\ \text{in cents} \end{bmatrix}$$

$$5(x + 19) \quad + \quad 10(x) \quad = \quad 290.$$

b. Solve for x.

$$5x + 95 + 10x = 290$$
$$15x = 195$$
$$x = 13$$

The solution set of the equation is $\{13\}$.

c. Thus the number of dimes, represented by x, is 13 and the number of nickels, $x + 19$, is 32. Do 13 dimes and 32 nickels amount to $2.90? Yes.

11. A man has $2.40 in change consisting of three more dimes than quarters. How many dimes and quarters does he have?

12. A collection of nickels and quarters has a value of $3.85. There are twice as many quarters as nickels. How many nickels and quarters are there?

13. A woman has $6.48 in pennies, nickels, and dimes. There are 102 coins in all and twice as many dimes as pennies. How many of each kind are there?

14. A post-office clerk sells some 8-cent and 11-cent stamps for a total of $2.80. The number of 8-cent stamps is three times the number of 11-cent stamps. How many of each were sold?

Mixture Problems

In solving mixture problems, the important idea is that the amount of a substance in a mixture is equal to the product of the amount of the mixture and the per cent of the substance in the mixture; and that the amount of the substance before and after mixing remains constant.

Example.   How many quarts of a 20% solution of acid should be added to 10 quarts of a 30% solution of acid to obtain a 25% solution?

Solution.   a. Let $x =$ a number of quarts of a 20% solution of acid. The mathematical model relating the *amount of pure acid* before and after combining the solutions can be obtained from the fact that the mathematical model is

$$\begin{bmatrix} \text{Pure acid in} \\ 20\% \text{ solution} \end{bmatrix} + \begin{bmatrix} \text{Pure acid in} \\ 30\% \text{ solution} \end{bmatrix} = \begin{bmatrix} \text{Pure acid in} \\ 25\% \text{ solution} \end{bmatrix}$$

$$0.20x \quad + \quad 0.30(10) \quad = \quad 0.25(x + 10).$$

b. Solve for x.

$$20x + 30(10) = 25(x + 10)$$
$$20x + 300 = 25x + 250$$
$$50 = 5x$$
$$10 = x$$

The solution set of the equation is $\{10\}$.

c. Thus, 10 quarts of the 20% solution are necessary to produce a 25% solution. In terms of the amount of pure acid, is $0.20(10) + 0.30(10) = 0.25(20)$? Yes.

15. How many quarts of a 20% solution of acid should be added to 30 quarts of a 50% solution of acid to obtain a 40% solution of acid?

16. How many quarts of a 40% salt solution must be added to 50 quarts of a 15% salt solution to obtain a 25% salt solution?

17. How many ounces of an alloy containing 50% aluminum must be melted with an alloy containing 70% aluminum to obtain 30 ounces of an alloy containing 55% aluminum?

18. How many pounds of an alloy containing 60% copper must be melted with an alloy containing 20% copper to obtain four pounds of an alloy containing 35% copper?

Interest Problems
━━━━━━━━ ━━━━━━━━

In solving interest problems, the basic idea is that the amount of interest earned in one year at simple interest equals the product of the rate of interest and the amount of money invested. For example, $1,000 invested for one year at $5\frac{1}{2}\%$ yields an interest of $(0.055)(1,000) = \$55$.

Example. Two investments produce an annual interest of $160. $1,000 more is invested at 5% than at 6%. How much is invested at each rate?

Solution. a. Let $x =$ an amount invested at 6%; then $x + 1,000 =$ the amount invested at 5%. The mathematical model relating the *interest* from each investment and the *total interest* received is

$$\begin{bmatrix} \text{Interest from} \\ 6\% \text{ investment} \end{bmatrix} + \begin{bmatrix} \text{Interest from} \\ 5\% \text{ investment} \end{bmatrix} = [\text{Total interest}].$$

$$0.06(x) \quad + \quad 0.05(x + 1,000) \quad = \quad 160.$$

b. Solve for x.

$$6x + 5(x + 1,000) = 16,000$$

$$6x + 5x + 5,000 = 16,000$$

$$11x = 11,000$$

$$x = 1,000$$

The solution set of the equation is $\{1,000\}$.

c. Thus, the amount invested at 6% is $1,000 and the amount invested at 5% is $x + \$1,000$ or $2,000. Is $0.06(1,000) + 0.05(2,000) = 160$? Yes.

19. Two investments produce an annual income of $472. One investment earns 4% and the other earns 5%. How much is invested at each rate if the amount invested at 5% is $6,200 more than the amount invested at 4%?

20. An amount of money is invested at 7% and $1,200 more than that amount is invested at $6\frac{1}{2}\%$. How much income is received from each investment if the total interest is $726?

21. An amount of $26,000 is invested, part at 6% and the remainder at 7%. Find the yearly interest on either part if the interest on each investment is the same.

22. An amount of money is invested at $4\frac{1}{2}$% and twice that amount is invested at 4%. How much is invested at each rate if the total income is $585?

Formula Problems

Some problems can be solved by simply substituting given information in a given formula and solving for one of the variables.

Example. Boyle's Law, in physics, states that the *product* of the pressure and the volume of a quantity of a gas remains constant at a constant temperature. The formula is $P_1 V_1 = P_2 V_2$, where P_1 and V_1 are the pressure and volume at a particular time and P_2 and V_2 are the pressure and volume at another time. If 200 cc of a gas at a pressure of 40 grams is reduced to a volume of 160 cc, what is the new pressure?

Solution. Substituting 200 for V_1, 40 for P_1, and 160 for V_2 in the formula gives

$$40(200) = P_2(160)$$

$$P_2 = 50.$$

Hence, the new pressure is 50 grams.

23. To determine the weight of a person over 5 feet tall, the formula $w = 5.4h - 220$ gives a good approximation. w is the weight in pounds and h is the height in inches. What is the approximate weight of a person whose height is 65 inches?

24. In certain electrical circuits, the formula $E = IR$ applies, where E is the voltage, I is the strength of the current in amperes, and R is the resistance in ohms. What is the strength of a current in a circuit whose resistance is 14 ohms if the voltage is 7 volts?

25. The formula for determining an individual's intelligence quotient (I.Q.) is I.Q. $= \dfrac{MA \cdot 100}{CA}$, where MA is the mental age as determined by testing and CA is the chronological age. What is the I.Q. of an individual whose mental age is 18 and whose chronological age is $15\frac{1}{2}$?

26. The present value of an amount of money invested at simple interest is given by the formula $A = p + prt$, where p is the principal, or amount invested, r is the rate, or per cent, and t is the number of years. What is the present value of $2,000 invested at a rate of 5% for 4 years?

B

Uniform Motion Problems

In solving uniform motion problems, the basic idea is that the distance traveled (at a uniform rate) is equal to the product of the rate of speed and the time traveled ($d = rt$). For example, if one travels 30 miles per hour for 4 hours, the distance $d = (30)(4) = 120$ miles. Furthermore, the model usually involves an equality between distances or between times.

Example. A plane travels 450 miles in the same time that a train travels 60 miles. If the plane goes 416 miles per hour faster than the train, find each rate.

Solution. a. Let r represent a rate for the train; then $r + 416$ represents the rate for the plane. Note that the *times* are the same and that

$$\text{time} = \frac{\text{distance}}{\text{rate}} .$$

A mathematical model can be determined from the fact that

[Time for plane] = [Time for train].

Express the time of each in terms of r.

$$\frac{450}{r + 416} = \frac{60}{r}$$

b. Solve for r.

$$450(r) = 60(r + 416) \quad (r \neq 0, -416)$$
$$450r = 60r + 24{,}960$$
$$390r = 24{,}960$$
$$r = 64$$

The solution set of the equation is {64}.

c. The train's rate, r, is 64 miles per hour; the plane's rate, $r + 416 = 480$ miles per hour. Does $\dfrac{450}{64 + 416} = \dfrac{60}{64}$? Yes.

27. A rocket travels 12,600 miles in the same time that a supersonic plane travels 3,600 miles. If the rate of the rocket is 2,700 miles per hour greater than the rate of the plane, find the rate of each.

28. Two rockets are propelled over a 5,600-mile test range. One rocket travels twice as fast as the other. The faster rocket covers the entire distance in two hours less time than the slower. How fast in miles per hour are the rockets traveling?

29. Two cars start together and travel in the same direction, one going three times as fast as the other. At the end of 2 hours, they are 100 miles apart. How fast is each traveling?

30. Two trains start together and travel in opposite directions, one going twice as fast as the other. At the end of four hours, they are 216 miles apart. How fast is each train traveling?

Inequality Problems

The basic principles used in setting up the preceding sets of problems also apply in this set, with the exception that inequality symbols are used instead of the equality sign.

Example. A student must have an average of 70% to 80% on five tests in a course to receive a "C." His grades on the first four tests were 83%, 76%, 66%, and 72%. What grade on the fifth test would guarantee him a "C" in the course?

Solution. a. Let x represent a grade (in per cent) on the fifth test. The mathematical model is

$$70 \leq \underbrace{\frac{83 + 76 + 66 + 72 + x}{5}}_{\text{average of 5 tests}} < 80.$$

b. Solve for x.

$$350 \leq 83 + 76 + 66 + 72 + x < 400$$

$$350 \leq 297 + x < 400$$

$$53 \leq x < 103$$

The solution set is $\{x \mid 53 \leq x < 103\}$.

c. Thus, any grade equal to or greater than 53 would guarantee a "C." Note that in this interpretation, the upper limit of 103% in the model was not pertinent to the stated problem because the per cent would never exceed 100.

31. As in the preceding example, what grade on the fifth test would give the student a "B" (80% to 90%) if his grades on the first four tests were 76%, 92%, 95%, and 60%?

32. What grade on the fourth test would give a student a "C" if his grades on the first three tests were 81%, 62%, and 63%?

33. Fahrenheit and Centigrade temperatures are related by the equation $F = \frac{9}{5}C + 32$. Within what range must the temperature be in Centigrade degrees for the Fahrenheit temperature to be between $-20°$ and $10°$?

34. Within what range must the temperature be in Fahrenheit degrees for the temperature in Centigrade degrees to be equal to or greater than $-10°$ and less than $50°$? (*Hint:* First solve $F = \frac{9}{5}C + 32$ for C in terms of F.)

CHAPTER SUMMARY

8.1 Sentences which involve only equality relationships are called **equations.** If the variable in an equation is replaced with a numeral and the resulting statement is true, the number represented by the numeral is called a **solution** or **root** of the equation and is said to **satisfy** the equation. The set of all numbers that satisfy an equation is called the **solution set** of the equation. Equations that are not true for all elements in the replacement set of the variables are called **conditional equations.** Equations that result in a true statement for every permissible replacement of the variable are called **identities.**

8.2 Equations or inequalities that have the same solution set are said to be **equivalent.** Methods used to generate equivalent equations include:

  a. Substituting an expression for an equal expression in one or both members of the given equation.

  b. Adding the same expression, representing a real number, to each member of the given equation.

  c. Multiplying each member by an expression representing a non-zero real number.

8.3 The distributive law is often useful to generate equivalent equations. If each member of an equation is multiplied by an expression containing a variable, the variable cannot assume a value that makes the expression equal to zero.

8.4   An equation in more than one variable can be "solved" for one variable in terms of the other variables by generating equivalent equations.

8.5   Conditional inequalities can have an infinite number of solutions. Methods used to generate equivalent inequalities include:

    a. Adding the same expression, representing a real number, to each member of an inequality to produce an equivalent inequality in which *the order remains the same.*

    b. Multiplying each member of an inequality by an expression representing a *positive* real number to produce an equivalent inequality in which *the order of the inequality remains the same.*

    c. Multiplying each member of an inequality by an expression representing a *negative* real number to produce an equivalent inequality in which *the order of the inequality is reversed.*

8.6–8.8   Word problems which state relationships between numbers or in which numbers are used as measures of physical quantities are solved by first expressing the relationship between the measures in terms of a mathematical model (equation or inequality) and then interpreting the solution set in terms of the word problem.

CHAPTER REVIEW

A

8.1   *Find the solution set of each equation by inspection. If the equation is an identity, so state.*

    1. $4 + x = 21$ 2. $15 = y - 7$

    3. $2(2x - 3) = 4x - 6$ 4. $\frac{1}{2}x = 24$

8.2   *State the axiom or theorem that implies that the second given equation is equivalent to the first equation.*

    5. $2x + 3x = 15$; $5x = 15$
    6. $y - 2 = 11$; $y - 2 + 2 = 11 + 2$
    7. $4x + 1 = 5$; $4x + 1 - 1 = 5 - 1$
    8. $5y = 35$; $y = 7$

8.2, 8.3 *Solve each equation.*

9. $3(2x + 7) = 21$

10. $x - (6 - 2x) = 3$

11. $6 - (x + 2)(x - 3) = 4 - x^2$

12. $(x - 2)^2 = x^2 + 4$

13. $\dfrac{x - 2}{x} = \dfrac{1}{3}$

14. $\dfrac{5}{x} - \dfrac{2}{x} = \dfrac{21}{5}$

8.4 *Solve each equation for the specified variable. Indicate any restrictions that must be placed on the other variables.*

15. $3a + 2x = x - a$ for x

16. $\dfrac{2}{3}y + c = a^2c$ for c.

17. $yx = 2y - 3$ for y.

18. $A = \dfrac{h}{2}(b + c)$ for h.

8.5 *Solve each inequality and graph the solution set on a number line.*

19. $2x < 10, x \in J$

20. $3x + 2 > x - 10, x \in R$

21. $\dfrac{4x - 7}{3} \geq 3, x \in R$

22. $\dfrac{x - 5}{2} \leq \dfrac{-5}{2}, x \in J$

8.6–8.8 *Solve each word problem.*

23. Four times the second of two consecutive integers less one-half the first is 18. Find the integers.

24. The sum of two integers is 83. Six times the smaller is equal to 8 more than the larger. Find the two integers.

25. The length of a rectangle is 4 feet less than two times the width. The perimeter is 46 feet. What are the dimensions?

26. One side of a triangle is three inches longer than another, and the third side is equal to four inches less than the sum of the first two. If the perimeter is 46 inches, find the lengths of the sides.

27. A man has $3.70 in change, consisting of 9 more dimes than quarters. How many of each kind of coin does he have?

28. A collection of dimes, quarters, and half-dollars has a value of $6.90. There are 3 more quarters than dimes, and half as many half-dollars as quarters. How many of each denomination are there?

29. How many ounces of fine powder worth 30 cents an ounce must be mixed with 50 ounces of course powder worth 12 cents an ounce so that the mixture can be sold for 20 cents an ounce?

30. How many gallons of rocket fuel containing 80% liquid oxygen must be mixed with 250 gallons containing 40% liquid oxygen to produce a mixture containing 70% liquid oxygen?

31. An amount of money is invested at 4% and $1600 more than that amount is invested at 5%. How much is invested at each rate if the annual interest is $800?

32. A man invests $\frac{1}{3}$ of his money at 6%, $\frac{1}{6}$ at 7%, and the remainder at 5.5%. His total income is $710. How much did he invest at each rate?

33. The formula for the pressure exerted by sea water is $P = 15\left(1 + \dfrac{d}{33}\right)$, where P is the pressure in pounds per square foot and d is the depth in feet below the surface. What is the pressure per square foot on a diving bell 132 feet below the surface?

34. The formula for converting temperature in degrees from the Fahrenheit scale to the Centigrade scale is $C = \frac{5}{9}(F - 32)$. What is the temperature on the Centigrade scale if the Fahrenheit temperature is 59°?

B

35. A man rode a bicycle for 10 miles and then hiked for 4 miles at a rate which was 3 miles per hour less than his rate riding. If he took the same time for each portion of his journey, find each rate.

36. The distance from the Earth to the Moon is about 240,000 miles. A space ship, moving in a direct path to the Moon, had an average velocity for the second half of the flight that was one-half the average velocity for the first half of the flight. If the entire flight took 3 days, find the average velocity in miles per hour for each half of the flight.

37. What grade on a fourth examination will give a student a B grade (80% to 90%) if his scores on the first three examinations were 72%, 81%, and 79%?

38. The mach number of a supersonic plane is determined by dividing the speed of the plane by the speed of sound. What is the range in speed of a plane if the mach number lies between 1.2 and 1.7? Assume the speed of sound to be 680 miles per hour.

chapters 6-8

Cumulative Review

CHAPTER 6

1. The product $5 \cdot 7$ is called the _____ form of 35.
2. When a number is written as the product of prime numbers only, it is in _____ factored form.
3. A polynomial of one term is a monomial, a polynomial of two terms is a binomial, and a polynomial of three terms is a _____.
4. In the expression $3x^4$, 4 is called the exponent and 3 is called the numerical _____.
5. The degree of a polynomial in one variable is the _____ exponent that appears in the polynomial.
6. If $P(x) = 2x^2 - 5x + 7$, then $P(3)$ equals $2(3)^2 - 5(3) + 7$ where x is replaced with _____.
7. By the closure laws for addition and multiplication, polynomials represent _____ _____ for real number replacements of the variable.
8. If $a \in R$, $m, n \in N$, then $a^m \cdot a^n = a^{m+n}$ and $a^m/a^n =$ _____.
9. By definition, a^0 $(a \neq 0)$ is equal to _____.
10. By definition, a^{-m} $(a \neq 0)$ is equal to _____.
11. A rational number expressed as the product of a number between 1 and 10 and a power of ten is said to be in _____ notation.
12. The _____ law justifies rewriting the product $2(x^2 + 3x + 2)$ as $2x^2 + 6x + 4$ or writing the polynomial $2x^2 + 6x + 4$ as the product $2(x^2 + 3x + 2)$.
13. The trinomial $ax^2 + bx + c$ is factorable if there are two factors of the product ac whose sum is _____.

CHAPTER 7

In Problems 14 to 21, assume denominators do not equal zero.

14. $\dfrac{P(x)}{Q(x)} = P(x) \cdot \dfrac{1}{?}$

15. $\dfrac{P(x) \cdot R(x)}{Q(x) \cdot R(x)} = \dfrac{?}{Q(x)}$

16. $\dfrac{P(x)}{R(x)} + \dfrac{Q(x)}{R(x)} = \dfrac{?}{R(x)}$

17. $\dfrac{P(x)}{Q(x)} = \dfrac{?}{-Q(x)}$

18. $\dfrac{-P(x)}{Q(x)} = -\dfrac{?}{-Q(x)}$

19. $\dfrac{P(x)}{Q(x)} - \dfrac{R(x)}{Q(x)} = \dfrac{?}{Q(x)}$

20. $\dfrac{P(x)}{Q(x)} \cdot \dfrac{R(x)}{S(x)} = \dfrac{?}{Q(x) \cdot S(x)}$

21. $\dfrac{P(x)}{Q(x)} \div \dfrac{R(x)}{S(x)} = \dfrac{P(x)}{Q(x)} \cdot \dfrac{?}{?}$

22. Write $\dfrac{x + \dfrac{2}{x}}{1 - \dfrac{1}{x}}$ as a single fraction in lowest terms.

CHAPTER 8

23. A number which satisfies an equation or inequality is called a _____ or root of the particular equation or inequality.

24. The set of all numbers that satisfy an equation or inequality is called its _____ _____.

25. Equations that are not true for all elements in the replacement set of the variable are called conditional equations; equations that are true for each permissible replacement of the variable are called _____.

26. Equations or inequalities which have the same solution set are called _____ equations or inequalities respectively.

27. Equations or inequalities are solved by inspection or by generating _____ equations or inequalities from which the solution set is evident.

28. Equivalent equations can be generated by adding the same expression to each member of a given equation, or by _____ each member by the same expression representing a nonzero real number.

29. An equivalent inequality in the (same/opposite) order can be generated by adding the same expression representing a real number to each member of a given inequality.

30. An equivalent inequality in the (same/opposite) order can be generated by multiplying each member of a given inequality by the same expression representing a *positive* real number.

31. An equivalent inequality in the (same/opposite) order can be generated

by multiplying each member of a given inequality by the same expression representing a *negative* real number.

32. Word problems are solved by first expressing the relationship between the given measures of physical quantities in terms of a _____ _____ and then interpreting the solution set in terms of the word problem.

9
chapter

Relations, Functions, and Their Graphs

The word "relationship" is quite common in everyday life. There are many examples such as the "doctor-patient" relationship, the "umpire-player" relationship, the "banker-borrower" relationship, and so on. All of these indicate that a relationship involves two sets of things and a means of determining if a member of one set is related in a prescribed way with a member of another set. In mathematics, the word "relation" is used in a very similar but more precise way.

In this chapter we consider the very important mathematical concepts of relations and functions associated with solutions and solution sets of equations and inequalities in two variables such as

$$2x + y = 7, \qquad \frac{1}{3}x = y - 4, \qquad \text{or} \qquad x - 3y \geq 8,$$

and with a means of representing these solutions graphically.

9.1 Solutions of Equations in Two Variables

An equation in two variables expresses a relation between the variables and serves to pair numbers in the replacement sets of the variables. For example, in the equation $y = 5x$, if $x, y \in R$, replacing x with the number 2 gives $y = 5(2) = 10$. This equation, therefore, pairs the number 2 with the number 10, where 2 is the replacement for x and 10 is the replacement for y. Obviously there are many such pairs of numbers. For each replacement of x, there is a value for y. If x is assigned the value 3, then $y = 15$; if $x = 4$, then $y = 20$; if $x = -4$, then $y = -20$; etc.

To simplify writing pairs of numbers that are values for two different variables, the numbers are expressed as **ordered pairs.** (See Section 1.4.) In any ordered pair, the number on the left is called the **first component** and the number on the right is called the **second component.** Thus (3, 2) is an ordered pair in which 3 is the first component and 2 is the second component. The ordered pair (3, 2) is *not* the same as the ordered pair (2, 3) because the pairs do not have the same first and second components.

If the components of an ordered pair represent replacements for the variables x and y, it is customary to write the replacement of x as the first component in the ordered pair, and the replacement of y as the second component. If the components represent replacements for the variables a and b, the replacement for a is generally the first component and the replacement for b is the second component. For any other variables, the order in which the replacements are made should be specified.

Examples.   a. An ordered pair (x, y) whose x component is 2 and whose y component is -7 is $(2, -7)$.

b. An ordered pair (a, b) whose a component is -4 and whose b component is 0 is $(-4, 0)$.

If the variables in an equation are replaced with the components of an ordered pair in the specified order and the resulting statement is true, the *ordered pair* is called a **solution** of the equation. The set of all such ordered pairs is called the **solution set** of the equation.

Example.   The ordered pair (2, 8) is a solution of $y = 2x + 4$ because replacing x with 2 and y with 8 gives $8 = 2(2) + 4$, which is a true statement.

Some other ordered pairs that are solutions of $y = 2x + 4$ in the above example can be found by assigning arbitrarily any real number as a replacement for x and then determining the corresponding value for y. For example:

$$\text{If } x \text{ is replaced with } 0, \quad y = 2(0) + 4 = 4.$$
$$\text{If } x \text{ is replaced with } 1, \quad y = 2(1) + 4 = 6.$$

Therefore, three ordered pairs which are solutions of $y = 2x + 4$ are

$$(0, 4), \quad (1, 6), \quad \text{and} \quad (2, 8).$$

It would appear from the preceding example that, if $x, y \in R$, such an equation has an **infinite number** of ordered pairs as solutions. Since it is impossible to list all the members of the solution set, we can use the set notation $\{(x, y) \mid y = 2x + 4, x \in R\}$ to describe the infinite set of ordered pairs making up the solution set of the equation $y = 2x + 4$. This notation is read: "the set of ordered pairs, (x, y), such that y equals 2 times x plus 4, x is an element in the set of real numbers."

In any equation involving two variables where the left-hand or right-hand member consists of one variable only, that variable is said to be expressed **explicitly** in terms of the other variable. For example, in the equations

$$y = 5x - 4, \quad y = x^2 + 3x - 2, \quad \text{and} \quad C = \pi d,$$

y is **explicitly related** to x, and C is explicitly related to d. In equations such as

$$3x + 5y = 6, \quad 8 = 3xy, \quad \text{and} \quad 2x^2 - 3y - 5 = 0,$$

x and y are said to be **implicitly related**.

Example. Write $x + 2y = 3$ as an equation in which y is expressed explicitly in terms of x.

Solution. $x + 2y = 3$

$$2y = -x + 3 \qquad \text{Theorem 8.1}$$

$$y = \frac{1}{2} \cdot (-x + 3) \qquad \text{Theorem 8.2}$$

$$y = \frac{-x + 3}{2} \qquad \text{Theorem 4.1}$$

Exercise 9.1

A

For each equation, find the solution with the specified first component.

Example. $y = x + 4$; $(2, \ ), (-3, \ ), (0, \ )$

Solution.   Replacing x with 2, -3, and 0, y is determined to be 6, 1, and 4 respectively.   Hence the solutions are $(2, 6)$, $(-3, 1)$, and $(0, 4)$.

1. $y = x + 1$; $(1, \ ), (0, \ ), (-2, \ )$

2. $y = x - 2$; $(2, \ ), (0, \ ), (-1, \ )$

3. $2x - y = 1$; $(-5, \ ), (0, \ ), (3, \ )$

4. $2x - 3y = 14$; $(-2, \ ), (1, \ ), (5, \ )$

5. $3x + y = 5$; $(0, \ ), (2, \ ), (4, \ )$

6. $x + 3y = 6$; $(-3, \ ), (0, \ ), (6, \ )$

7. $y = 5 + 2x$; $(-2, \ ), (0, \ ), (2, \ )$

8. $y = -7 - 3x$; $(-3, \ ), (0, \ ), (3, \ )$

9. $y = \dfrac{2}{x + 1}$; $(-2, \ ), (0, \ ), (5, \ )$

10. $y = \dfrac{4}{1 - 2x}$; $(-4, \ ), (1, \ ), (3, \ )$

For each equation, find the solution with the specified second component.   Hint: Replace y with the second component and find the associated value of x.

11. $3x - y = 4$; $(\ , -4), (\ , 2), (\ , 0)$

12. $x - 2y = 11$; $(\ , -1), (\ , 0), (\ , 1)$

13. $y = \dfrac{3}{x + 1}$; $(\ , 1), (\ , 3), (\ , -3)$

14. $y = \dfrac{-2}{x - 3}$; $(\ , 1), (\ , 2), (\ , -1)$

Find solutions of each equation for the specified replacement set for x.

Example. $y = x + 5$, $x \in \{1, 3, 5\}$

Solution on page 264.

Solution.   Replacing x with 1, 3, and 5 in turn, y is found to be 6, 8, and 10, respectively. Hence, the solutions are (1, 6), (3, 8), and (5, 10).

15. $x + y = 3,\quad x \in \{-2, -1, 0, 1, 2\}$

16. $3x - y = 10,\quad x \in \{-5, -3, -1, 1, 3, 5\}$

17. $4x + y = -1,\quad x \in \left\{\dfrac{1}{2}, \dfrac{1}{4}, \dfrac{1}{8}, \dfrac{1}{16}\right\}$

18. $x + 2y = -6,\quad x \in \left\{-\dfrac{1}{3}, 0, \dfrac{1}{3}\right\}$

19. $xy = 8,\quad x \in \{1, 2, 3, 4\}$

20. $xy = -4,\quad x \in \{1, 2, 3, 4\}$

Solve each equation explicitly for y in terms of x.

Example. $3x + \dfrac{y}{2} = 5$

Solution. $\dfrac{y}{2} = -3x + 5$ \qquad\qquad Theorem 8.1

$\qquad\qquad y = 2(-3x + 5)$ \qquad Theorem 8.2

$\qquad\qquad y = -6x + 10$ \qquad\quad Distributive law

21. $x + y = 6$ \hfill 22. $x - 2y = 5$

23. $x = \dfrac{y}{3} + 4$ \hfill 24. $\dfrac{x}{4} = -y + 6$

25. $2x - 2y = -3$ \hfill 26. $3x + 2y = 7$

27. $\dfrac{y + 3}{7} = x$ \hfill 28. $\dfrac{2y - 4}{5} = -2x$

B

Solve each equation explicitly for y. Assume no denominator equals zero.

29. $3x + \dfrac{y}{4} = 6$ \hfill 30. $2x = \dfrac{y}{3} + 6$

31. $x = \dfrac{3}{y}$ \hfill 32. $x = \dfrac{4}{y - 2}$

33. $yx^2 = y - 3$

34. $yx^3 + y = 7$

35. $3x = \dfrac{y - 1}{y}$

36. $\dfrac{x}{4} = \dfrac{y + 2}{y - 3}$

What definition(s), axiom(s), or theorem(s) justify that the second equation in the following pairs is equivalent to the first equation? $x, y \in R$.

37. $x - 3y = 5$,

    $x - 3y - x = 5 - x$

38. $y - 2x = 7$,

    $y - 2x + 2x = 7 + 2x$

39. $4y = x - 6$,

    $\dfrac{1}{4}(4y) = \dfrac{1}{4}(x - 6)$

40. $\dfrac{y}{2} = -4x - 10$,

    $2\left(\dfrac{y}{2}\right) = 2(-4x - 10)$

41. $xy = 2$,

    $y = \dfrac{2}{x} \quad (x \neq 0)$

42. $x = \dfrac{y}{5} - 2$,

    $y = 5x + 10$

43. $\dfrac{y + 2}{8} = \dfrac{x}{3}$,

    $y = \dfrac{1}{3}(8x - 6)$

44. $\dfrac{2y - 5}{x} = -3, \quad (x \neq 0)$

    $y = \dfrac{5 - 3x}{2}$

45. In the equation, $y = x + 7$, how many values of y can you determine for *each* real number replacement for x? Find a value for y associated with a value of 5 for x.

46. In the inequality $y > x + 7$, how many values of y can you determine for *each* real number replacement for x? Find several values of y associated with a value of 5 for x.

9.2 Relations

The notions of a relation and a function are most basic and very useful in mathematics. Essentially a relation or a function is an association between the elements of two sets, a function being a *special* kind of relation.

    Consider the possibilities that arise when two dice are tossed, one red and one green. The numbers represented on the red die and the green die can be written as elements of

$$R = \{1, 2, 3, 4, 5, 6\}$$

and
$$G = \{1, 2, 3, 4, 5, 6\},$$
respectively. The possibilities that arise on a single toss of two dice are elements of

$$\{(1, 1), (1, 2), \cdots, (2, 1), (2, 2), \cdots, (3, 1), (3, 2), \cdots,$$
$$(4, 1), (4, 2), \cdots, (5, 1), (5, 2), \cdots, (6, 1), (6, 2), \cdots, (6, 6)\}$$

where the first component of each ordered pair is an element of R and each second component is an element of G. In this case each element of R is paired with each of the six elements of G. This particular kind of association is an example of a relation and suggests the following.

Definition 9.1.   A relation is an association between the elements of two sets.

Observe, as in the above example, that *one* way of exhibiting the association (or correspondence) is through the use of an ordered pair and that the relation itself can be considered to be *a set of ordered pairs*.

In the example above, the relation is a set of ordered pairs and is the Cartesian product, $R \times G$ or $G \times R$ (See Section 1.4). However, as we shall see, this is only one way in which a relation between two sets can be formed.

The set of first components in a set of ordered pairs is called the **domain** of the relation and the set of second components is called the **range**. For example, in the relation shown in Figure 9.1, the domain is $\{1, 2\}$ and the range is $\{3, 4, 5\}$.

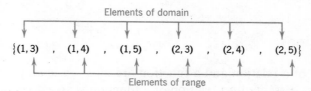

Elements of domain

$$\{(1, 3)\ ,\ (1, 4)\ ,\ (1, 5)\ ,\ (2, 3)\ ,\ (2, 4)\ ,\ (2, 5)\}$$

Elements of range

Figure 9.1

Example.   If $A = \{x \mid x \in R\}$ and $B = \{y \mid y \in R\}$, write an expression for the relation $A \times B$. What is the domain? What is the range?

Solution. $A \times B = \{(x, y) \mid x, y \in R\}$. The domain is R; the range is R.

We cannot list all of the ordered pairs in the relation in the preceding example, but set notation makes it possible for us to specify how the relation is formed. In general, we shall consider $R \times R$ the set from which ordered pairs in any relation are selected. A relation can be defined in this universe by specifying a set of real numbers for the domain and a *rule* for associating each element in the domain with a real number in the range.

Example. Specify the domain, the rule for association, and the set of ordered pairs of the relation $\{(x, y) \mid y = 2x + 1, x \in \{2, 3\}\}$.

Solution. In this case, $\{2, 3\}$ is the domain, and $y = 2x + 1$ is the rule for association.

$$\text{If } x = 2, \text{ then } y = 2(2) + 1 = 5.$$
$$\text{If } x = 3, \text{ then } y = 2(3) + 1 = 7.$$

Thus, $\{(x, y) \mid y = 2x + 1, x \in \{2, 3\}\} = \{(2, 5), (3, 7)\}$.

The relation in the preceding example can also be specified as

$$\{(x, 2x + 1) \mid x \in \{2, 3\}\}.$$

In this case, the second component of each ordered pair is found in the same manner as before; y and $2x + 1$ are symbols that represent the same number for each value of x.

The pairings of elements in a relation can also be viewed as a mapping of one set of elements onto another. Figure 9.2 illustrates different types of mappings, where in each case A is the domain and B is the range.

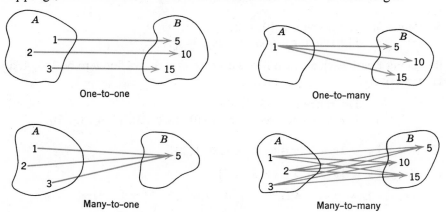

One-to-one   One-to-many

Many-to-one   Many-to-many

Figure 9.2

Exercise 9.2

A

Specify the domain and range for each relation.

Examples.   a. $\{(0, 1), (0, 2), (4, 1), (4, 2)\}$
                 b. $\{(-2, -2), (-2, -1), (0, -2), (0, -1), (2, -2), (2, -1)\}$

Solutions.   a. Domain is $\{0, 4\}$; range is $\{1, 2\}$.
                 b. Domain is $\{-2, 0, 2\}$; range is $\{-2, -1\}$.

1. $\{(1, 4), (1, 5)\}$

2. $\{(2, -1), (2, 0)\}$

3. $\{(0, 1), (0, 2), (2, 1), (2, 2)\}$

4. $\{(3, 0), (9, 0), (3, -1), (9, -1)\}$

5. $\{(1, 0), (2, 0), (3, 0), (1, 2), (2, 2), (3, 2)\}$

6. $\{(2, 1), (4, 1), (8, 1), (2, 3), (4, 3), (8, 3)\}$

7. $\{(-5, 4), (-5, 5), (-5, 6), (-5, 7), (-5, 8), (-5, 9)\}$

8. $\{(0, 2), (1, 2), (2, 2), (3, 2), (4, 2), (5, 2)\}$

Specify (a) the domain, (b) the rule for association, (c) the range, and (d) the set of ordered pairs for each relation.

Example. $\{(x, y) \mid y = x + 2, x \in \{0, 1, 2, 3\}\}$

Solution.   a. The domain is $\{0, 1, 2, 3\}$.
                b. The rule for association is $y = x + 2$.

Replacing x with each element in the domain in turn, we find y to be 2, 3, 4, and 5, respectively.

                c. The range is $\{2, 3, 4, 5\}$.
                d. The set of ordered pairs is $\{(0, 2), (1, 3), (2, 4), (3, 5)\}$.

9. $\{(x, y) \mid y = 2x - 3, x \in \{-2, -1, 0, 1, 2\}\}$

10. $\{(x, y) \mid y = 3x + 6, x \in \{1, 3, 5, 7\}\}$

11. $\{(x, y) \mid 2x - y = -4, x \in \{0, 1, 4, 9\}\}$

12. $\{(x, y) \mid 3x + y = 5, x \in \{3, 6, 9, 12\}\}$

13. $\{(x, y) \mid x + 2y = 1, x \in \{-3, -1, 1, 3\}\}$

14. $\{(x, y) \mid x - 3y = 2, x \in \{-10, -7, -4, -1\}\}$

15. $\{(x, x + 3) \mid x \in \{-2, -1, 0, 1\}\}$ 16. $\{(x, x - 2) \mid x \in \{0, 1, 2, 3\}\}$

17. $\{(x, 2x - 1) \mid x \in \{2, 4, 6, 8\}\}$ 18. $\{(x, 3x + 2) \mid x \in \{1, 3, 5, 7\}\}$

19. $\left\{\left(x, \dfrac{-2}{x}\right) \,\middle|\, x \in \{-3, -1, 1, 3\}\right\}$ 20. $\left\{\left(x, \dfrac{3}{x}\right) \,\middle|\, x \in \{-4, -2, 2, 4\}\right\}$

21. $\left\{\left(x, \dfrac{2}{x - 1}\right) \,\middle|\, x \in \{2, 3, 4, 5\}\right\}$ 22. $\left\{\left(x, \dfrac{4}{x + 3}\right) \,\middle|\, x \in \{-2, -1, 0, 1\}\right\}$

B

23. $\{(x, y) \mid y = \sqrt{x + 4}, x \in \{0, 1, 2, 3\}\}$

24. $\{(x, y) \mid y = \sqrt{x - 9}, x \in \{9, 10, 11, 12\}\}$

25. $\{(x, \sqrt{x - 2}) \mid x \in \{3, 4, 5, 6\}\}$

26. $\{(x, \sqrt{x + 5}) \mid x \in \{-3, -2, -1, 0\}\}$

27. $\{(x, y) \mid y = \sqrt{x^2 - 4}, x \in \{-4, -2, 2, 4\}\}$

28. $\{(x, y) \mid y = \sqrt{16 - 2x^2}, x \in \{-\sqrt{8}, -2, 0, 2, \sqrt{8}\}\}$

29. $\{(x, \sqrt{9 - x^2}) \mid x \in \{0, 1, 2, 3\}\}$

30. $\{(x, \sqrt{x^2 - 25}) \mid x \in \{5, 6, 7, 8\}\}$

31. $\{(x, y) \mid y = |x|, x \in \{-2, -1, 0, 1, 2\}\}$

32. $\{(x, y) \mid y = |x - 2|, x \in \{-3, -2, -1, 0, 1, 2, 3\}\}$

*For what replacement values of x are the right-hand members of each equation
undefined?*

33. $y = \dfrac{x - 2}{x - 3}$ 34. $y = \dfrac{5}{6 - x}$ 35. $y = \dfrac{2x - 7}{3 - 2x}$

36. $y = \dfrac{1 - x}{5 + 2x}$ 37. $y = \sqrt{9 - x}$ 38. $y = \sqrt{9 + x}$

9.3 Functions

In Section 9.2 we mentioned that a function was a special kind of relation
In particular, it is a relation such that each element in the domain is

associated with *one and only ane element in the range.* For example, in the equation

$$y = x + 7,$$

for each real number replacement of x there corresponds one and only one real number y. Thus, if x is assigned a value of 5, the associated value for y is 12. Hence, this equation defines a function. However, in the inequality

$$y > x + 7,$$

we can find an infinite number of values for y corresponding to a given value of x. Thus, if x is assigned a value 5, some solutions of the inequality are (5, 13), (5, 14), (5, 15), and, in fact, any ordered pair in which the first component x equals 5 and in which the second component y is greater than 12. Hence, this inequality does not define a function. Hereafter we shall work for the most part with functions and formalize the definition for convenient reference.

Definition 9.2.   A function is a relation in which each element in the domain is associated with only one element in the range.

Examples.   a. The relation   {(1, 2), (3, 4), (5, 6), (7, 8)}   is a function because for each first component there is one and only one second component.

b. The relation {$(x, y) \mid y = x - 3$} is a function because for every $x \in R$, $x - 3$ is a unique real number. Thus, for every x, there is one and only one y.

Another example of a function is the set of ordered pairs generated by the equation $C = \pi d$ (the circumference of a circle equals π times the length of a diameter, where π is an irrational number ≈ 3.14), which pairs a circumference of a circle with the length of its diameter. Since we want the elements of each ordered pair (d, C) to be meaningful in a practical sense, we specify the domain of the function, the replacement value for d, to be the set of *positive* real numbers. The variable, d, representing an element in the domain, is often referred to as an **independent variable,** while the variable, C, representing an element in the range is called a **dependent variable.** These names take on meaning in a context when any choice in the set of positive numbers may be made to replace d, while the value of C "depends upon" the choice that has been made for d.

In $C = \pi d$ we specified the replacements for d to be the set of positive real numbers because the rule was related to a geometric concept. If the domain is not specified or implied from the rule relating to the measures of things, we shall consider the domain and range to consist of the set of real numbers or a subset for which the rule is meaningful. For example, in the function $\{(x, y) \mid y = 3x + 2\}$, the replacement values of x will be elements in the set of real numbers because for each $x \in R$, there is a $y \in R$. In the function

$$\left\{(x, y) \mid y = \frac{5}{x - 2}\right\},$$

the domain is a *subset* of the set of real numbers, because we must exclude the value of 2 for x, since the replacement of 2 for x gives

$$y = \frac{5}{2 - 2} = \frac{5}{0},$$

which is undefined in the set of real numbers. In the function

$$\{(x, y) \mid y = \sqrt{x - 4}\},$$

the domain is a *subset* of the set of real numbers, since it will contain only those real numbers greater than or equal to 4, $(x \geq 4)$, since for any $x < 4$, y is not an element of R.

Examples. Specify the domain of the function defined by each equation where x and y are elements of R.

a. $y = \dfrac{7}{x + 5}$
b. $y = \sqrt{x + 2}$

Solutions. a. If $x = -5$, the right-hand member is undefined. Therefore the domain is $\{x \mid x \in R, x \neq -5\}$.

b. If $x < -2$, then $x + 2 < 0$, and $\sqrt{x + 2}$ does not represent a real number. Therefore the domain is $\{x \mid x \in R, x \geq -2\}$.

Exercise 9.3

A

Find the element in the range of the function $\{(x, y) \mid y = 3x - 2\}$ associated with the given element, x, in the domain.

Examples.   a. -2 **b.** 4

Solutions

a. Replacing x with -2 yields   b. Replacing x with 4 yields

$$y = 3(-2) - 2$$
$$= -6 - 2 = -8.$$

$$y = 3(4) - 2$$
$$= 12 - 2 = 10.$$

1. 2   2. 5   3. 6   4. 3   5. $\dfrac{1}{3}$

6. $\dfrac{2}{3}$ 7. -2 8. -6 9. 0   10. $-\dfrac{1}{3}$

Find the element in the range of the function $\{(x, 2x + 1)\}$ associated with the given element, x, in the domain.

11. 5   12. 7   13. $\dfrac{1}{2}$ 14. $\dfrac{3}{4}$

15. -3 16. -4 17. $-\dfrac{1}{4}$ 18. 0

Find the element in the domain of the function $\{(x, y) \mid y = 3x - 2\}$ associated with the given element, y, in the range.

Example.   6

Solution.   Replacing y with 6 in the equation $y = 3x - 2$ yields

$$6 = 3x - 2,$$
$$8 = 3x,$$

from which we have x equal to 8/3.

19. 4   20. 10   21. -2 22. -5

23. 0   24. $\dfrac{5}{3}$ 25. $\dfrac{1}{3}$ 26. $-\dfrac{2}{3}$

Find the element in the domain of the function $\{(x, 2x + 1)\}$ associated with the given element in the range.

Example. 5

Solution. Setting 5 equal to $2x + 1$ yields

$$5 = 2x + 1,$$
$$4 = 2x,$$
$$2 = x.$$

27. 3 28. 12 29. $\dfrac{1}{5}$ 30. $\dfrac{3}{2}$

31. $-\dfrac{1}{2}$ 32. 0 33. -1 34. -7

Specify the domain of the function defined by each equation where x and y are elements of R.

Examples. a. $y = \dfrac{1}{x - 3}$ b. $y = \sqrt{x - 2}$

Solutions. a. If $x = 3$, the right-hand member is undefined. Therefore the domain is $\{x \mid x \in R,\ x \neq 3\}$.

b. If $x < 2$, then $x - 2 < 0$, and $\sqrt{x - 2}$ does not represent a real number. Therefore the domain is $\{x \mid x \in R,\ x \geq 2\}$.

35. $y = \dfrac{1}{x}$ 36. $y = \dfrac{1}{x - 4}$ 37. $y = \dfrac{3}{x + 3}$

38. $y = \dfrac{5}{x + 7}$ 39. $y = \dfrac{7}{2x - 1}$ 40. $y = \dfrac{4}{3x - 2}$

41. $y = \sqrt{x}$ 42. $y = \sqrt{x + 4}$ 43. $y = \sqrt{x - 6}$

44. $y = \sqrt{3 - x}$ 45. $y = \sqrt{6 - 2x}$ 46. $y = \sqrt{12 - 6x}$

9.4 Function Notation

We have frequently used a single symbol, usually a letter, as the name for a set. Functions, being sets, can be designated in the same way. The symbol f is generally used for this purpose. Thus we may speak of the function f defined by a specific equation. If the discussion includes a consideration of more than one function, other letters such as g, h, F, P, or some other symbol may be used.

Recall that in Chapter 6 we used the symbol $P(x)$ to denote a polynomial, and the symbol $P(a)$ to designate the particular value of the polynomial when x is replaced by some real number a. In a similar manner, the symbol for the function can be used in conjunction with the variable representing an element in the domain, to represent the associated element in the range. For example, $f(x)$ is read: "f of x" or "the value of f at x," and is the element in the *range* of f associated or paired with the element x in the *domain*. Thus we can discuss functions defined by an equation such as

$$y = x - 7, \quad f(x) = x - 7, \quad g(x) = x - 7, \quad \text{etc.,}$$

where the symbol $f(x)$, $g(x)$, etc., is playing *exactly* the same role as y.

The notation $f(x)$ is especially useful because, by replacing x with a specific number a in the domain, the notation $f(a)$ will then denote the paired, or corresponding, element in the range.

Examples.   If $f(x) = x^2 - 2$, then

$$f(2) = (2)^2 - 2 = 4 - 2 = 2,$$
$$f(0) = (0)^2 - 2 = 0 - 2 = -2,$$
$$f(a) = (a)^2 - 2 = a^2 - 2,$$
$$f(a - 1) = (a - 1)^2 - 2 = a^2 - 2a + 1 - 2 = a^2 - 2a - 1.$$

It is frequently convenient to use $f(x)$, $g(x)$, $Q(x)$, etc., simply as symbols representing expressions in x. Recall that we have already used such symbols whenever we wanted to discuss equations, such as $P(x) = Q(x)$, obtained by setting one expression in x equal to another.

Exercise 9.4

A

If $f(x) = x - 3$, find each of the following.

Example. $f(2)$.

Solution.   Replacing x with 2,

$$f(2) = 2 - 3$$
$$= -1.$$

1. $f(0)$ 2. $f(-1)$ 3. $f(-3)$

4. $f(9)$ 5. $f\left(\frac{1}{2}\right)$ 6. $f(a)$

If $g(x) = x^2 - 3$, find each of the following.

7. $g(0)$ 8. $g(1)$ 9. $g(-1)$

10. $g(a)$ 11. $g(a + 1)$ 12. $g(a - 1)$

Find the range of the function whose rule and domain are as given. Write the function as a set of ordered pairs.

Example. $f(x) = \sqrt{x + 2}$; domain is $\{-1, 0, 1, 2\}$

Solution. Replacing x with $-1, 0, 1$, and 2 in turn yields

$$f(-1) = \sqrt{-1 + 2} = \sqrt{1} = 1, \qquad f(0) = \sqrt{0 + 2} = \sqrt{2},$$
$$f(1) = \sqrt{1 + 2} = \sqrt{3}, \qquad \text{and} \qquad f(2) = \sqrt{2 + 2} = \sqrt{4} = 2.$$

Thus the range is $\{1, \sqrt{2}, \sqrt{3}, 2\}$; $f = \{(-1, 1), (0, \sqrt{2}), (1, \sqrt{3}), (2, 2)\}$.

13. $f(x) = 2x - 1$; domain is $\{-1, 0, 2, 3\}$

14. $f(x) = 3 - x$; domain is $\{2, 4, 6, 8\}$

15. $g(x) = x^2 + 3$; domain is $\{-2, 0, 1, 2\}$

16. $g(x) = 6 - x^2$; domain is $\{-3, -2, -1, 0\}$

17. $G(x) = \frac{1}{x}$; domain is $\left\{\frac{1}{4}, \frac{1}{2}, \frac{3}{4}, 1\right\}$

18. $G(x) = \frac{-2}{x}$; domain is $\left\{\frac{1}{2}, \frac{1}{4}, \frac{1}{8}, \frac{1}{16}\right\}$

19. $P(x) = \sqrt{x - 4}$; domain is $\{5, 7, 9, 11\}$

20. $P(x) = \sqrt{5 - x}$; domain is $\{-1, 0, 1, 2\}$

21. $f(x) = \sqrt{9 - x^2}$; domain is $\{0, 1, 2, 3\}$

22. $f(x) = \sqrt{x^2 - 6}$; domain is $\{3, 4, 5, 6\}$

B

23. If $h(x) = x^2 - 1$, find each of the following.

    a. $h(x + 2)$ b. $h(x)$ c. $h(x + 2) - h(x)$

24. If $f(x) = 2x^2 - x + 4$, find each of the following.

    a. $f(x + 3)$ b. $f(x)$ c. $f(x + 3) - f(x)$

25. If $f(x) = x^2 - 3x + 7$, find $\dfrac{f(x + h) - f(x)}{h}$.

26. If $f(x) = x^3 - 2x^2 + x - 9$, find $\dfrac{f(x + h) - f(x)}{h}$.

9.5 Graph of a Function; Linear Functions

We shall now consider the association that exists between *ordered pairs of numbers* and the *points in a geometric plane*. Recall that we discussed the construction of a number line in Chapter 2 and that we called points on the line the graphs of numbers and called the numbers the coordinates of the points. The point corresponding to 0 is called the origin. In Figure 9.3a, the coordinate of the point marked A is -2. A number line can also be constructed in a vertical position as shown in Figure 9.3b. If two

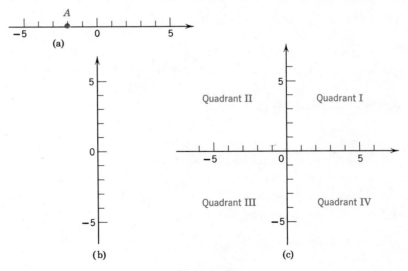

Figure 9.3

number lines are drawn perpendicular to each other at their origins, as in Figure 9.3c, they form a **rectangular coordinate system.** A rectangular coordinate system is also called a **Cartesian coordinate system.** These perpendicular number lines are called **axes** (singular, **axis**). The horizontal line is usually called the **x-axis** and the vertical line is usually called the **y-axis.** The x-axis and the y-axis divide the plane into four regions called **quadrants,** as shown.

Any point in the plane can be associated with a unique ordered pair of real numbers. For example, the point P, in Figure 9.4, is in Quadrant I

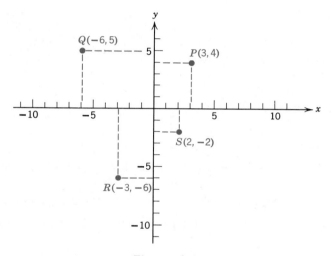

Figure 9.4

and can be associated with the ordered pair (3, 4), where 3 represents the perpendicular distance of the point to the right of the y-axis and 4 represents the perpendicular distance of the point above the x-axis. The point designated by Q in Quadrant II can be associated with the ordered pair (−6, 5), where −6 represents the perpendicular distance of the point to the left of the y-axis and 5 represents the perpendicular distance of the point above the x-axis. The points R and S can be associated with the ordered pairs (−3, −6) and (2, −2) respectively.

The components of an ordered pair associated with a point in the plane are called the **coordinates** of the point. The first component of the ordered pair associated with a point in the plane is called the **abscissa** of the point. The second component is called the **ordinate** of the point.

Example.   Name the quadrants in which the points *L*, *M*, *N*, and *P* lie, and give the coordinates of the points in the figure.

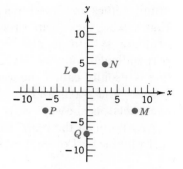

Solution.   *L*, Quadrant II, (−2, 4); *M*, Quadrant IV, (8, −3); *N*, Quadrant I, (3, 5); *P*, Quadrant III, (−7, −3); *Q*, or any other point lying on an axis, is said to lie in no quadrant, (0, −7).

In Definition 9.2 a function was defined in part, as a set of ordered pairs. Consequently, a function—or at least a part of a function—can now be displayed as a set of points in the plane. The restriction "part of a function" is used here since the domain or range or both may be the infinite set of real numbers, which we cannot show on a finite straight line. For example, consider the function defined by

$$f(x) = x + 2.$$

Solutions to this equation can be obtained by assigning arbitrary values to *x* and obtaining the corresponding values for $f(x)$. Thus six solutions to this equation are

$$(-10, -8), (-7, -5), (-2, 0), (0, 2), (3, 5), \text{ and } (7, 9).$$

Locating these points on a coordinate system, we have Figure 9.5a. These

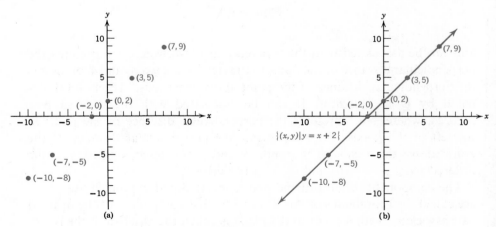

Figure 9.5

points appear to lie on a straight line and, in fact, they do. It can be shown that the coordinates of any point on the line in Figure 9.5b constitute a solution of the first-degree equation in two variables, $y = x + 2$, and conversely, every solution of $y = x + 2$ corresponds to a point on this line where the variable y is used here in the same way as $f(x)$. The line is referred to as the graph of the function defined by $y = x + 2$, or, alternately, as the graph of the solution set of the equation, or simply as the graph of the equation. Obviously, only part of the graph can be displayed. Thus an arrowhead is placed at either end to indicate that the graph, or line, continues in both directions indefinitely.

The graphs of all first-degree equations in two variables are straight lines, therefore such equations are also called **linear equations** and the functions defined by these equations are called **linear functions.** Since a straight line can be determined by any two points on the line, the graph of a linear function can be drawn by finding two ordered pairs in the solution set of the defining equation. A third ordered pair is generally used as a check.

Example.   Graph the function $\{(x, y) \mid y = 6 - x\}$.

Solution.   Select two or three values for x, say 0, 3, and 8, and find associated values for y by use of the defining equation $y = 6 - x$. Graph the three ordered pairs in the solution set, (0, 6), (3, 3), and (8, -2). Draw the straight line which contains all the points whose coordinates are elements in the solution set.

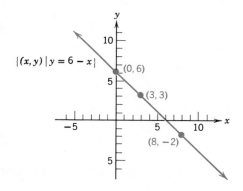

Exercise 9.5

A

State the quadrant or the axis in which the point with given coordinates is located.

1. (1, 6)   2. (3, 4)   3. (-4, 1)   4. (2, -5)

5. (8, -3)   6. (-1, 7)   7. $\left(-\dfrac{1}{2}, -3\right)$ 8. $\left(-4, -\dfrac{2}{3}\right)$

9. $(-\sqrt{5}, 2)$ 10. $(\sqrt{3}, -5)$ 11. $\left(\dfrac{1}{4}, \sqrt{6}\right)$ 12. $\left(\sqrt{5}, \dfrac{1}{3}\right)$

13. $(5, 0)$ 14. $(0, -2)$ 15. $(0, 4)$ 16. $(-3, 0)$

Graph each function by selecting any three values for x.

Example. $\{(x, y) \mid y = x + 1\}$

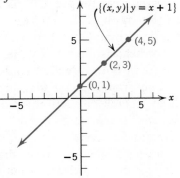

Solution. Select arbitrary values for x, say 0, 2, and 4, and find the associated values for y, which are 1, 3, and 5. Three ordered pairs in the function are $(0, 1)$, $(2, 3)$, and $(4, 5)$. The points are plotted and the graph is completed.

17. $\{(x, y) \mid y = 2x - 3\}$ 18. $\{(x, y) \mid x - y = 4\}$

19. $\{(x, y) \mid y - x = 6\}$ 20. $\{(x, y) \mid 3x - 2y = 10\}$

21. $\{(x, y) \mid 3x + 2y = 5\}$ 22. $\{(x, y) \mid 4x - 5y = 7\}$

23. $\{(x, x + 3)\}$ 24. $\{(x, x + 5)\}$

25. $\{(x, x - 2)\}$ 26. $\{(x, x - 7)\}$

27. $\{(x, 2x + 1)\}$ 28. $\{(x, 2x - 5)\}$

29. $\{(x, 3x - 1)\}$ 30. $\{(x, 3x + 2)\}$

9.6 Graphing Using Intercepts

Since any number can be used for one of the variables to obtain ordered pairs in the solution set of the equation, it makes sense to choose a value that will make it easy to find the other component in the ordered pair. Thus, in the example on page 279, we chose 0 for x to find the corresponding value of 6 for y. We could also have chosen 0 for y to find the corresponding value of 6 for x. This method of choosing 0 for x to obtain one ordered pair and then 0 for y to obtain a second ordered pair is called the **intercept method** for graphing first-degree equations because the ordered pairs are the coordinates of the points at which the graph crosses,

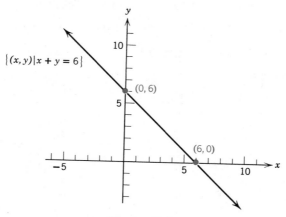

Figure 9.6

or intersects, the x- and y-axes as shown in Figure 9.6. In general, the graph of a linear equation will intersect both coordinate axes, unless it is parallel to one of them, in points designated by $(a, 0)$ and $(0, b)$. In this context, a is called the **x-intercept** and b is called the **y-intercept.**

Sometimes it is not possible to use the intercept method for graphing a first degree equation, and sometimes, while possible, it is not very satisfactory. If the graph intersects, or crosses, the axes at or near the origin, we only have one point or the two points are too close together to be of much use in drawing the graph. It is then necessary to graph at least one other point at a distance far enough removed from the origin to establish

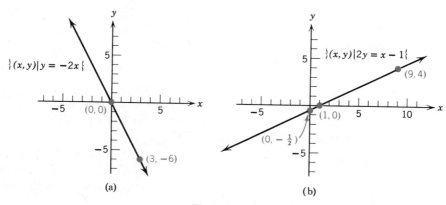

Figure 9.7

the line with accuracy. For example, consider the graphs of the functions

$$\{(x, y) \mid y = -2x\} \qquad \text{and} \qquad \{(x, y) \mid 2y = x - 1\}$$

in Figures 9.7a and 9.7b on page 281.

Exercise 9.6

A

Graph each function by the intercept method.

Example. $\{(x, y) \mid 2x + 3y = 6\}$

Solution. If x is replaced by 0, then $y = 2$.
If y is replaced by 0, then $x = 3$. Thus two
ordered pairs in the function are $(0, 2)$ and
$(3, 0)$. The points are plotted and the graph
is completed.

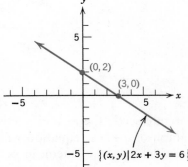

1. $\{(x, y) \mid 3x - 4y = 12\}$

2. $\{(x, y) \mid 2x + 5y = 10\}$

3. $\{(x, y) \mid y + 3x = 6\}$

4. $\{(x, y) \mid y - 3x = 5\}$

5. $\{(x, y) \mid 3x - y + 9 = 0\}$ 6. $\{(x, y) \mid 4x - 3y - 7 = 0\}$

7. $\{(x, y) \mid x = 2y + 11\}$ 8. $\{(x, y) \mid 2x = 3y - 2\}$

9. $\{(x, y) \mid 2y = 3(x - 2)\}$ 10. $\{(x, y) \mid 3x = 5(4 - 2y)\}$

11. $\{(x, 2x + 4)\}$ 12. $\{(x, 3x - 2)\}$

13. $\left\{\left(x, \dfrac{2}{3}x - 1\right)\right\}$ 14. $\left\{\left(x, \dfrac{1}{2}x + 5\right)\right\}$

B

15. Graph the solution sets of $y = x + 1$, $y = 2x + 1$, and $y = 3x + 1$ on the
same set of axes. How are the "steepnesses" of the lines related to the
numerical coefficients of x in the equations?

16. Graph the solution sets of $x + y = 5$ and $2x - 3y = 0$ on the same set of
axes. Estimate the coordinates of the point at which the lines cross each
other, the point of intersection. How are the coordinates of this point
related to the two equations?

9.7 Special Cases of Linear Equations

Observe that an equation such as

$$y = 2$$

when written as

$$0x + y = 2,$$

can be considered a linear equation in two variables, in which the co-efficient of x is zero. For each real number replacement for x, therefore, this equation assigns a value of 2 to y. This means any ordered pair of the form $(x, 2)$ is a member of the solution set of the equation $y = 2$. In particular,

$$(1, 2), (2, 2), (3, 2), \text{ and } (4, 2)$$

are all solutions of the equation. In general, the equation

$$y = b \qquad (b \in R)$$

assigns to each x the same value of y and its graph is a straight line perpendicular to the y-axis as shown in Figure 9.8a for the case $y = 2$. Functions

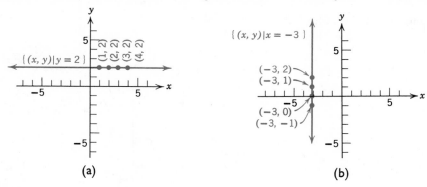

(a)

(b)

Figure 9.8

defined by such equations are called **constant functions** because the same value of y is assigned to each x. Now note that an equation such as

$$x = -3$$

can be considered as a linear equation in two variables when written as

$$x + 0y = -3$$

in which the coefficient of y is zero. Here, only one value is permissible for x, namely -3, whereas any replacement value in the set of real numbers may be assigned to y. Thus any ordered pair of the form $(-3, y)$ is a solution of this equation. For example,

$$(-3, 2), (-3, 1), (-3, 0), \text{ and } (-3, -1),$$

are all solutions of the equation $x = -3$. In general, the equation

$$x = a \qquad (a \in R)$$

assigns to each y the same value of x, and its graph is a straight line perpendicular to the x-axis as shown in Figure 9.8b for the case $x = -3$.

Because the first component of every solution of an equation of the form $x = a$ is the same number, such equations do not define functions in accordance with Definition 9.2.

Exercise 9.7

Graph each relation.

Examples.   a. $\{(2, 3), (4, 3), (6, 3), (8, 3)\}$
                 b. $\{(4, 1), (4, 3), (4, 5), (4, 7)\}$

Solutions.   a.   b.

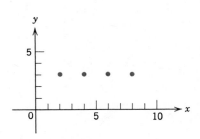

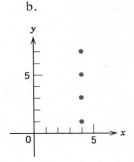

1. $\{(2, 1), (3, 1), (4, 1), (5, 1)\}$

2. $\{(2, -1), (4, -1), (6, -1), (8, -1)\}$

3. $\{(-3, 2), (-1, 2), (1, 2), (3, 2)\}$

4. $\{(-4, -2), (-2, -2), (2, -2), (4, -2)\}$

5. $\{(2, -1), (2, 1), (2, 3), (2, 5)\}$

6. $\{(1, 0), (1, 2), (1, 4), (1, 6)\}$

7. $\{(-3, -2), (-3, 0), (-3, 2), (-3, 4)\}$
8. $\{(4, -3), (4, -2), (4, -1), (4, 0)\}$

Examples.   a. $\{(x, y) \mid 0x + y = 3, x \in \{2, 4, 6, 8\}\}$
        b. $\{(x, y) \mid x + 0y = 4, y \in \{1, 3, 5, 7\}\}$

Solutions.   a.   b.

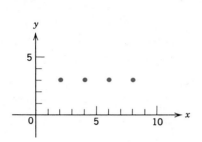
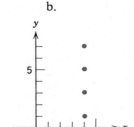

9. $\{(x, y) \mid 0x + y = 5, x \in \{1, 3, 5, 7\}\}$
10. $\{(x, y) \mid 0x + 2y = 6, x \in \{0, 3, 6, 9\}\}$
11. $\{(x, y) \mid 0x + 3y = 9, x \in \{2, 4, 6, 8\}\}$
12. $\{(x, y) \mid 0x + 4y = 16, x \in \{1, 2, 3, 4\}\}$
13. $\{(x, y) \mid x + 0y = -3, y \in \{1, 2, 3, 4\}\}$
14. $\{(x, y) \mid x + 0y = -1, y \in \{-1, 0, 1, 2\}\}$
15. $\{(x, y) \mid 2x + 0y = 10, y \in \{-3, -1, 1, 3\}\}$
16. $\{(x, y) \mid 3x + 0y = 12, y \in \{-4, -2, 0, 2\}\}$

Examples.   a. $\{(x, y) \mid 0x + 2y = 6, x \in R\}$
        b. $\{(x, y) \mid 3x + 0y = 9, y \in R\}$

Solutions.   a.   b.

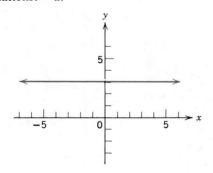
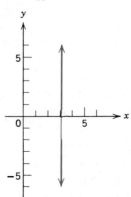

17. $\{(x, y) \mid 0x + y = -2, x \in R\}$ 18. $\{(x, y) \mid 0x + y = 5, x \in R\}$

19. $\{(x, y) \mid 0x + 3y = 12, x \in R\}$ 20. $\{(x, y) \mid 0x + 2y = 4, x \in R\}$

21. $\{(x, y) \mid x + 0y = -4, y \in R\}$ 22. $\{(x, y) \mid x + 0y = -2, y \in R\}$

23. $\{(x, y) \mid 5x + 0y = 15, y \in R\}$ 24. $\{(x, y) \mid 3x + 0y = -5, y \in R\}$

Graph each equation.

Examples.   a. $y = -2$ b. $x = -3$

Solutions.   a.   b.

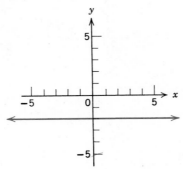   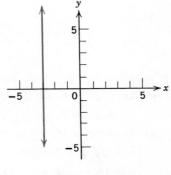

25. $y = 3$ 26. $y = 6$ 27. $y = -4$

28. $y = -3$ 29. $x = 7$ 30. $x = -5$

9.8   Special Property of Linear Functions

In Section 9.5, we observed that the graph of a linear function is a straight line and that such a line can be determined by two points. Now consider the points in Figure 9.9 associated with the ordered pairs $(2, 4)$ and $(5, 9)$. The first point is located 2 units to the right of the origin and the second point is located 5 units to the right of the origin. Thus the horizontal distance between the two points is $5 - 2$ or 3 units. Similarly, the vertical distance between the two points is $9 - 4$, or 5 units. In general, for two points in the plane named by $P_1(x_1, y_1)$ and $P_2(x_2, y_2)$, shown in Figure 9.10, the horizontal distance between the two points is $x_2 - x_1$

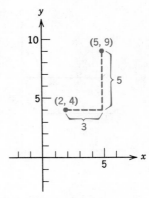

Figure 9.9

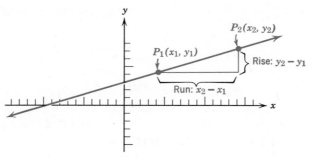

Figure 9.10

and the vertical distance is $y_2 - y_1$. The ratio of the vertical distance, called the **rise,** to the horizontal distance, called the **run,** is called the **slope, m,** of the line containing the two points P_1 and P_2.

Definition 9.3. The slope of the line containing two points $P_1(x_1, y_1)$ and $P_2(x_2, y_2)$ is

$$m = \frac{y_2 - y_1}{x_2 - x_1} \quad (x_2 \neq x_1).$$

Example. Determine the slope of the straight line containing the points associated with the ordered pairs $(2, 2)$ and $(4, 5)$.

Solution. Using the points labeled A as (x_1, y_1) and B as (x_2, y_2),

$$y_2 - y_1 = 5 - 2 = 3,$$

and

$$x_2 - x_1 = 4 - 2 = 2.$$

Thus,

$$m = \frac{y_2 - y_1}{x_2 - x_1} = \frac{3}{2}.$$

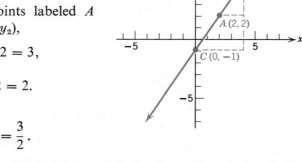

If the points labeled B and C had been used, the change in y is $5 - (-1)$, or 6, and the change in x is $4 - 0$, or 4. Thus, the slope $m = 6/4 = 3/2$, as before. You can show that the slope as determined by the points labeled A and C is also $3/2$.

Now observe that if we consider A as (x_2, y_2) and B as (x_1, y_1) in the example above, we obtain

$$m = \frac{y_2 - y_1}{x_2 - x_1} = \frac{2 - 5}{2 - 4} = \frac{-3}{-2} = \frac{3}{2},$$

and the slope is the same as before. This suggests, and correctly so, that the coordinates of two points on a line can be taken in either order to find the slope of the line. However, the differences of the corresponding components must be found in the same order:

$$m = \frac{y_2 - y_1}{x_2 - x_1} = \frac{y_1 - y_2}{x_1 - x_2}.$$

The concept discussed in the above example is that the slope of a straight line, determined by using the coordinates of any two points on the line, is constant. However, if the line is perpendicular to the x axis, the x coordinates of any two points are the same and the slope of such a line is undefined.

Some graphs of linear functions rise or fall more rapidly than others. Consider the graphs of $\{(x, y) \mid y = x\}$ and $\{(x, y) \mid y = 2x\}$ shown in Figure 9.11 as the lines L_1 and L_2, respectively. The coordinates of two solutions are shown on each graph.

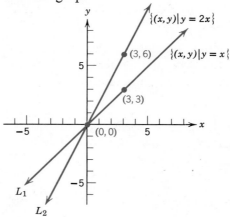

Figure 9.11

How do their slopes compare? Using the points $(0, 0)$ and $(3, 3)$ on L_1,

$$m_1 = \frac{3 - 0}{3 - 0} = \frac{3}{3} = 1,$$

and using the points $(0, 0)$ and $(3, 6)$ on L_2,

$$m_2 = \frac{6 - 0}{3 - 0} = \frac{6}{3} = 2.$$

Thus the slope of L_2 is greater than the slope of L_1. This means that the change in values of y on L_2 is greater than the change in values of y on L_1 for the *same* change in values of x.

Consider the graph of $\{(x, y) \mid 2y + x = 4\}$ in Figure 9.12 in which

$$m_3 = \frac{2 - 0}{0 - 4} = -\frac{2}{4} = -\frac{1}{2}.$$

The slope is negative because values of y decrease with increasing values of x.

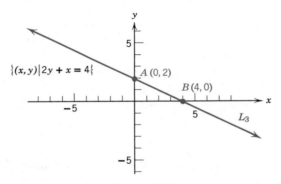

Figure 9.12

If the equation, $2y + x = 4$, of the graph in Figure 9.12 is solved explicitly for y, we obtain

$$2y = -x + 4,$$

$$y = -\frac{1}{2}x + 2.$$

Notice that the equations above

$$y = x, \qquad y = 2x, \qquad \text{and} \qquad y = -\frac{1}{2}x + 2,$$

whose graphs are shown in Figures 9.11 and 9.12, have slopes 1, 2, and $-\frac{1}{2}$, respectively. Note that these slopes are also the numerical coefficients of the variable x, where y is expressed explicitly in terms of x.

Now observe that if we replace x with 0 in the equation $y = -\frac{1}{2}x + 2$, we obtain $y = 2$, the y-intercept of the line. This example suggests the following theorem.

Theorem 9.1. The graph of an equation of the form

$$y = mx + b$$

has a slope m and y-intercept b.

This theorem is a consequence of Definition 9.3. If we let b equal the y-intercept ($x = 0$) of any line, and substitute b for y and 0 for x in the slope formula, $m = (y_2 - y_1)/(x_2 - x_1)$, with (x, y) any other point on the line, we obtain

$$, = \frac{y - b}{x - 0}.$$

If this equation is solved for y in terms of the other variables, we first obtain

$$mx = y - b$$

from which

$$y = mx + b,$$

the desired form. This equation is called the **slope-intercept form** of the equation of a line.

Examples. Determine the slope of the graph of each equation.

a. $y = -2x + 5$ b. $y = \dfrac{3}{4}x - 7$ c. $2y - 4x = -10$

Solutions. a. The equation is in the form $y = mx + b$. The coefficient of x is -2. Thus, the slope is -2.

b. Slope is $\dfrac{3}{4}$.   c. Solve for y in terms of x.

$$2y - 4x = -10$$
$$2y = 4x - 10$$
$$y = 2x - 5$$

Slope is 2.

Exercise 9.8

A

Find the slope of the line passing through the two points associated with the given ordered pairs.

Example. $(-4, 3), (1, -9)$

Solution. $m = \dfrac{y_2 - y_1}{x_2 - x_1} = \dfrac{-9 - 3}{1 - (-4)} = \dfrac{-12}{5} = -\dfrac{12}{5}.$

1. $(1, 1), (3, 4)$ 2. $(1, 2), (5, 1)$ 3. $(2, 2), (5, 4)$

4. $(1, -1), (9, 3)$ 5. $(-2, 5), (2, 11)$ 6. $(-1, -3), (-3, 7)$

7. $(3, 2), (-1, 0)$ 8. $(1, 0), (-1, 0)$ 9. $(-9, -4), (7, -1)$

10. $(-7, -1), (5, 1)$ 11. $(8, -2), (-4, -6)$ 12. $(-3, -2), (3, -6)$

Write each equation in slope-intercept form and specify the slope of its graph.

Example. $x - 3y = 6$

Solution. Solve the equation for y in terms of x.

$$x - 3y = 6$$
$$-3y = -x + 6$$
$$y = \frac{1}{3}x - 2.$$

Compare this equation with the slope-intercept form $y = mx + b$; the slope is $1/3$, the coefficient of x.

13. $x + y = 3$ 14. $x - y = 2$ 15. $2x + y = 4$

16. $x + 2y = -3$ 17. $2x + 3y = 5$ 18. $x - 3y = 8$

19. $3x - y = 2$ 20. $3x + 2y = 0$ 21. $5x - 4y = 0$

22. $-x = 5y - 2$ 23. $x = 3 - 5y$ 24. $x + 4 = 4y$

B

Without graphing, determine if the three points with the specified coordinates lie on the same line. (Hint: How are the slopes of line segments between pairs of points related?)

25. $(1, 6)$, $(-3, -2)$, $(4, 12)$ 26. $(8, 4)$, $(-1, 7)$, $(2, 6)$

27. Write the equation of the line with the same slope as the graph of $x - 2y = 4$ and passing through the origin. Graph the functions defined by the equations.

28. Write the equation of the line with the same slope as the graph of $x + 3y = 6$ and passing through the origin. Graph the functions defined by the equations.

29. Write the equation of the line with the same slope as the graph of $2x + y = 8$ and passing through the point $(1, 1)$. Graph the functions defined by the equations.

30. Write the equation of the line with the same slope as the graph of $3x - y = 3$ and passing through the point $(-1, -1)$. Graph the functions defined by the equations.

Specify the slope of the graph of each equation if such a slope exists.

31. $y = -2$ 32. $x = -3$ 33. $x = 5$ 34. $y = 4$

9.9 Graphs of Inequalities in Two Variables

Inequalities involving two variables such as

$$x + y < 7, \qquad 2x - 7 \geq y, \qquad \text{and} \qquad y \leq \frac{2x}{3} + 7,$$

also have ordered pairs of numbers as solutions. Thus $(2, 1)$ is a member of the solution set of $x + y < 7$, because when x is replaced by 2 and y by 1, the resulting statement, $2 + 1 < 7$, is a true statement. Furthermore, $(2, 3)$ is also a member of the solution set, because $2 + 3 < 7$ is also a true statement. These examples illustrate the fact that inequalities do *not* define functions, because each element of the replacement set of the variable x (in this case, 2) is *not* associated with a *unique* (one and only one) element in the replacement set of the variable y. The number 2 was first paired with 1, and then with 3.

As another type of example, compare the equation $y = 2x + 1$ with the inequality $y < 2x + 1$. If x is replaced by 3 in each, in one case the resulting equation $y = 7$ determines a *single* value for y, namely 7 (see Figure 9.13a), but in the other case the resulting inequality, $y < 7$, will be true for an *infinite* number of values of y (see Figure 9.13b). Some of the members of

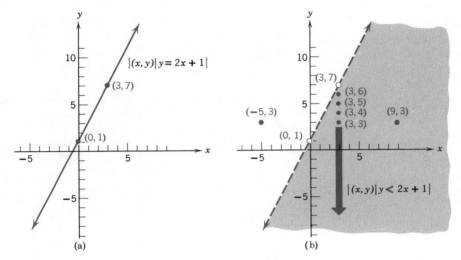

Figure 9.13

the solution set of $y < 2x + 1$ are (3, 6), (3, 5), (3, 4), and (3, 3). In fact, for $x = 3$, any real number replacement for y which is less than 7 will satisfy the inequality.

The shaded portion of Figure 9.13b shows the graph of the solution set of $y < 2x + 1$. Note that (9, 3) lies in the shaded portion because (9, 3) is a solution of $y < 2x + 1$. Note also that (−5, 3) does not lie in the shaded portion because (−5, 3) is not a solution, since replacing x with −5 and y with 3 yields $3 < 2(−5) + 1$ or $3 < −9$, which is a false statement. In Figure 9.13b, the line, which is the graph of the function $\{(x, y) \mid y = 2x + 1\}$, is shown by dashes to indicate that this line is not part of the graph of the inequality. The ordered pair (3, 7) represents a point on the line, but (3, 7) is not a solution of $y < 2x + 1$, since $7 < 2(3) + 1$, or $7 < 7$, is not a true statement.

Consider the relation

$$\{(x, y) \mid x + y − 2 > 0\}.$$

When the defining inequality is rewritten equivalently in the form

$$y > 2 - x,$$

we see that solutions of the inequality are such that for each x-replacement, y must be greater than $2 - x$.

The graph of the equation $y = 2 - x$ is the straight line shown in Figure 9.14a. To graph the inequality $y > 2 - x$, we need only observe that any point above this line has x, y-coordinates that satisfy the inequality $y > 2 - x$. Consequently, the solution set of $y > 2 - x$ corresponds to the entire region, or *half-plane*, above the line shown in Figure 9.14b.

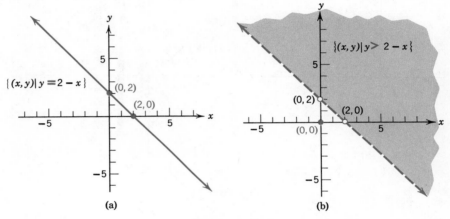

(a)   (b)

Figure 9.14

One way to determine in which half-plane the graph of the solution of the inequality lies is to note whether or not the coordinates of the origin, $(0, 0)$ belong to the solution set. If $(0, 0)$ is a member of the solution set, then the half-plane in which $(0, 0)$ lies represents the complete solution set of the inequality. In the example above, the coordinates of the origin do not satisfy the inequality $y > 2 - x$ because $0 > 2 - 0$, or $0 > 2$, is not a true statement. Therefore the half-plane which does not include the origin is shaded. Another point should be used to determine the half-plane if the graph of the associated equation passes through the origin.

Sometimes, a relation includes the cases where the members are equal as well as the cases where they are unequal. Thus, $y \le 3 - x$ is read "y is less than or equal to $3 - x$," and includes in its solution set the solution sets of both $y = 3 - x$ and $y < 3 - x$. The graph of

$$\{(x, y) \mid y \le 3 - x\}$$

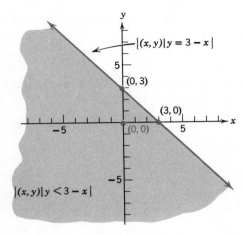

Figure 9.15

is shown in Figure 9.15 where the graph of

$$\{(x, y) \mid y = 3 - x\}$$

is a *solid* line indicating that the line is a part of the graph. Note that the coordinates of the origin, $(0, 0)$, satisfy $y \leq 3 - x$ ($0 \leq 3 - 0$, or $0 \leq 3$, is a true statement), and therefore the half-plane which includes the origin is shaded.

Exercise 9.9

A

State whether the given ordered pair belongs to the solution set of the given inequality.

Example. $y > x - 3$; $(2, 5)$

Solution.   Replacing x with 2 and y with 5 gives $5 > 2 - 3$, or $5 > -1$. Since this is a true statement, the ordered pair $(2, 5)$ is a solution of $y > x - 3$.

1. $y < x$; $(1, 0)$
2. $y > x$; $(-2, 5)$
3. $y < x + 3$; $(2, 2)$
4. $y > x - 5$; $(4, 6)$
5. $x + y < -2$; $(-2, 0)$
6. $x + y > -1$; $(-3, 0)$
7. $2x + y > 1$; $(1, 7)$
8. $2x - y < 3$; $(4, 2)$
9. $x < 4y - 1$; $(0, 4)$
10. $x < 3y + 5$; $(-2, 0)$
11. $0 \leq 4x + y$; $(0, 0)$
12. $2 > x + 2y$; $(0, 0)$

Graph each relation. First graph the corresponding equation and then shade the appropriate region.

Example. $\{(x, y) \mid y < 2x + 2\}$

Solution.   Figure a shows the graph of the corresponding equation $y = 2x + 2$. Figure b shows the graph of $\{(x, y) \mid y < 2x + 2\}$.

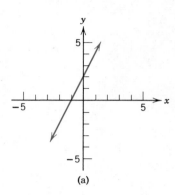

(a)

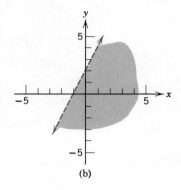

(b)

13. $\{(x, y) \mid y < 4 - 2x\}$

14. $\left\{(x, y) \mid y < \dfrac{1}{3}x - 5\right\}$

15. $\left\{(x, y) \mid y > 3x + \dfrac{2}{3}\right\}$

16. $\{(x, y) \mid y > 2x - 7\}$

17. $\{(x, y) \mid 3x + 2y \le 14\}$

18. $\{(x, y) \mid 6x + 4y \ge 12\}$

19. $\left\{(x, y) \left| \dfrac{1}{4}x - \dfrac{1}{3}y < \dfrac{5}{12}\right.\right\}$

20. $\left\{(x, y) \left| x + \dfrac{3}{4}y < 6\right.\right\}$

21. $\{(x, y) \mid y + 3x - 6 > 0\}$

22. $\{(x, y) \mid y - x + 5 < 0\}$

23. $\{(x, y) \mid x - y - 4 \le 0\}$

24. $\{(x, y) \mid 2x + y - 3 \ge 0\}$

25. $\{(x, y) \mid y > x\}$

26. $\{(x, y) \mid y \le x\}$

27. $\{(x, y) \mid y \le 2\}$

28. $\{(x, y) \mid y > -3\}$

29. $\{(x, y) \mid x > -3\}$

30. $\{(x, y) \mid x \le 4\}$

CHAPTER SUMMARY

9.1   If the variables in an equation in two variables are replaced by the components of an ordered pair and the resulting statement is true, the ordered pair is called a **solution** of the equation. The set of all such ordered pairs is called the **solution set** of the equation.

9.2 A **relation** is an association or correspondence between the elements of two sets; a set of ordered pairs. The set of first components in a set of ordered pairs is called the **domain** of the relation and the set of second components is called the **range.** A relation can be defined in a universe by specifying a set of numbers, the domain, and a rule (usually an equation or inequality) for associating each element in the domain with an element in the range.

9.3 A **function** is a special kind of relation in which each element in the domain is associated with *only one* element in the range.

9.4 Functions are usually designated by symbols such as f, g, h, etc.; if x represents an element in the domain, then $f(x)$, $g(x)$, $h(x)$, etc., represent elements in the range of the respective function.

9.5 Ordered pairs of numbers can be associated with points in a geometric plane through the use of a **rectangular coordinate system.** The components of the ordered pair associated with a point in the plane are called the **coordinates** of the point, the first component is the **abscissa,** the second component is the **ordinate.** Solutions (ordered pairs) of a first-degree equation in two variables correspond to points on a *straight line.* Thus functions defined by first-degree equations in two variables are called **linear functions.**

Linear functions can be graphed by plotting ordered pairs in the solution set of the defining equations. The ordered pairs can be found by assigning arbitrary values to one variable and determining the corresponding values of the second variable.

9.6 Linear functions can also be graphed by using the **intercept method,** in which the value of 0 is assigned to each variable in turn to find the corresponding value of the second variable.

9.7 An equation of the form $y = b$ can be viewed as the equation in two variables $0x + y = b$ and has solutions of the form (x, b). An equation of the form $x = a$ can be viewed as $x + 0y = a$ and has solutions of the form (a, y).

9.8 The **slope** of the line containing two points $P_1(x_1, y_1)$ and $P_2(x_2, y_2)$ is given by

$$m = \frac{y_2 - y_1}{x_2 - x_1} \quad (x_2 \neq x_1).$$

If a linear equation in two variables, x and y, is written in the **slope intercept form**

$$y = mx + b,$$

m is the slope and b is the y-intercept of the graph of its equation.

9.9 Inequalities in two variables have ordered pairs as solutions but do not define functions, because each element in the domain is associated with *more than one element* in the range. In general, their graphs consist of the set of points in a half-plane.

CHAPTER REVIEW

9.1 *For each equation, find solutions with the specified first components.*

1. $y = 3x - 2$; $(-4, \ \ )$, $(0, \ \ )$, $(4, \ \ )$

2. $2y - 3x = 11$; $(-2, \ \ )$, $(5, \ \ )$, $(7, \ \ )$

3. $y = 3$; $(-\sqrt{2}, \ \ )$, $(0, \ \ )$, $(5, \ \cdot)$

4. $2x - y = 6$; $(-3, \ \ )$, $(0, \ \ )$, $(4, \ \ )$

For each equation, find solutions with the specified second components.

5. $y = 4x - 3$; $(\ \ , -3)$, $(\ \ , 2)$, $(\ \ , 0)$

6. $2x - 5y = 13$; $(\ \ , 1)$, $(\ \ , 0)$, $(\ \ , -1)$

7. $x = -4$; $(\ \ , 0)$, $(\ \ , -3)$, $(\ \ , \sqrt{6})$

8. $2y + x = 4$; $(\ \ , 0)$, $(\ \ , -1)$, $(\ \ , 2)$

Find solutions of each equation for the specified replacement set for x.

9. $x - y = 7$; $x \in \{-1, 0, 1\}$ $\qquad$ 10. $x - 3y = -2$; $x \in \{0, 1, 2\}$

11. $3x - y = 10$; $x \in \left\{-\dfrac{1}{3}, 0, \dfrac{1}{3}\right\}$

12. $2x + y = 6$; $x \in \{-1, -2, -3\}$

Solve for y in terms of x.

13. $2x = \dfrac{-y}{4} + 3$ $\qquad\qquad\qquad$ 14. $\dfrac{2y - 3}{2} = -x$

9.2 *Specify the domain and range for each relation.*

15. $\{(-1, 2), (-1, 3), (1, 2), (1, 3)\}$

16. $\{(3, 0), (4, 0), (5, 0), (6, 0)\}$

17. $\{(x, y) \mid y = 3x + 1, x \in \{5, 7, 9, 11\}\}$

18. $\left\{(x, y) \mid y = \dfrac{2}{4 - x}, x \in \{-2, -1, 0, 1\}\right\}$

9.3 *Specify the domain of the function defined by each equation.*

19. $y = \dfrac{-2}{4 - x}$ 20. $y = \sqrt{3 - x}$

9.4 *If $g(x) = 2x^2 - x + 4$, find each of the following.*

21. $g(-1)$ 22. $g(0)$ 23. $g(3)$ 24. $g(a + 1)$

9.5, 9.6 *Graph each function.*

25. $\{(x, y) \mid 4x - y = 5\}$ 26. $\{(x, y) \mid 3x - 2y = 7\}$

27. $\{(x, 5x + 1)\}$ 28. $\{(x, 2x + 5)\}$

9.7 *Graph each set of ordered pairs.*

29. $\{(0, 3), (1, 3), (2, 3), (3, 3)\}$

30. $\{(5, -2), (5, -3), (5, -4), (5, -5)\}$

31. $\{(x, y) \mid 0x - 2y = 4, x \in \{0, 1, 2, 3\}\}$

32. $\{(x, y) \mid 2x - 0y = -6, y \in \{1, 3, 5, 7\}\}$

33. $\{(x, y) \mid y = 1\}$

34. $\{(x, y) \mid x = 2\}$

9.8 *Find the slope of the line passing through two points associated with the given ordered pairs.*

35. $(-2, 1), (8, -7)$ 36. $(6, 2), (-1, 4)$

Write each equation in slope-intercept form and specify the slope of the graph of the equation.

37. $2x - 3y = \dfrac{1}{4}$ 38. $-3x = 4y + 1$

9.9 *State whether the given ordered pair belongs to the solution set of the given inequality.*

39. $y < x - 7$; $(4, -6)$ 　　　　　　　40. $x > y + 3$; $(1, -1)$

Graph each relation.

41. $\left\{(x, y) \mid 2x - \dfrac{3}{4}y > -5\right\}$ 　　　　42. $\{(x, y) \mid x + 3 \geq 0\}$

10
chapter

Systems of Linear Equations and Inequalities

In Chapter 9 we discussed the solution sets of first-degree linear equations in two variables. These sets are composed of an infinite number of ordered pairs of numbers. It is often necessary to consider *pairs* of such equations and to determine whether or not their solution sets contain any ordered pair in common. Such common ordered pairs can be approximated by graphical methods or determined precisely by algebraic methods.

10.1 Graphical Solutions of Systems of Equations

The graph of $\{(x, y) \mid 3y - x = 9\}$ is a straight line, as shown in Figure 10.1. The coordinates of each point on this line are the components of a solution of the equation $3y - x = 9$. Also shown in Figure 10.1 is the graph of $\{(x, y) \mid 2y + x = 1\}$. The coordinates of each point on this line are the components of a solution of the equation $2y + x = 1$. The point that lies on *both* lines, shown as having the coordinates -3 and 2, is called the **point of intersection** of the lines.

Any ordered pair that is a solution of *both* $3y - x = 9$ and $2y + x = 1$ must have components that are the coordinates of a point that lies on both

301

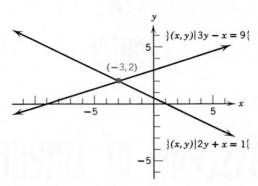

Figure 10.1

lines. The ordered pair $(-3, 2)$ is a solution of $3y - x = 9$, because if x is replaced by -3 and y by 2, the equation

$$3(2) - (-3) = 9,$$

or

$$6 + 3 = 9,$$

is a true statement. Also $(-3, 2)$ is a solution of $2y + x = 1$, since

$$2(2) + (-3) = 1,$$

or

$$4 - 3 = 1,$$

is a true statement. Two or more equations considered together, as in the example above, form a **system of equations,** and any ordered pair which is a solution of both equations is a solution of the system.

Recall that the intersection of two sets A and B, symbolized by $A \cap B$, is the set of all elements that are in both A and B. Hence, the coordinates of the point of intersection of the graphs of two linear equations are the components of the ordered pair in the intersection of the solution sets of the equations. Thus, in the preceding example,

$$\{(x, y) \mid 3y - x = 9\} \cap \{(x, y) \mid 2y + x = 1\} = \{(-3, 2)\}.$$

In many cases, the coordinates of a point of intersection and therefore the solution of a system can only be *approximated* by graphical methods.

Example. Verify by direct substitution in the two equations that the following statement is true. Also graph the system.

$$\{(x, y) \mid 2x - 5y = 0\} \cap \{(x, y) \mid 2x - y = 8\} = \{(5, 2)\}.$$

Solution. Replace x and y with 5 and 2 respectively in the two equations

$$2x - 5y = 0 \quad \text{and} \quad 2x - y = 8$$

Then,

$$2(5) - 5(2) = 0$$

and

$$2(5) - (2) = 8$$

are both true statements. Thus,

$$\{(x, y) \mid 2x - 5y = 0\} \cap \{(x, y) \mid 2x - y = 8\} = \{(5, 2)\}.$$

If the solution sets of two equations do not have an ordered pair in common, the equations are said to be **inconsistent**. For example, from an inspection of the graph in Figure 10.2, it appears that

$$\{(x, y) \mid 2x + y = 6\} \cap \{(x, y) \mid 2x + y = 10\} = \varnothing.$$

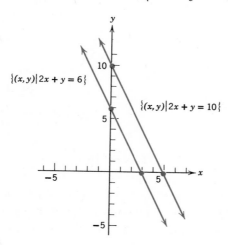

Figure 10.2

The slopes of their graphs are equal and the lines are said to be parallel. To verify that their slopes are equal, we solve each equation explicitly for y and obtain

$$y = -2x + 6 \quad \text{and} \quad y = -2x + 10. \tag{1}$$

Observe that the slope of each line is -2, the coefficient of x, and that the two graphs have different y-intercepts, namely 6 and 10.

If the graphs of two equations are actually the same line, the solution sets have an infinite number of ordered pairs in common. We shall refer to such equations as **dependent equations.** For example, the equations

$$2x - y = 6 \quad \text{and} \quad 4x - 2y = 12 \tag{2}$$

whose graphs are shown in Figure 10.3 are dependent. If we solve each equation explicitly for y in terms of x, we obtain

$$y = 2x - 6 \quad \text{and} \quad y = 2x - 6.$$

Notice that the slope of each line is 2 and that the y-intercept of each is -6.

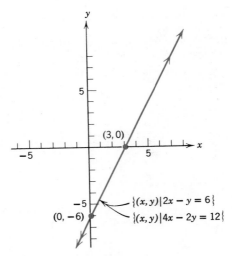

Figure 10.3

Observe that the equations in System (1), which are *inconsistent*, have the same slopes and different intercepts and that the ratios of the coefficients of the y-variable, the x-variable, and of the constant terms, respectively, are in the relationship

$$\frac{1}{1} = \frac{-2}{-2} \neq \frac{6}{10}.$$

Further notice that the equations in System (2), which are *dependent*, have the same slope and the same intercept and that the ratios of the coefficients of the x-variable, the y-variable, and of the constant terms, respectively, are in the relationship

$$\frac{2}{4} = \frac{-1}{-2} = \frac{6}{12}.$$

These examples suggest the following theorem.

Theorem 10.1. The equations in the system

$$a_1 x + b_1 y = c_1$$
$$a_2 x + b_2 y = c_2$$

are inconsistent if

$$\frac{a_1}{a_2} = \frac{b_1}{b_2} \neq \frac{c_1}{c_2},$$

and are dependent if

$$\frac{a_1}{a_2} = \frac{b_1}{b_2} = \frac{c_1}{c_2}.$$

If the equations in a system are neither inconsistent nor dependent, they are **independent equations.**

You may have observed that the above discussion of systems implies that if the solution of a system of two linear equations in two variables is finite and non-empty, then it contains exactly one ordered pair.

Exercise 10.1

A

Verify by direct substitution that the given ordered pair satisfies each equation. Graph each system.

Example
$$\{(x,\ y)\ |\ 3x + 2y = 7\} \cap \{(x,\ y)\ |\ x + y = 3\} = \{(1, 2)\}.$$

Solution. Replace x and y with 1 and 2 respectively in

$$3x + 2y = 7 \quad \text{and} \quad x + y = 3.$$

Then,

$$3(1) + 2(2) = 7$$

and

$$(1) + (2) = 3.$$

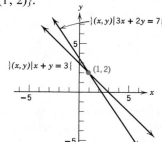

1. $\{(x, y) \mid x + y = 4\} \cap \{(x, y) \mid 2x - 3y = 3\} = \{(3, 1)\}$

2. $\{(x, y) \mid x + y = 0\} \cap \{(x, y) \mid 3x + 2y = 1\} = \{(1, -1)\}$

3. $\{(x, y) \mid y = x - 3\} \cap \{(x, y) \mid y = 2x + 4\} = \{(-7, -10)\}$

4. $\{(x, y) \mid 2x - 3y = 6\} \cap \{(x, y) \mid x + 3y = 3\} = \{(3, 0)\}$

5. $\{(x, y) \mid 2x - y = 0\} \cap \{(x, y) \mid x = -3\} = \{(-3, -6)\}$

6. $\{(x, y) \mid 3x - 4y = -5\} \cap \{(x, y) \mid y = 2\} = \{(1, 2)\}$

7. $\{(x, y) \mid 5x + 2y = -4\} \cap \{(x, y) \mid 6x - y = 2\} = \{(0, -2)\}$

8. $\{(x, y) \mid 2x - y = 4\} \cap \{(x, y) \mid x + 3y = 23\} = \{(5, 6)\}$

Determine if the equations in each system are inconsistent or dependent.

Examples

a. $6x + 4y = 12$
   $3x + 2y = 18$

b. $2x - 3y = 5$
   $6x - 9y = 15$

Solutions

a. Since $\dfrac{6}{3} = \dfrac{4}{2} \neq \dfrac{12}{18}$, by

Theorem 10.1 the equations are inconsistent.

b. Since $\dfrac{2}{6} = \dfrac{-3}{-9} = \dfrac{5}{15}$, by

Theorem 10.1 the equations are dependent.

9. $2x - y = 8$
   $4x - 2y = 12$

10. $x - 3y = 4$
    $3x - 9y = 12$

11. $2x - 3y = 5$
    $-4x + 6y = -10$

12. $2x - 4y = 7$
    $x - 2y = 1$

13. $3x - 4y = -7$
    $-6x + 8y = 15$

14. $3x + y = -4$
    $6x + 2y = -8$

15. $\dfrac{1}{3} x - \dfrac{2}{3} y = 2$

    $x - 2y = 6$

16. $x = \dfrac{1}{4} y - 3$

    $4x = y - 5$

Determine if the equations in each system are inconsistent or dependent. If they are not inconsistent or dependent, estimate the element of the solution set of the system by graphical means.

Example. $x + y = 5$

$2x - y = 4$

Solutions. Since $\dfrac{1}{2} \neq \dfrac{1}{-1} \neq \dfrac{5}{4}$, by Theorem
10.1 the equations are independent. From the
figure, the solution set is estimated as $\{(3, 2)\}$.

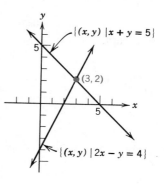

17. $x - y = 5$
   $2x + 3y = 5$

18. $x + y = -4$
   $3x - 2y = -7$

19. $3x + 2y = 7$ 20. $x + 2y = 0$ 21. $3x + y = -4$
   $x - y = -1$ $-3x + y = -7$ $6x + 2y = -8$

22. $15x - 2y = 5$ 23. $-3x + 8y = 5$ 24. $8x - 4y = 4$
   $30x - 4y = 10$ $-x + y = 0$ $3x - 2y = 3$

25. $2x - 3y = -1$ 26. $x + 3y = 3$ 27. $2x + 3y = 0$
   $4x - 6y = 24$ $2x + 6y = 8$ $5x - 2y = 19$

28. $x + 3y = 4$ 29. $x + 3y = 8$ 30. $2x - y = 6$
   $x - 3y = -2$ $x = 2$ $y = 4$

B

31. State the conditions under which the equations in the system

$$a_1x + b_1y = c_1$$
$$a_2x + b_2y = c_2$$

   are inconsistent.

32. State the conditions under which the equations in the system

$$a_1x + b_1y = c_1$$
$$a_2x + b_2x = c_2$$

   are dependent.

10.2 Solution of Systems by Analytic Methods

There are many important uses for graphing in mathematics. However,
graphing is not always the best way to solve systems of linear equations.
In the first place, graphing takes time, and second, you frequently have to
estimate the coordinates of points on the graph.

We now consider a much more efficient method using the following theorem.

Theorem 10.2. Any ordered pair (x, y) that satisfies the equations

$$a_1x + b_1y + c_1 = 0 \tag{1}$$
$$a_2x + b_2y + c_2 = 0 \tag{2}$$

also satisfies the equation

$$s \cdot (a_1x + b_1y + c_1) + t \cdot (a_2x + b_2y + c_2) = 0 \tag{3}$$

for any real numbers s and t.

This is true because the replacement of the variables in equation (3) above with the components of any ordered pair (x, y) that satisfies both Equations (1) and (2) results in

$$s \cdot (0) + t \cdot (0) = 0,$$

which is clearly true for any real numbers s and t. The left-hand member of Equation (3) is called a **linear combination** of the left-hand members of Equations (1) and (2).

Theorem 10.2 can be used to identify any ordered pairs which satisfy both equations in a system of equations such as (1) and (2). Consider the system

$$3x - y + 9 = 0 \tag{4}$$
$$x + 2y - 4 = 0. \tag{5}$$

We can multiply each member of (4) by 1 and each member of (5) by -3 to obtain

$$3x - y + 9 = 0 \tag{6}$$
$$-3x - 6y + 12 = 0. \tag{7}$$

The sum of the left-hand members of (6) and (7) is a linear combination of the left-hand members of (4) and (5) and by Theorem 10.2 the result is

$$0x - 7y + 21 = 0, \tag{8}$$

or

$$y = 3, \tag{9}$$

must contain in its solution set any solution common to the solution sets of the two original Equations (4) and (5). But any solution to $y = 3$ must be of the form $(x, 3)$—that is, it must have a y-component of 3. Thus, substituting 3 for y in either (4) or (5), the required x-component can be determined. If Equation (4) is used, we have

$$3x - (3) + 9 = 0 \quad \text{or} \quad x = -2,$$

and, if Equation (5) is used, we have

$$x + 2(3) - 4 = 0 \quad \text{or} \quad x = -2.$$

Since the ordered pair $(-2, 3)$ satisfies both Equations (4) and (5), the solution set is

$$\{(x, y) \mid 3x - y + 9 = 0\} \cap \{(x, y) \mid x + 2y - 4 = 0\} = \{(-2, 3)\}.$$

Note that the multipliers of Equations (4) and (5), 1 and -3, were chosen so that the resulting linear Equation (8) was free of the variable x. If multipliers 2 and 1 are used for Equations (4) and (5), respectively, the resulting equation will be free of y. The same solution set, $\{(-2, 3)\}$, is obtained.

The foregoing method proceeds in the same manner when the constant terms are written as the right-hand members of the two equations, as

$$a_1 x + b_1 y = c_1 \tag{1'}$$
$$a_2 x + b_2 y = c_2, \tag{2'}$$

because by Theorem 8.1, each equation can be written equivalently with the right member equal to 0. We shall simply cite Theorem 10.2 as our justification when forming a linear combination of the members in (1) and (2) or (1') and (2'). Furthermore, if the equations in a system do not happen to be written in a convenient form for solution, they should first be transformed into equivalent equations of the form $ax + by = c$. A system in which each equation is written in this form is said to be in **standard form.**

Exercise 10.2

A

Write each system in standard form.

Examples

a. $x = 5 + y$
   $y + 2x = 6$

b. $3y = 14 - x$
   $3y = 4 - 2x$

c. $x = 4y - 3$
   $y = 3x + 6$

Solutions on page 310.

Solutions

a. $x - y = 5$
   $2x + y = 6$

b. $x + 3y = 14$
   $2x + 3y = 4$

c. $x - 4y = -3$
   $-3x + y = 6$

1. $x - 2y = 6$
   $y - 3x = 4$

2. $y = 3x - 4$
   $x = 2 - y$

3. $2x = y + 4$
   $y + 2 = x$

4. $x = 3y - 4$
   $y = x$

5. $\dfrac{1}{2}x - 2 = y$

   $x = 3 + \dfrac{2}{3}y$

6. $y - \dfrac{1}{4}x = 3$

   $x + \dfrac{3}{2} = \dfrac{1}{4}y$

Solve.

Example. $4x = 3y + 2$
   $y + 2 = -4x$

Solution.   Write the system in standard form.

$$4x - 3y = 2 \qquad (1)$$
$$4x + y = -2 \qquad (2)$$

Multiply both members of Equation (2) by -1.

$$4x - 3y = 2$$
$$-4x - y = 2$$

By Theorem 10.2,

$$-4y = 4, \quad \text{from which} \quad y = -1.$$

Replace y with -1 in Equation (1) or (2). Here we shall use Equation (1).

$$4x - 3(-1) = 2$$
$$4x + 3 = 2$$
$$4x = -1$$

$$x = -\frac{1}{4}$$

Thus, $\{(x, y) \mid 4x = 3y + 2\} \cap \{(x, y) \mid y + 2 = -4x\} = \left\{\left(-\dfrac{1}{4}, -1\right)\right\}.$

7. $x - y = 3$
   $x + y = 5$

8. $2x + y = -6$
   $x - y = -3$

9. $-3y + x = 1$
   $1 - x = 2y$

10. $4 = 2x - y$
    $-y = x + 7$

11. $x = 3y$
    $3y + 2x = 45$

12. $x = 4y - 6$
    $x = 2y + 12$

13. $-3x + y = 1$
    $2x + y = 1$

14. $y = -3$
    $x + y = 4$

15. $4x = 6y + 26$
    $-6y = x + 1$

16. $2x - y = -12$
    $-4x - y = 12$

17. $\dfrac{1}{6}x - \dfrac{1}{4}y = -\dfrac{7}{12}$
    $\dfrac{5}{6}x + \dfrac{1}{4}y = \dfrac{19}{12}$

18. $\dfrac{1}{3}x + \dfrac{1}{3}y = 0$
    $\dfrac{1}{2}x + \dfrac{1}{3}y = \dfrac{15}{6}$

Example. $3x - 2y = 3$ (1)
$\qquad\qquad -5x + 3y = -4$ (2)

Solution. Multiply each member of Equations (1) and (2) by 3 and 2, respectively.

$$9x - 6y = 9$$
$$-10x + 6y = -8$$

By Theorem 10.2,

$$-x = 1 \text{ from which } x = -1.$$

Replace x with -1 in Equation (1) or (2). We shall use Equation (1).

$$3(-1) - 2y = 3$$
$$-3 - 2y = 3$$
$$-2y = 6$$
$$y = -3.$$

Thus, $\{(x, y) \mid 3x - 2y = 3\} \cap \{(x, y) \mid -5x + 3y = -4\} = \{(-1, -3)\}.$

19. $5 - x = 3y$
    $2x = y + 3$

20. $5 = 2y + 3x$
    $5 - y = -x$

21. $x - 3y = 2$
    $y = -4$

22. $2x - 5y = -4$
    $x = 3$

23. $5y = 7x - 37$
    $23 - 3x = 5y$

24. $9y = 7x + 3$
    $8 + 2x = -y$

25. $7x = -3y - 57$
    $5x = 8y + 10$

26. $2x = 5y - 41$
    $5x = -4y + 13$

27. $4x - 3y = 27$
    $17 = 3x + y$

28. $3x + 8y = -1$
    $35 = 9x + 5y$

29. $2x + 3y = 800$
    $5x - 2y = 100$

30. $3x - 2y = 26,000$
    $4x + 5y = 4,000$

B

Solve for x and y in terms of a and b.

31. $x + y = a$
    $x - y = b$

32. $ax + by = 1$
    $bx + ay = 1$

33. Find values of a and b so that the graph of the equation $ax + by - 4 = 0$ passes through the points whose coordinates are $(-2, -5)$ and $(4, 4)$. (*Hint:* First replace x and y with -2 and -5, respectively, and then with 4 and 4 to obtain a system in the variables a and b.)

34. The use of Theorem 10.2 to solve the system

$$4x - 5y = 3$$
$$-8x + 10y = -6$$

results in the statement $0 + 0 = 0$. Explain.

35. The use of Theorem 10.2 to solve the system

$$3x + 6y = 9$$
$$6x + 12y = 27$$

results in the statement $0 + 0 = 9$. Explain.

10.3 Word Problems Using Two Variables

Recall that you learned how to solve word problems in Chapter 8 by using one equation in one variable. Since you have all the tools necessary to solve a system of two linear independent equations in two variables, we shall now consider the solution of word problems by the use of such systems. In general, the use of a system of equations as a mathematical model simplifies the solution of these problems. First, we shall see how word sentences which state conditions on two variables can be represented by equations in two variables.

Examples. Represent the following sentences symbolically:

a. The sum of one number and twice a second number is 20.
b. One number equals four times a second number.
c. The difference of two numbers is 14.

Solutions. a. $x + 2y = 20$.   b. $x = 4y$.
c. Assuming that $x > y$, we can represent this sentence in three different forms,

$$x - y = 14, \quad x = y + 14, \quad \text{or} \quad x - 14 = y.$$

It is convenient, though not always necessary, to use *two* variables to solve word problems when *two* conditions are imposed on the variables. These conditions can be represented by two equations. A solution meeting the conditions of the word problem must be a solution of both equations. For example, the sentence "The sum of two numbers is twelve and their difference is eight" imposes two conditions on two numbers. The first condition, that their sum is twelve, can be represented by the equation

$$x + y = 12.$$

The second condition, that their difference is eight, can be represented by the equation

$$x - y = 8,$$

in which form x now represents the larger of the two numbers. The solution set of the system

$$x + y = 12$$
$$x - y = 8$$

is found to be $\{(10, 2)\}$. That 10 and 2 is the correct solution to the problem can be verified by noting that the sum of the numbers is 12 and their difference is 8, the conditions on the numbers stated in the word problem.

Example. A 20-foot board is cut into two pieces, one of which is 2 feet longer than the other. How long is each piece?

Solution. Let x and y represent a number of feet for each piece. In this

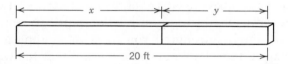

case, a figure is helpful in setting up the equations. The conditions of the problem are restated as the system

$$x + y = 20$$
$$x - y = 2.$$

You can verify the lengths of the pieces to be 11 feet and 9 feet.

Example. How many quarts of a 15% solution of acid should be added to 30 quarts of a 50% solution of acid to obtain a 40% solution?

Solution. Let x represent a number of quarts of the 15% solution to be added, and y a number of quarts of the mixture, the 40% solution.

Solution continued on page 314.

Write a mathematical model. Use one equation to represent the relation between the number of quarts being used.

$$x + 30 = y \tag{1}$$

Use a second equation to represent the relation between the amount of *pure* acid in the solution added, the original solution, and the final mixture.

$$0.15(x) + 0.50(30) = 0.40(y) \tag{2}$$

Equivalently, Equations (1) and (2) can be written as

$$-40x + 40y = 1200, \tag{1'}$$
$$15x - 40y = -1500. \tag{2'}$$

By Theorem 10.2

$$-25x = -300,$$

from which

$$x = 12,$$

and by substituting 12 for x in Equation (1),

$$y = 42.$$

The solution set of the systems is $\{(12, 42)\}$. Hence, 12 quarts of the 15% solution must be added.

These preceding examples are typical of the types in which two equations in two variables can be used to express relationships between measures of things. We have simply introduced you to possible applications of this type of mathematical model. We note here again that the work with the mathematical model, in this case the system of equations, is quite independent of the physical problem. *A solution to a system of equations is first obtained and then this solution is interpreted for the particular word problem.*

Exercise 10.3

A

Solve each word problem by using a system of equations as the mathematical model.

General Problems

Example. If 1/5 of a certain integer is added to 1/3 of the next consecutive integer, the result is 19. Find the integers.

Solution. Let x represent an integer, and y represent the next consecutive integer.

Write the mathematical model (system of equations) which describes the conditions on the variables and then solve the system.

$$\frac{1}{5}(x) + \frac{1}{3}(y) = 19 \tag{1}$$

$$y - x = 1 \tag{2}$$

Equivalently,

$$3x + 5y = 285 \tag{1'}$$

$$-x + y = 1, \tag{2'}$$

and

$$3x + 5y = 285 \tag{1''}$$

$$-3x + 3y = 3. \tag{2''}$$

By Theorem 10.2

$$8y = 288$$

from which

$$y = 36,$$

and by substituting 36 for y in Equation (2)

$$x = 35.$$

Thus, the solution set of the system is $\{(35, 36)\}$, and the desired integers are 35 and 36.

1. The sum of two numbers is 18 and one of the numbers is 6 less than twice the other. Find the numbers.

2. Find two consecutive even integers whose sum is 90.

3. The denominator of a certain fraction is 15 more than the numerator and the fraction is equal to 7/12. Find the numerator and the denominator.

4. The sum of the digits of a two-digit number is 9. The number formed by reversing the digits is 27 less than the original number. Find the digits of the original number. (*Hint:* If t represents a tens digit and u a units digit, then $10t + u$ represents the number.)

5. A 54-foot rope is cut into two pieces so that one piece is 20 feet longer than the other. How long is each piece?

6. At a recent election, the winning candidate received 85 votes more than his opponent. How many votes did each candidate receive if there were 911 votes cast?

Coin Problems

Recall from Section 8.8 that the value of a number of coins of the same denomination is equal to the product of the value of one coin and the number of coins.

Example.   A collection of nickels and dimes has a value of $2.55. How many nickels and dimes are in the collection if there are 3 more dimes than nickels?

Solution.   Let *d* represent a number of dimes and *n* a number of nickels.
    Write the mathematical model. Use one equation to represent the relation between the *number* of coins, and a second equation to represent the relation between the *value* of the dimes in cents and the *value* of the nickels in cents to the *value* of the entire collection in cents. Then solve the system of equations.

$$d - n = 3 \qquad (1)$$
$$10d + 5n = 255 \qquad (2)$$

Equivalently,

$$5d - 5n = 15 \qquad (1')$$
$$10d + 5n = 255 \qquad (2')$$

By Theorem 10.2,

$$15d = 270$$

from which

$$d = 18,$$

and by substituting 18 for *d* in Equation (1)

$$n = 15.$$

The solution set of the system is $\{(18, 15)\}$; the number of dimes in the collection is 18, the number of nickels, 15.

7. A man has $4.30 in change consisting of 8 more nickels than quarters. How many quarters and nickels does he have?

8. A man has $355 in ten-dollar and five-dollar bills. There are ten more ten-dollar bills than five-dollar bills. How many of each kind does he have?

9. Admission fees at a football game were $1.25 for adults and $.55 for children. The receipts were $530.40 for 454 paid admissions. How many adults and children attended the game?

10. Three pounds of bacon and 4 pounds of coffee cost $4.32. One pound of bacon and 5 pounds of coffee cost $3.97. How much are the costs of a pound of bacon and a pound of coffee?

Mixture Problems

Recall from Section 8.8 that the amount of a substance in a mixture is equal to the product of the per cent of the substance in the mixture and the amount of the mixture.

11. How many ounces each of a 12% salt solution and a 30% salt solution must be mixed to obtain 45 ounces of a 24% solution?

12. A grocer blends walnuts worth 65 cents a pound with walnuts worth 90 cents a pound to obtain 30 pounds of a mixture worth 80 cents a pound. How many pounds of 65-cent walnuts were used?

13. How many pounds each of a 50% copper alloy and a 75% copper alloy must be mixed to obtain $87\frac{1}{2}$ pounds of a 60% alloy?

14. How many quarts each of pure alcohol and of an antifreeze mixture that is 30 per cent alcohol must be mixed to produce 24 quarts of a solution that is 65 per cent alcohol?

Interest Problems

Recall from Section 8.8 that the amount of interest earned in one year at simple interest equals the product of the rate of interest and the amount of money invested.

15. A sum of $3,000 is invested, part at 5% and the remainder at 6%. Find the amount invested at each rate if the yearly income from the two investments is $169.

16. A man invests $12,000, part at 4% and part at $5\frac{1}{2}$%. If the simple interest for 2 years is $1,080, how much was invested at $5\frac{1}{2}$%?

17. A man has twice as much money invested at $6\frac{1}{2}$% as he has invested at 7%. If his yearly income from the investments is $300, how much does he have invested at each rate?

18. A man has three times as much money invested at 4% as he has invested at $4\frac{3}{4}$%. If his yearly income from the investments is $268, how much does he have invested at each rate?

B

Uniform Motion Problems

Recall from Section 8.8 that the distance traveled at a constant rate is equal to the product of the rate and the time traveled.

Example. Two cars start together and travel in the same direction, one going twice as fast as the other. At the end of three hours, they are 93 miles apart. How fast is each traveling?

Solution. Let x represent a rate for the faster car, y a rate for the slower car.
Write one equation expressing the fact that the greater rate is two times the other rate.

$$x = 2y \tag{1}$$

Write a second equation expressing the fact that the difference of the distances traveled is 93 miles (distance equals rate × time).

$$3x - 3y = 93 \tag{2}$$

The system can be written equivalently,

$$-3x + 6y = 0 \tag{1'}$$
$$3x - 3y = 93. \tag{2'}$$

By Theorem 10.2,

$$3y = 93,$$

from which

$$y = 31,$$

and by substituting 31 for y in Equation (1) above

$$x = 62.$$

The solution set is $\{(62, 31)\}$. The faster car is traveling 62 miles per hour; the slower, 31 miles per hour.

19. Two airplanes left towns A and B, which are 2,400 miles apart, at the same time and proceeded toward each other along parallel routes. At the time they passed each other, the plane from A had traveled 250 miles farther than the plane from B. How many miles from A were the planes when they passed each other?

20. An express train travels 238 miles in the same time that a freight train travels 175 miles. If the express goes 18 miles per hour faster than the freight, find the rate of each.

21. A freight train leaves town A for town B. Two hours later, a passenger train also leaves town A for town B, traveling at an average rate $1\frac{1}{2}$ times the rate of the freight train. Find the rates of each if the passenger train passes the freight train 180 miles from town A.

22. Two motorists 140 miles apart and driving at different constant speeds would meet in 1 hour if they drove towards each other, and in 7 hours if they drove in the same direction. Find their rates.

Lever Problems

In solving lever problems, the basic formula involves forces that are applied on both sides of a fixed point, called the fulcrum, as shown in the diagram. If

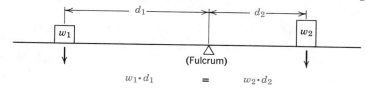

$$w_1 \cdot d_1 \quad = \quad w_2 \cdot d_2$$

a force w_1 is applied to one side of a lever at a distance d_1 from the fulcrum, and a force w_2 is applied on the other side of the fulcrum at a distance d_2 from the fulcrum, then the lever will be balanced if and only if $w_1 d_1 = w_2 d_2$.

Example. Two weights balance when placed 6 feet and 3 feet from the fulcrum of a lever. If the positions of the weights are interchanged, the larger one must be decreased by $4\frac{1}{2}$ ounces if the balance is to be maintained. Find the weights.

Solution. Let x represent a weight in ounces for the larger weight, and y a weight in ounces for the smaller. The mathematical model representing the first statement of the problem is

$$6y = 3x. \tag{1}$$

An equation for the second statement is

$$6\left(x - \frac{9}{2}\right) = 3y. \tag{2}$$

The system can be written equivalently as

$$\begin{array}{ll} 3x - 6y = 0 \\ 6x - 3y = 27 \end{array} \quad \text{or} \quad \begin{array}{ll} -2x + 4y = 0 & (1') \\ 2x - y = 9. & (2') \end{array}$$

By Theorem 10.2,

$$3y = 9$$

from which

$$y = 3,$$

and by substituting 3 for y in Equation (1) above, we obtain

$$x = 6.$$

The solution set is $\{(6, 3)\}$. The weights are 6 ounces and 3 ounces.

23. Two weights balance when placed 2 feet and 5 feet from the fulcrum of a lever. If the positions of the weights are interchanged, the smaller one must be increased by 105 pounds if the balance is to be maintained. Find the weights.

24. Two weights balance when placed 4 feet and 3 feet from the fulcrum of a lever. If the lighter weight is increased by 60 pounds, the other weight must then be moved to a distance of 8 feet from the fulcrum in order to maintain a balance. Find the weights.

25. A 20-pound weight is placed at one end of a 10-foot lever and a 5-pound weight is placed at the other. Where should the fulcrum be placed in order to obtain a balance? (*Hint:* Let x and y represent the two distances on either side of the fulcrum.)

26. Where should the fulcrum be placed on a 12-foot lever if a weight of 920 pounds at one end is to be lifted by a force of 40 pounds at the other end?

10.4 Systems of Inequalities

Graphs are sometimes used to represent solutions of systems of inequalities. The solution set of such a system is determined by the coordinates of the points that are common to the graphs of every inequality in the system. For example, the solution set of a system of inequalities, such as

$$y < x + 2$$
$$y > 4 - x,$$

would consist of the coordinates of all those points common to the solution set of each inequality. Symbolically, the solution set is represented as

$$\{(x, y) \mid y < x + 2\} \cap \{(x, y) \mid y > 4 - x\}.$$

The solution set is represented in the three parts of Figure 10.4. The graph of the inequality $y < x + 2$ is shown in Figure 10.4a. The graph of the inequality $y > 4 - x$ is shown in Figure 10.4b. The combined graphs of $\{(x, y) \mid y < x + 2\}$ and $\{(x, y) \mid y > 4 - x\}$ are shown in Figure 10.4c. The heavy-shaded region is the set of all points whose coordinates are components of ordered pairs in the solution set.

As another example, consider the system

$$y > 2$$
$$x < 3.$$

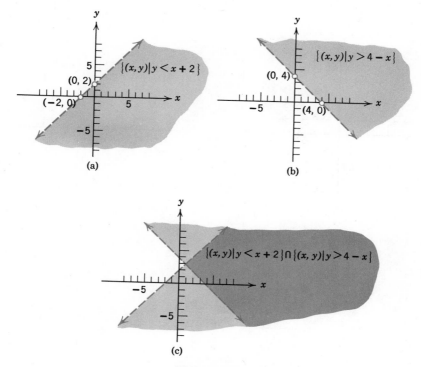

Figure 10.4

The graph of the solution set of $y > 2$, shown as the shaded region in Figure 10.5a, consists of all points above the graph of the linear equation $y = 2$. The graph of the solution set of $x < 3$, shown as the shaded region in Figure 10.5b, consists of all points to the left of the graph of the equation $x = 3$. The solution set of the system, shown as the heavy-shaded region in Figure 10.5c, consists of all points whose coordinates are components of ordered pairs common to the solution sets of $y > 2$ and $x < 3$. This solution set can be written

$$\{(x, y) \mid y > 2\} \cap \{(x, y) \mid x < 3\}.$$

As a final example, consider a system containing an equation and an inequality such as

$$y = x + 2$$

$$y > -\frac{1}{3}x + 1,$$

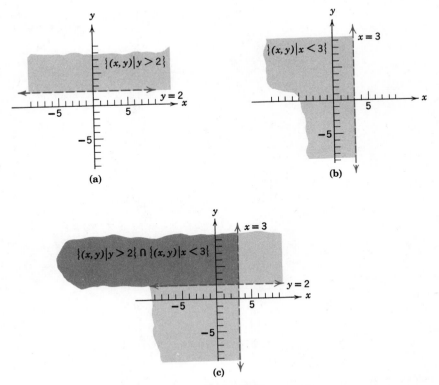

Figure 10.5

The graph of the solution set of $y = x + 2$ consists of all points on the line in Figure 10.6a. The graph of the solution set of $y > -\frac{1}{3}x + 1$, shown as the shaded region in Figure 10.6b, consists of all points above the graph of the equation $y = -\frac{1}{3}x + 1$. The solution set of the system consists of the coordinates of all points common to both graphs. In this case these common points exist only on the portion of the graph of $y = x + 2$ which lies in the graph of the solution set of $y > -\frac{1}{3}x + 1$, as shown in Fig. 10.6c. This solution set can be written

$$\{(x, y) \mid y = x + 2\} \cap \{(x, y) \mid y > -\tfrac{1}{3}x + 1\}.$$

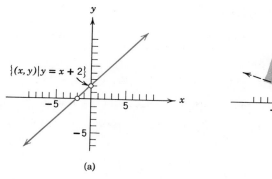

(a)

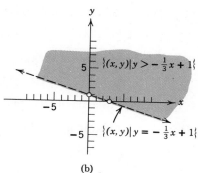

(b)

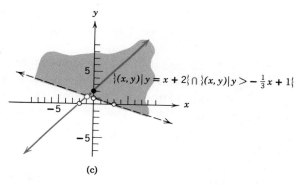

(c)

Figure 10.6

Exercise 10.4

A

Graph each system in Problems 1 to 12. Indicate, by proper shading or by labeling, the region in the plane representing the solution set of the system.

1. $y \geq x + 2$
   $y \geq 4 - x$

2. $y \geq 3$
   $x \geq 5$

3. $y \geq x$
   $y + x \geq 0$

4. $3y - x \geq 2$
   $x \leq -4$

5. $y + 2x < 6$
   $y > 3$

6. $x > 4$
   $x + y \geq 6$

7. $x = 4$
   $x + y < 3$

8. $y = 3$
   $2y - x < 8$

9. $\{(x, y) \mid x + 3y < 5\} \cap \{(x, y) \mid y - x = 1\}$

10. $\{(x, y) \mid 2x - y \le 4\} \cap \{(x, y) \mid x + y = 2\}$

11. $\{(x, y) \mid y < 3x + 1\} \cap \{(x, y) \mid x - y < 4\}$

12. $\{(x, y) \mid y > 2x + 4\} \cap \{(x, y) \mid x + y \ge 4\}$

13. The region representing the set

$$\{(x, y) \mid y > 2\} \cap \{(x, y) \mid x > -2\} \cap \{(x, y) \mid x + y < 1\}$$

is the interior of a small triangle. Graph the corresponding system.

14. Estimate the coordinates of the three vertices of the triangle in Problem 13.

15. Graph the set

$$\{(x, y) \mid x + 3y < 6\} \cap \{(x, y) \mid 2x + y > 4\} \cap \{(x, y) \mid y > 0\}.$$

16. Estimate the coordinates of the three vertices of the triangle in Problem 15.

CHAPTER SUMMARY

10.1 A pair of linear equations when considered together are referred to as a **system of linear equations.** Any ordered pair that is a solution of each of two equations in two variables has components that are the coordinates of a point that lies on the graphs of both equations. Such a common point is called the **point of intersection** of their graphs.

If the graphs of the solution sets of two linear equations are *parallel lines*, the equations do not have any ordered pairs in common and the equations are **inconsistent.** If the graphs of the solution sets of the two linear equations are actually the *same line*, the equations have an infinite number of ordered pairs in common and they are **dependent.** The equations in the system

$$a_1 x + b_1 y = c_1$$
$$a_2 x + b_2 y = c_2$$

are inconsistent if $\dfrac{a_1}{a_2} = \dfrac{b_1}{b_2} \ne \dfrac{c_1}{c_2}$ and dependent if $\dfrac{a_1}{a_2} = \dfrac{b_1}{b_2} = \dfrac{c_1}{c_2}$.

Otherwise, they are independent.

10.2 The coordinates of a point of intersection can only be *approximated* by graphical methods, but can be determined *exactly* by algebraic methods. One method of solving a system of two linear equations in two variables is based on the theorem:

Any ordered pair (x, y) that satisfies the equations

$$a_1x + b_1y + c_1 = 0$$
$$a_2x + b_2y + c_2 = 0$$

also satisfies the equation

$$s \cdot (a_1x + b_1y + c_1) + t \cdot (a_2x + b_2y + c_2) = 0,$$

where $s, t \in R$.

10.3 A system of equations is often easier to use as a mathematical model to solve word problems than is a single equation. In general such problems involve *two conditions* on *two variables*.

10.4 The solution set of a system of linear inequalities consists of the ordered pairs corresponding to coordinates of all points common to the graphs of each inequality in the system. Regions containing such points can be shown graphically by proper shading techniques.

CHAPTER REVIEW

A

10.1 *Verify that each statement is true.*

1. $\{(x, y) \mid 2x - 3y - 4 = 0\} \cap \{(x, y) \mid y = 5x + 3\} = \{(-1, -2)\}$

2. $\{(x, y) \mid y = 4x - 4\} \cap \{(x, y) \mid 3x + y = 3\} = \{(1, 0)\}$

Determine if the equations in each system are inconsistent, or dependent, or if the system has a unique (only one) solution. Estimate by graphical methods the solution set of each system that has a unique solution.

3. $x - 4y + 8 = 0$ 4. $3x + 2y = 5$ 5. $x - 8y = 12$
   $2y = -x + 10$ $y = 4x + 19$ $9 + 8y = x$

6. $5x - y = 0$ 7. $3y - 4x = -2$ 8. $x = 2y - 5$
   $3y = -9x - 24$ $8x = 6y + 1$ $6y - 3x = 15$

10.2 *Solve by algebraic methods.*

9. $3x - 2y + 13 = 0$ 10. $4y - 3x = -5$ 11. $\dfrac{x}{5} - \dfrac{y}{3} = 0$
   $y - 4 = -6x$ $x - 7 = -20y$
                                                    $2x - y = 7$

12. $\dfrac{y}{6} = -\dfrac{x}{2}$

    $7y + 2x = 19$

13. $\dfrac{2x}{3} + \dfrac{y}{2} = \dfrac{-19}{6}$

    $\dfrac{x + 3}{4} = \dfrac{-y}{10}$

14. $\dfrac{x}{5} - \dfrac{4y}{3} = \dfrac{-32}{5}$

    $\dfrac{y - 2}{3} = \dfrac{x}{6}$

10.3 *Solve using two variables.*

15. The sum of two rational numbers is 7 and the difference between two times the first and the second is -3. Find the numbers.

16. A man has $7.40 in change consisting of 3 more quarters than dimes. How many quarters and dimes does he have?

17. A boat travels 105 miles in the same time that a freight train travels 225 miles. If the rate of the freight train is 24 miles per hour greater than the rate of the boat, find the rate of each.

10.4 *Graph each system. Indicate, by proper shading or by labeling, the region in the plane representing the solution set of the system.*

18. $3x + 2y > 7$

    $x \le y - 1$

19. $x + 5y \le 0$

    $x - 3 + y \le 0$

20. Graph $\{(x, y) \mid x + y \le 1\} \cap \{(x, y) \mid 6y > 2x - 8\}$.

chapters 9 & 10

Cumulative Review

CHAPTER 9

1. In any ordered pair, the number on the left is called the first component, and the number on the right is called the _____ _____.
2. The ordered pair $(-1, 5)$ (is/is not) the same as the ordered pair $(5, -1)$.
3. If the variables in an equation of two variables are replaced by the components of an ordered pair in a specified order and the resulting statement is true, the ordered pair is called a _____ of the equation.
4. The set of *all* ordered pairs whose elements are solutions of an equation is called the _____ _____ of the equation.
5. A linear equation in two variables has a(n) (finite/infinite) number of ordered pairs as solutions.
6. A relation or a function is a(n) _____ between the elements of two sets.
7. A relation can be defined as any set of _____ _____.
8. The set of first components in a set of ordered pairs is called the _____ of the relation; the set of second components is called the range.
9. A function is a relation in which each element in the domain is associated with one and _____ _____ element in the range.
10. A variable representing an element in the domain is often referred to as an independent variable, while the variable representing an element in the range is called a(n) _____ variable.
11. What is the domain of the function $\{(x, y) \mid y = \sqrt{x - 9}, x \in R\}$?
12. The symbol $f(x)$ represents the element in the _____ of f associated or paired with the element x in the domain.
13. If $g(x) = x^3 - 2x^2 + 7$, what is the value of $g(-1)$?
14. Any point in the geometric plane can be associated with a unique _____ _____ of real numbers.
15. Functions defined by first-degree equations are sometimes called first-degree functions or more generally _____ functions.

16. The graphs of the solution sets of all first-degree equations in two variables are _____ _____.

17. The method of graphing solution sets of first-degree equations in the variables x and y by choosing 0 for x to obtain one ordered pair and then 0 for y to obtain a second ordered pair is called the _____ method.

18. What ordered pairs would be determined in using the intercept method for graphing the solution set of the equation $2x - y - 15 = 0$?

19. The ratio of the change in the value of y with the change in the value of x between any two points on a straight line is called the _____ of the line.

20. If a linear equation in two variables, x and y, is written in the form $y = ax + b$, $a, b \in R$, the value of a will be the _____ of the straight line which is the graph of the solution set of the equation.

21. Determine the slope of the graph of $4x + 5y + 7 = 0$.

22. Because each element in the domain is usually associated with *more than one element* in the range, inequalities in two variables (do/do not) in general define functions.

23. The graph of the solution set of a linear *equation* in two variables is a straight line. The graph of the solution set of a linear *inequality* in two variables is, in general, the _____ on one side of a straight line.

CHAPTER 10

24. Pairs of first-degree linear equations in two variables which are considered together are referred to as a _____ of linear equations.

25. A point that is common to the graphs of two equations is called the point of _____.

26. If the solution sets of two linear equations do not have an ordered pair in common, the equations are _____ and the graphs of the equations are parallel lines.

27. If the solution sets of two linear equations have an infinite number of ordered pairs in common, the equations are _____ and the graphs of the equations are the same line.

28. The equations $\begin{array}{l} a_1x + b_1y = c_1 \\ a_2x + b_2y = c_2 \end{array}$ are _____ if $\dfrac{a_1}{a_2} = \dfrac{b_1}{b_2} \neq \dfrac{c_1}{c_2}$.

29. The equations $\begin{array}{l} a_1x + b_1y = c_1 \\ a_2x + b_2y = c_2 \end{array}$ are _____ if $\dfrac{a_1}{a_2} = \dfrac{b_1}{b_2} = \dfrac{c_1}{c_2}$.

30. The left-hand member of the equation

$$s \cdot (a_1x + b_1y + c_1) + t \cdot (a_2x + b_2y + c_2) = 0,$$

where s and t are elements of R, is a _____ _____ of the left-hand members of the equations

$$a_1x + b_1y + c_1 = 0,$$
$$a_2x + b_2y + c_2 = 0.$$

31. If the left-hand member of $y - 2 = 0$ is a linear combination of the left-hand members of a system of two equations, then the y-component of a solution of the system is _____.

32. A system of linear equations in which each equation is written as an equivalent equation in the form $ax + by = c$ is said to be in _____ _____.

33. It is convenient to use two variables to solve word problems when _____ conditions are imposed on two variables.

34. The solution set of a system of inequalities consists of the coordinates of the points that are _____ to the graphs of every inequality in the system.

11
chapter

Roots and Radicals

Many problems require mathematical models that are non-linear equations or inequalities. For example, suppose we want to find the length of the hypotenuse of a right triangle. The Pythagorean theorem states that the square of the length of the hypotenuse is equal to the sum of the squares of the lengths of the other two sides. The mathematical model that can be used to solve this problem is $c^2 = a^2 + b^2$ (see Figure 11.1). This equation

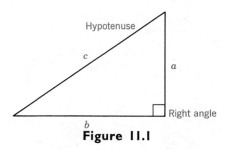

Figure 11.1

is non-linear because it contains variables of degree greater than one. The solutions of such equations are sometimes irrational numbers which are conveniently represented in radical form. In this chapter we shall study several theorems that will facilitate our work with radical notation.

11.1 Roots in Radical Form

Recall how $\sqrt{a}$ was defined in Section 5.1.

> **Definition 5.1.** For all $a \in R$ and $a \geq 0$, $\sqrt{a}$ is the non-negative number such that $\sqrt{a} \cdot \sqrt{a} = a$.

Since, for $a > 0$, $\sqrt{a}$ is a positive real number, then $-\sqrt{a}$ is a negative real number. If $a < 0$, there is no real square root because the square of a real number is always positive.*

Examples. a. $\sqrt{36} = 6$ b. $-\sqrt{49} = -7$

Consider the product

$$\sqrt{4} \cdot \sqrt{9} = 2 \cdot 3 = 6.$$

The same result can be obtained by writing

$$\sqrt{4} \cdot \sqrt{9} = \sqrt{4 \cdot 9} = \sqrt{36} = 6.$$

This example suggests the following theorem.

> **Theorem 11.1.** If $a, b \in R$, $a, b \geq 0$, then
> $$\sqrt{a}\sqrt{b} = \sqrt{ab} \quad \text{and} \quad \sqrt{ab} = \sqrt{a}\sqrt{b}.$$

Examples. a. $\sqrt{3}\sqrt{5} = \sqrt{3 \cdot 5} = \sqrt{15}$ b. $\sqrt{30x} = \sqrt{30}\sqrt{x}$ $(x \geq 0)$

Now consider the quotient

$$\frac{\sqrt{100}}{\sqrt{4}} = \frac{10}{2} = 5.$$

The same result can be obtained by writing

$$\frac{\sqrt{100}}{\sqrt{4}} = \sqrt{\frac{100}{4}} = \sqrt{25} = 5.$$

---

* Square roots of negative real numbers are called **imaginary numbers**. You shall have an opportunity to study these numbers in Chapter 12.

This example suggests the following theorem.

Theorem 11.2.   If $a, b \in R$, $a \geq 0$, and $b > 0$, then

$$\frac{\sqrt{a}}{\sqrt{b}} = \sqrt{\frac{a}{b}} \quad \text{and} \quad \sqrt{\frac{a}{b}} = \frac{\sqrt{a}}{\sqrt{b}}.$$

Examples.   a. $\dfrac{\sqrt{7}}{\sqrt{2}} = \sqrt{\dfrac{7}{2}}$ b. $\sqrt{\dfrac{3}{5}} = \dfrac{\sqrt{3}}{\sqrt{5}}$

From Definition 5.1 and the theorems in this section, a radical expression can be written in a variety of forms. We shall designate certain forms involving the square root symbol as being in the "simplest" form if they meet the following conditions.

1. *The radicand contains no polynomial factor raised to a power equal to or greater than 2.*

Examples.   a. $\sqrt{50}$ b. $\sqrt{18x^3y^2}$ $(x, y \geq 0)$

$$= \sqrt{5^2 \cdot 2} \qquad\qquad = \sqrt{3^2x^2y^2 \cdot 2x}$$

$$= \sqrt{5^2} \cdot \sqrt{2} \qquad\qquad = \sqrt{3^2x^2y^2}\sqrt{2x} \qquad \text{Theorem 11.1}$$

$$= 5\sqrt{2} \qquad\qquad\quad = 3xy\sqrt{2x} \qquad\qquad \text{Definition 5.1}$$

2. *No radical expressions are contained in denominators of fractions.*

Examples.   a. $\dfrac{5}{\sqrt{2}}$ b. $\dfrac{3}{\sqrt{x}}$ $(x > 0)$

$$= \frac{5\sqrt{2}}{\sqrt{2}\sqrt{2}} \qquad\qquad = \frac{3\sqrt{x}}{\sqrt{x}\sqrt{x}} \qquad \text{Fundamental principle of fractions}$$

$$= \frac{5\sqrt{2}}{2} \qquad\qquad = \frac{3\sqrt{x}}{x} \qquad\qquad \text{Definition 5.1}$$

3. *The radicand does not contain a fraction.*

Examples. a. $\sqrt{\dfrac{2}{3}}$ b. $\sqrt{\dfrac{3}{5x}}$ $(x > 0)$

$\qquad = \dfrac{\sqrt{2}}{\sqrt{3}}$ $= \dfrac{\sqrt{3}}{\sqrt{5x}}$ Theorem 11.2

$\qquad = \dfrac{\sqrt{2}\sqrt{3}}{\sqrt{3}\sqrt{3}}$ $= \dfrac{\sqrt{3}\sqrt{5x}}{\sqrt{5x}\sqrt{5x}}$ Fundamental principle of fractions

$\qquad = \dfrac{\sqrt{2}\sqrt{3}}{3}$ $= \dfrac{\sqrt{3}\sqrt{5x}}{5x}$ Definition 5.1

$\qquad = \dfrac{\sqrt{6}}{3}$ $= \dfrac{\sqrt{15x}}{5x}$ Theorem 11.1

We have designated certain forms of radical expressions as being "simplest" simply as a matter of convenience in order to practice changing the forms of radical expressions and to develop the ability to recognize equal radical expressions. There are times, however, when the forms we have designated as simplest are not the ones that are most convenient to use.

Exercise 11.1

In this exercise, assume that all variables represent positive real numbers.

A

Simplify each expression if possible.

Examples. a. $\sqrt{25} = 5$ b. $\sqrt{7}$ (simplest form)   c. $-\sqrt{4y^2} = -2y$

1. $\sqrt{9}$ 2. $\sqrt{16}$ 3. $\sqrt{81}$ 4. $\sqrt{64}$

5. $-\sqrt{3}$ 6. $\sqrt{17}$ 7. $\sqrt{\dfrac{4}{9}}$ 8. $-\sqrt{\dfrac{9}{25}}$

9. $-\sqrt{x^2 y^2}$ 10. $\sqrt{4x^2 y^4}$ 11. $\sqrt{\dfrac{1}{x^4 y^6}}$ 12. $\sqrt{\dfrac{1}{y^2}}$

13. $\sqrt{\dfrac{y^2}{x^4}}$ 14. $\sqrt{\dfrac{9y^2}{16x^6}}$

Write each expression as an equal expression in radical form.

Examples.   a. $8 = \sqrt{64}$ b. $-3x = -\sqrt{9x^2}$

15. 9   16. 1   17. 6   18. -4

19. $\dfrac{2}{5}$ 20. $\dfrac{-3}{8}$ 21. $2x$ 22. $-3xy^2$

23. $\dfrac{1}{x}$ 24. $\dfrac{-x}{2y^2}$ 25. $\dfrac{1}{6y^3}$ 26. $\dfrac{5x^2}{7y^5}$

State the definition, axiom, or theorem that justifies each statement.

Example.   a. $\sqrt{6xy^2} = \sqrt{6x}\sqrt{y^2}$ Theorem 11.1

                 b. $= \sqrt{6x}\,y$ Definition 5.1

                 c. $= y\sqrt{6x}$ Commutative law of multiplication

27. $\sqrt{25} = 5$ 28. $\sqrt{81} = 9$

29. a. $\sqrt{18} = \sqrt{9}\sqrt{2}$ 30. a. $\sqrt{75x} = \sqrt{25}\sqrt{3x}$

    b. $= 3\sqrt{2}$ b. $= 5\sqrt{3x}$

31. a. $\sqrt{2x^2} = \sqrt{2}\sqrt{x^2}$ 32. a. $\sqrt{4xy^2} = \sqrt{4y^2x}$

    b. $= \sqrt{2}\,x$ b. $= \sqrt{4y^2}\sqrt{x}$

    c. $= x\sqrt{2}$ c. $= 2y\sqrt{x}$

Simplify each expression.

33. $\sqrt{16}$ 34. $\sqrt{144}$ 35. $\sqrt{9x}$ 36. $\sqrt{49y}$

37. $\sqrt{20}$ 38. $\sqrt{45}$ 39. $\sqrt{8x}$ 40. $\sqrt{27y^3}$

41. $\sqrt{18x^2}$ 42. $\sqrt{50y^2}$ 43. $\sqrt{32x^2y^3}$ 44. $\sqrt{54xy^2}$

State the definition, axiom, or theorem that justifies each statement.

Example.   a. $\dfrac{3}{\sqrt{2}} = \dfrac{3\sqrt{2}}{\sqrt{2}\sqrt{2}}$ Fundamental principle of fractions

                 b. $= \dfrac{3\sqrt{2}}{2}$ Definition 5.1

45. a. $\dfrac{2}{\sqrt{3}} = \dfrac{2\sqrt{3}}{\sqrt{3}\sqrt{3}}$

    b. $= \dfrac{2\sqrt{3}}{3}$

46. a. $\dfrac{x}{\sqrt{y}} = \dfrac{x\sqrt{y}}{\sqrt{y}\sqrt{y}}$

    b. $= \dfrac{x\sqrt{y}}{y}$

47. a. $\dfrac{5}{2\sqrt{5}} = \dfrac{5\sqrt{5}}{2\sqrt{5}\sqrt{5}}$

    b. $= \dfrac{5\sqrt{5}}{2 \cdot 5}$

    c. $= \dfrac{\sqrt{5}}{2}$

48. a. $\dfrac{2x}{\sqrt{x}} = \dfrac{2x\sqrt{x}}{\sqrt{x}\sqrt{x}}$

    b. $= \dfrac{2x\sqrt{x}}{x}$

    c. $= 2\sqrt{x}$

Simplify each expression.

49. $\dfrac{1}{\sqrt{4}}$

50. $\dfrac{6}{\sqrt{9}}$

51. $\dfrac{1}{\sqrt{2}}$

52. $\dfrac{2}{\sqrt{3}}$

53. $\dfrac{6}{\sqrt{3}}$

54. $\dfrac{14}{\sqrt{7}}$

55. $\dfrac{8}{3\sqrt{2}}$

56. $\dfrac{15}{2\sqrt{5}}$

57. $\dfrac{6}{-2\sqrt{3}}$

58. $\dfrac{-15}{3\sqrt{3}}$

59. $\dfrac{xy}{\sqrt{y}}$

60. $\dfrac{6xy}{\sqrt{2x}}$

State the definition, axiom, or theorem that justifies each statement.

Example. a. $\sqrt{\dfrac{2}{3}} = \dfrac{\sqrt{2}}{\sqrt{3}}$ Theorem 11.2

    b. $= \dfrac{\sqrt{2}\sqrt{3}}{\sqrt{3}\sqrt{3}}$ Fundamental principle of fractions

    c. $= \dfrac{\sqrt{2}\sqrt{3}}{3}$ Definition 5.1

    d. $= \dfrac{\sqrt{6}}{3}$ Theorem 11.1

61. a. $\sqrt{\dfrac{3}{4}} = \dfrac{\sqrt{3}}{\sqrt{4}}$

    b. $= \dfrac{\sqrt{3}}{2}$

62. a. $5\sqrt{\dfrac{3x}{25}} = \dfrac{5\sqrt{3x}}{\sqrt{25}}$

    b. $= \dfrac{5\sqrt{3x}}{5}$

    c. $= \sqrt{3x}$

63. a. $\sqrt{\dfrac{2}{5}} = \dfrac{\sqrt{2}}{\sqrt{5}}$

   b. $\phantom{\sqrt{\dfrac{2}{5}}} = \dfrac{\sqrt{2}\sqrt{5}}{\sqrt{5}\sqrt{5}}$

   c. $\phantom{\sqrt{\dfrac{2}{5}}} = \dfrac{\sqrt{2}\sqrt{5}}{5}$

   d. $\phantom{\sqrt{\dfrac{2}{5}}} = \dfrac{\sqrt{10}}{5}$

64. a. $\sqrt{\dfrac{3x}{2y}} = \dfrac{\sqrt{3x}}{\sqrt{2y}}$

   b. $\phantom{\sqrt{\dfrac{3x}{2y}}} = \dfrac{\sqrt{3x}\sqrt{2y}}{\sqrt{2y}\sqrt{2y}}$

   c. $\phantom{\sqrt{\dfrac{3x}{2y}}} = \dfrac{\sqrt{3x}\sqrt{2y}}{2y}$

   d. $\phantom{\sqrt{\dfrac{3x}{2y}}} = \dfrac{\sqrt{6xy}}{2y}$

Simplify each expression.

65. $\sqrt{\dfrac{2}{9}}$

66. $\sqrt{\dfrac{3}{49}}$

67. $3\sqrt{\dfrac{1}{3}}$

68. $4\sqrt{\dfrac{3}{2}}$

69. $6\sqrt{\dfrac{2x}{3}}$

70. $2x\sqrt{\dfrac{3}{x}}$

71. $\dfrac{3y}{4}\sqrt{\dfrac{x}{y}}$

72. $\dfrac{xy}{3}\sqrt{\dfrac{2y}{x}}$

B

Simplify each expression.

Example. $\dfrac{\sqrt{18}\sqrt{12x}}{\sqrt{6x}} = \sqrt{\dfrac{18 \cdot 12x}{6x}} = \sqrt{36} = 6$

73. $\dfrac{\sqrt{2}\sqrt{3}}{\sqrt{6}}$

74. $\dfrac{\sqrt{16}\sqrt{3x}}{\sqrt{8x}}$

75. $\dfrac{\sqrt{81}\sqrt{5x}}{\sqrt{27x}}$

76. $\dfrac{2\sqrt{3}\sqrt{6y}}{\sqrt{2y}}$

77. $\dfrac{\sqrt{5y}\sqrt{xy}}{\sqrt{5x}}$

78. $\dfrac{\sqrt{3xy}\sqrt{4x}}{\sqrt{6y}}$

11.2 Sums and Differences

By the symmetric law of equality and the commutative law of multiplication, the distributive law

$$a(b + c) = ab + ac$$

can be written as

$$ba + ca = (b + c)a$$

where a, b, $c \in R$. Thus we can write sums and differences containing identical radical expressions as a single term. For example,

$$2\sqrt{5} + 7\sqrt{5} = (2 + 7)\sqrt{5} = 9\sqrt{5}$$

and

$$5\sqrt{y} - 3\sqrt{y} = 5\sqrt{y} + (-3\sqrt{y}) = [5 + (-3)]\sqrt{y} = 2\sqrt{y}.$$

Since $\sqrt{x}$ equals $1 \cdot \sqrt{x}$, the sum

$$3\sqrt{x} + \sqrt{x} = (3 + 1)\sqrt{x} = 4\sqrt{x}.$$

In some cases, sums or differences which do not contain identical radical expressions in each term can be simplified first and then written as a single term. For example,

$$
\begin{aligned}
7\sqrt{2} + 5\sqrt{18} &= 7\sqrt{2} + 5\sqrt{3^2 \cdot 2} \\
&= 7\sqrt{2} + 5\sqrt{3^2}\sqrt{2} \\
&= 7\sqrt{2} + 5 \cdot 3\sqrt{2} \\
&= 7\sqrt{2} + 15\sqrt{2} \\
&= 22\sqrt{2}.
\end{aligned}
$$

Exercise 11.2

In this exercise, assume that all variables represent positive real numbers.

A

Justify each statement by an axiom, definition, or theorem.

Example. a. $2\sqrt{3} + \sqrt{27} = 2\sqrt{3} + \sqrt{9}\sqrt{3}$ Theorem 11.1

b. $= 2\sqrt{3} + 3\sqrt{3}$ Definition 5.1

c. $= (2 + 3)\sqrt{3}$ Distributive law

d. $= 5\sqrt{3}$ Basic numeral for $2 + 3$

1. a. $5\sqrt{2} + 3\sqrt{2} = (5 + 3)\sqrt{2}$ 2. a. $3\sqrt{x} + 2\sqrt{x} = (3 + 2)\sqrt{x}$

    b. $= 8\sqrt{2}$ b. $= 5\sqrt{x}$

3. a. $3\sqrt{5} + \sqrt{20}$

    $= 3\sqrt{5} + \sqrt{4}\sqrt{5}$

   b. $= 3\sqrt{5} + 2\sqrt{5}$

   c. $= (3 + 2)\sqrt{5}$

   d. $= 5\sqrt{5}$

4. a. $\sqrt{18} + 3\sqrt{2}$

    $= \sqrt{9}\sqrt{2} + 3\sqrt{2}$

   b. $= 3\sqrt{2} + 3\sqrt{2}$

   c. $= (3 + 3)\sqrt{2}$

   d. $= 6\sqrt{2}$

5. a. $\sqrt{27} + \sqrt{48}$

    $= \sqrt{9}\sqrt{3} + \sqrt{16}\sqrt{3}$

   b. $= 3\sqrt{3} + 4\sqrt{3}$

   c. $= (3 + 4)\sqrt{3}$

   d. $= 7\sqrt{3}$

6. a. $2\sqrt{50} + 3\sqrt{8}$

    $= 2\sqrt{25}\sqrt{2} + 3\sqrt{4}\sqrt{2}$

   b. $= 2 \cdot 5\sqrt{2} + 3 \cdot 2\sqrt{2}$

   c. $= 10\sqrt{2} + 6\sqrt{2}$

   d. $= 16\sqrt{2}$

Write each of the following as an expression of one term.

7. $\sqrt{5} + 3\sqrt{5}$

8. $7\sqrt{2} - 3\sqrt{2}$

9. $3\sqrt{x} - 7\sqrt{x}$

10. $5\sqrt{y} + 2\sqrt{y}$

11. $3\sqrt{y} - 2\sqrt{y}$

12. $7\sqrt{x} + 2\sqrt{x}$

13. $\sqrt{8} + 5\sqrt{2}$

14. $\sqrt{75} - \sqrt{3}$

15. $\sqrt{49x} + 2\sqrt{25x}$

16. $3\sqrt{16y} - \sqrt{9y}$

17. $\sqrt{8xy} + \sqrt{18xy}$

18. $\sqrt{50xy} - 2\sqrt{32xy}$

Justify each statement by an axiom, definition, or theorem.

Example. a. $\dfrac{2 + \sqrt{8}}{4} = \dfrac{2 + \sqrt{4}\sqrt{2}}{4}$ Theorem 11.1

   b. $= \dfrac{2 + 2\sqrt{2}}{4}$ Definition 5.1

   c. $= \dfrac{2(1 + \sqrt{2})}{4}$ Distributive law

   d. $= \dfrac{1 + \sqrt{2}}{2}$ Fundamental principle of fractions

19. a. $\dfrac{\sqrt{32}}{12} = \dfrac{\sqrt{16}\sqrt{2}}{12}$

    b. $\quad = \dfrac{4\sqrt{2}}{12}$

    c. $\quad = \dfrac{\sqrt{2}}{3}$

20. a. $\dfrac{\sqrt{18x^2}}{6x} = \dfrac{\sqrt{9x^2}\sqrt{2}}{6x}$

    b. $\quad = \dfrac{3x\sqrt{2}}{6x}$

    c. $\quad = \dfrac{\sqrt{2}}{2}$

21. a. $\dfrac{6 + \sqrt{18}}{6} = \dfrac{6 + \sqrt{9}\sqrt{2}}{6}$

    b. $\quad = \dfrac{6 + 3\sqrt{2}}{6}$

    c. $\quad = \dfrac{3(2 + \sqrt{2})}{6}$

    d. $\quad = \dfrac{2 + \sqrt{2}}{2}$

22. a. $\dfrac{-3 + \sqrt{18}}{3} = \dfrac{-3 + \sqrt{9}\sqrt{2}}{3}$

    b. $\quad = \dfrac{-3 + 3\sqrt{2}}{3}$

    c. $\quad = \dfrac{3(-1 + \sqrt{2})}{3}$

    d. $\quad = -1 + \sqrt{2}$

Reduce each fraction to lowest terms.

23. $\dfrac{6 + 4\sqrt{3}}{2}$

24. $\dfrac{3 + 3\sqrt{5}}{3}$

25. $\dfrac{-2 + \sqrt{8}}{2}$

26. $\dfrac{6 - \sqrt{54}}{3}$

27. $\dfrac{12 - \sqrt{8}}{4}$

28. $\dfrac{21 + \sqrt{18}}{6}$

Write each sum or difference as a single fraction. $x, y \neq 0$.

Example. a. $\dfrac{2}{3} + \dfrac{5\sqrt{3}}{6} = \dfrac{(2)2}{(2)3} + \dfrac{5\sqrt{3}}{6}$ Fundamental principle of fractions

    b. $\quad = \dfrac{4}{6} + \dfrac{5\sqrt{3}}{6}$ Basic numeral

    c. $\quad = \dfrac{4 + 5\sqrt{3}}{6}$ Theorem 4.4

29. $\dfrac{3}{5} + \dfrac{\sqrt{2}}{5}$

30. $\dfrac{3}{2} - \dfrac{\sqrt{3}}{2}$

31. $\dfrac{2\sqrt{5}}{x} - \dfrac{3}{x}$

32. $\dfrac{\sqrt{2}}{y} + \dfrac{\sqrt{3}}{y}$

33. $\dfrac{1}{2} - \dfrac{\sqrt{3}}{8}$

34. $\dfrac{\sqrt{5}}{3} + \dfrac{1}{6}$

35. $\dfrac{3}{5} + \dfrac{\sqrt{3}}{2}$

36. $\dfrac{3}{4} - \dfrac{\sqrt{2}}{3}$

37. $\dfrac{3\sqrt{3}}{2} - \dfrac{\sqrt{2}}{3}$

38. $\dfrac{2\sqrt{5}}{3} - \dfrac{2}{5}$

39. $\dfrac{\sqrt{2}}{5} - 1$

40. $3 + \dfrac{\sqrt{3}}{4}$

41. $\dfrac{2\sqrt{3}}{3x} + \dfrac{\sqrt{7}}{x}$

42. $\dfrac{\sqrt{5}}{y} - \dfrac{2\sqrt{3}}{7y}$

43. $\dfrac{2\sqrt{3}}{y} - 1$

44. $4 + \dfrac{3\sqrt{5}}{2y}$

45. $\dfrac{\sqrt{3}}{2y} - 2$

46. $\dfrac{6}{x} + 3\sqrt{2}$

B

Simplify each expression.

Example.   a. $5\sqrt{8} - 7\sqrt{4} + 13\sqrt{2} = 5\sqrt{4}\sqrt{2} - 7(2) + 13\sqrt{2}$
$$= 5 \cdot 2\sqrt{2} - 14 + 13\sqrt{2}$$
$$= 10\sqrt{2} - 14 + 13\sqrt{2}$$
$$= 23\sqrt{2} - 14$$

47. $\sqrt{6} + 2\sqrt{24} - \sqrt{54}$

48. $2\sqrt{20} - \sqrt{80} + \sqrt{45}$

49. $\sqrt{27} + 2\sqrt{12} - 3\sqrt{48}$

50. $2\sqrt{32} - \sqrt{36} - \sqrt{49}$

51. $3\sqrt{144} - 4\sqrt{24} + \sqrt{49}$

52. $3\sqrt{3} + 2\sqrt{18} - \sqrt{12}$

53. $2\sqrt{27} - 3\sqrt{12} + 2\sqrt{8}$

54. $4\sqrt{24} - 6\sqrt{3} + 11\sqrt{6}$

55. $2\sqrt{12x} - 3\sqrt{18x} - 3\sqrt{2x}$

56. $\sqrt{64y^2} - 2\sqrt{24y^2} - \sqrt{54y^2}$

57. $\sqrt{4xy^3} - \sqrt{xy^3} + 2\sqrt{xy^3}$

58. $2\sqrt{xy^4} + 3\sqrt{4xy^4} - 4\sqrt{36xy^4}$

59. $\dfrac{3\sqrt{5} + \sqrt{15}}{\sqrt{5}}$

60. $\dfrac{4\sqrt{2} + \sqrt{8}}{\sqrt{2}}$

61. $\dfrac{7\sqrt{3} - 2\sqrt{27}}{\sqrt{3}}$

62. $\dfrac{\sqrt{7} - \sqrt{28}}{\sqrt{7}}$

63. $\dfrac{5\sqrt{x^3} + \sqrt{x}}{\sqrt{x}}$

64. $\dfrac{\sqrt{x} - 2\sqrt{x^3}}{2\sqrt{x}}$

11.3 Products and Quotients

A direct application of the distributive law permits us to write certain products which contain parentheses as equivalent expressions without parentheses. For example,

$$\sqrt{2}\,(\sqrt{3} + \sqrt{5}) = \sqrt{2} \cdot \sqrt{3} + \sqrt{2} \cdot \sqrt{5}$$
$$= \sqrt{6} + \sqrt{10},$$

and

$$\sqrt{3x}\,(\sqrt{3} + \sqrt{x}) = \sqrt{3x}\sqrt{3} + \sqrt{3x}\sqrt{x}$$
$$= \sqrt{3^2 x} + \sqrt{3x^2}$$
$$= \sqrt{3^2}\sqrt{x} + \sqrt{3}\sqrt{x^2}$$
$$= 3\sqrt{x} + \sqrt{3}\,x.$$

In Section 11.1 you rewrote quotients such as $1/\sqrt{2}$, in which the divisor contains a radical expression, by applying the fundamental principle of fractions and multiplying the numerator and denominator by the radical expression contained in the denominator. Thus,

$$\frac{1}{\sqrt{2}} = \frac{\sqrt{2}}{\sqrt{2}\sqrt{2}} = \frac{\sqrt{2}}{2}.$$

This process is called **rationalizing the denominator** of the fraction because the result is a fraction in which the denominator does not contain a *radical*.

We can also rationalize denominators of fractions with binomial denominators in which radicals occur in one or both of the two terms. First recall that

$$(b - c)(b + c) = b^2 - c^2,$$

where the expression $b^2 - c^2$ contains no linear term. Each of the two factors, $(b - c)$ and $(b + c)$, of such a product exhibiting this property is called the **conjugate** of the other. Now, if we multiply the numerator and denominator of a fraction of the form

$$\frac{a}{b + \sqrt{c}} \quad (c > 0, b + \sqrt{c} \neq 0)$$

by the conjugate of the denominator, the resulting denominator will be free of radicals. That is,

$$\frac{(b - \sqrt{c})}{(b + \sqrt{c})(b - \sqrt{c})} = \frac{ab - a\sqrt{c}}{b^2 - c} \quad (b - \sqrt{c} \neq 0).$$

Examples

a. $\dfrac{1}{3 + \sqrt{2}} = \dfrac{1(3 - \sqrt{2})}{(3 + \sqrt{2})(3 - \sqrt{2})}$

$\qquad = \dfrac{3 - \sqrt{2}}{9 - 2}$

$\qquad = \dfrac{3 - \sqrt{2}}{7}$

b. $\dfrac{\sqrt{y} - 1}{\sqrt{y} - 2} = \dfrac{(\sqrt{y} - 1)(\sqrt{y} + 2)}{(\sqrt{y} - 2)(\sqrt{y} + 2)}$

$\qquad = \dfrac{y + \sqrt{y} - 2}{y - 4}$

$\qquad (y \geq 0, y \neq 4)$

Again we rationalize denominators in this section primarily to develop the ability to recognize equivalent radical expressions. However, there are times when quotients in which the *divisor* (or denominator) contains a radical are preferred.

Exercise 11.3

In this exercise, assume that all variables represent positive real numbers.

A

State the axiom, definition, or theorem that justifies each statement.

Example. a. $\sqrt{2}\,(\sqrt{3} - \sqrt{2}) = \sqrt{2}\sqrt{3} - \sqrt{2}\sqrt{2}$ — Distributive law

b. $\qquad\qquad\qquad = \sqrt{6} - \sqrt{2}\sqrt{2}$ — Theorem 11.1

c. $\qquad\qquad\qquad = \sqrt{6} - 2$ — Definition 5.1

1. a. $\sqrt{3}\,(\sqrt{5} - 1) = \sqrt{3}\sqrt{5} - \sqrt{3}$

 b. $\qquad\qquad\quad = \sqrt{15} - \sqrt{3}$

2. a. $\sqrt{x}\,(y + \sqrt{y}) = \sqrt{xy} + \sqrt{x}\sqrt{y}$

 b. $\qquad\qquad\quad = \sqrt{xy} + \sqrt{xy}$

 c. $\qquad\qquad\quad = y\sqrt{x} + \sqrt{xy}$

Write each expression as an equal expression without parentheses in simplest form.

3. $2(\sqrt{3} + 1)$

4. $3(2 + \sqrt{2})$

5. $4(\sqrt{2} - 3)$

6. $2(3 - \sqrt{5})$

7. $\sqrt{3}(2 + \sqrt{5})$

8. $\sqrt{2}(4 - \sqrt{3})$

9. $\sqrt{2}(3 - \sqrt{2})$ 10. $\sqrt{5}(\sqrt{5} + 1)$ 11. $\sqrt{2}(\sqrt{6} - 4)$

12. $\sqrt{3}(\sqrt{6} - 3)$ 13. $\sqrt{5}(\sqrt{20} - \sqrt{5})$ 14. $\sqrt{6}(\sqrt{6} + \sqrt{3})$

State the axiom, definition, or theorem that justifies each statement.

15. a. $(\sqrt{2} + 1)(\sqrt{2} + 3) = \sqrt{2}\sqrt{2} + 3\sqrt{2} + 1\sqrt{2} + 1 \cdot 3$

 b. $\qquad\qquad\qquad = \sqrt{2}\sqrt{2} + 4\sqrt{2} + 1 \cdot 3$

 c. $\qquad\qquad\qquad = \sqrt{2}\sqrt{2} + 4\sqrt{2} + 3$

 d. $\qquad\qquad\qquad = 2 + 4\sqrt{2} + 3$

 e. $\qquad\qquad\qquad = 5 + 4\sqrt{2}$

16. a. $(1 + \sqrt{2})(3 - \sqrt{2}) = 1 \cdot 3 + 2\sqrt{2} - \sqrt{2}\sqrt{2}$

 b. $\qquad\qquad\qquad = 3 + 2\sqrt{2} - \sqrt{2}\sqrt{2}$

 c. $\qquad\qquad\qquad = 3 + 2\sqrt{2} - 2$

 d. $\qquad\qquad\qquad = 3 - 2 + 2\sqrt{2}$

 e. $\qquad\qquad\qquad = 1 + 2\sqrt{2}$

Write each product without parentheses and in simplest form.

17. $(2 + \sqrt{5})(3 - \sqrt{5})$ 18. $(1 + \sqrt{2})(2 - \sqrt{3})$

19. $(\sqrt{x} - 2)(\sqrt{x} + 4)$ 20. $(3 + \sqrt{y})(4 - \sqrt{y})$

21. $(\sqrt{y} + 4)(\sqrt{y} - 4)$ 22. $(3 + \sqrt{x})(3 - \sqrt{x})$

23. $(\sqrt{y} + 4)^2$ 24. $(3 + \sqrt{x})^2$

25. $(\sqrt{2} - \sqrt{3})(\sqrt{2} + 4\sqrt{3})$ 26. $(\sqrt{2} - \sqrt{6})(3\sqrt{2} + \sqrt{6})$

27. $(\sqrt{3} - \sqrt{2})(\sqrt{3} + \sqrt{2})$ 28. $(\sqrt{5} - \sqrt{7})(\sqrt{5} + \sqrt{7})$

29. $(\sqrt{x} - \sqrt{5})^2$ 30. $(2\sqrt{7} - \sqrt{y})^2$

Write each quotient as an equal expression in which the denominator has been rationalized. Assume no denominator equals zero.

Example. $\dfrac{1}{\sqrt{3} - 2} = \dfrac{1(\sqrt{3} + 2)}{(\sqrt{3} - 2)\,(\sqrt{3} + 2)} = \dfrac{\sqrt{3} + 2}{3 + 2\sqrt{3} - 2\sqrt{3} - 4}$

$\qquad\qquad = \dfrac{\sqrt{3} + 2}{3 - 4} = \dfrac{\sqrt{3} + 2}{-1} = -\sqrt{3} - 2$

31. $\dfrac{1}{\sqrt{2} - 1}$

32. $\dfrac{1}{3 - \sqrt{3}}$

33. $\dfrac{3}{1 - \sqrt{x}}$

34. $\dfrac{x}{2 - \sqrt{x}}$

35. $\dfrac{4 - \sqrt{3}}{\sqrt{3} + 1}$

36. $\dfrac{\sqrt{5} - 3}{2 - \sqrt{5}}$

37. $\dfrac{\sqrt{3}}{\sqrt{7} - \sqrt{3}}$

38. $\dfrac{2\sqrt{2}}{\sqrt{3} + \sqrt{5}}$

Write each quotient as an equal expression in which the numerator has been rationalized.

Examples.   a. $\dfrac{\sqrt{2}}{5} = \dfrac{\sqrt{2} \cdot \sqrt{2}}{5 \cdot \sqrt{2}} = \dfrac{2}{5\sqrt{2}}$

b. $\dfrac{3 - \sqrt{2}}{4} = \dfrac{(3 - \sqrt{2})(3 + \sqrt{2})}{4(3 + \sqrt{2})}$

$= \dfrac{9 + 3\sqrt{2} - 3\sqrt{2} - 2}{12 + 4\sqrt{2}} = \dfrac{7}{12 + 4\sqrt{2}}$

39. $\dfrac{\sqrt{3}}{3}$

40. $\dfrac{\sqrt{7}}{14}$

41. $\dfrac{1 + \sqrt{2}}{3}$

42. $\dfrac{\sqrt{5} - 2}{4}$

43. $\dfrac{\sqrt{3} - \sqrt{2}}{\sqrt{2}}$

44. $\dfrac{\sqrt{6} - \sqrt{2}}{\sqrt{6}}$

11.4   Distance Between Two Points

The Pythagorean theorem (page 330) can be used to find the distance between two points in the coordinate plane. For example, consider the points (2, 3) and (5, 7) shown in Figure 11.2. We can form a right triangle

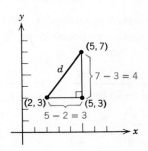

Figure 11.2

by drawing a line through (2, 3) parallel to the x-axis, and a line through (5, 7) parallel to the y-axis. These lines intersect in the point (5, 3) forming a right triangle. Now observe that the length of the line segment parallel to the x-axis is $5 - 2 = 3$, and that the length of the line segment parallel to the y-axis is $7 - 3 = 4$. The Pythagorean theorem can also be stated in the form

$$d = \sqrt{a^2 + b^2},$$

which leads to

$$d = \sqrt{3^2 + 4^2} = \sqrt{9 + 16}$$
$$= \sqrt{25} = 5$$

The preceding example suggests a method for finding the distance between any two points $P_1(x_1, y_1)$ and $P_2(x_2, y_2)$ in the coordinate plane as shown in Figure 11.3. Lines drawn through P_1 and P_2 parallel to the x-axis and y-axis, respectively, intersect in point $R(x_2, y_1)$ forming a right

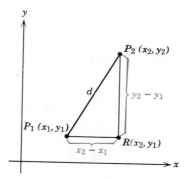

Figure 11.3

triangle. The length of the line segment parallel to the x-axis is equal to $x_2 - x_1$ (the difference of the x coordinates of P_1 and P_2), and the length of the line segment parallel to the y-axis is $y_2 - y_1$ (the difference of the y coordinates). Then by the Pythagorean theorem

$$d = \sqrt{(x_2 - x_1)^2 + (y_2 - y_1)^2}.$$

This equation is called the **distance formula.**

Example. Find the distance between the points $(1, -5)$ and $(-2, 3)$.

Solution. If we consider $(1, -5)$ to be P_1 and $(-2, 3)$ to be P_2, then

Solution continued on page 346.

$x_1 = 1$, $x_2 = -2$, $y_1 = -5$, and $y_2 = 3$. Substituting these values in the distance formula gives

$$d = \sqrt{(-2 - 1)^2 + [3 - (-5)]^2}$$
$$= \sqrt{(-3)^2 + 8^2} = \sqrt{9 + 64} = \sqrt{73}.$$

We can leave the result in this form, or find an approximation for this irrational number in the table inside the back cover. This value we find to be 8.544.

If in the preceding example, we designate $(-2, 3)$ as P_1 and $(1, -5)$ as P_2 then $x_1 = -2$, $x_2 = 1$, $y_1 = 3$, and $y_2 = -5$. Substituting these values in the distance formula, gives

$$d = \sqrt{[1 - (-2)]^2 + (-5 - 3)^2}$$
$$= \sqrt{3^2 + (-8)^2} = \sqrt{9 + 64} = \sqrt{73},$$

the same value we obtained above. This suggests that the order in which two points are considered in the distance formula has no effect on the result.

If two points $P_1(x_1, y_1)$ and $P_2(x_2, y_2)$ have the same y coordinates, the distance between the two points is the difference of the x coordinates. Substituting in the distance formula we obtain

$$d = \sqrt{(x_2 - x_1)^2 + (y - y)^2}$$
$$= \sqrt{(x_2 - x_1)^2 + 0} = \sqrt{(x_2 - x_1)^2}.$$

In a similar manner, if two points have the same x coordinate, the distance between them is given by

$$d = \sqrt{(y_2 - y_1)^2}.$$

Exercise 11.4

A

Find the distance between the given pairs of points.

Examples.   a. $(2, -4)$ and $(7, -4)$ b. $(-9, 2)$ and $(3, 5)$

Solutions. a. Because the points have the same y coordinate,

$$d = \sqrt{(x_2 - x_1)^2} = \sqrt{(2 - 7)^2} = \sqrt{25} = 5$$

b. $d = \sqrt{(x_2 - x_1)^2 + (y_2 - y_1)^2} = \sqrt{(-9 - 3)^2 + (2 - 5)^2}$

$$= \sqrt{(-12)^2 + (-3)^2} = \sqrt{144 + 9} = \sqrt{153}$$

1. $(4, 1)$ and $(4, -7)$
2. $(-5, 1)$ and $(-5, 2)$
3. $(7, 5)$ and $(1, 5)$
4. $(-6, 3)$ and $(-1, 3)$
5. $(0, 8)$ and $(2, 7)$
6. $(4, 0)$ and $(7, 3)$
7. $(6, 3)$ and $(8, 1)$
8. $(3, 8)$ and $(5, 7)$
9. $(-4, 8)$ and $(7, 2)$
10. $(2, -3)$ and $(1, 4)$
11. $(2, -1)$ and $(-1, 5)$
12. $(-4, 1)$ and $(4, -5)$
13. $(-7, -6)$ and $(6, -9)$
14. $(2, -7)$ and $(-5, -4)$
15. $(0, 0)$ and $(-4, 5)$
16. $(-8, 9)$ and $(0, 0)$

Find the perimeter of each triangle having the given points as vertices.

17. $(0, 0)$, $(3, 0)$, and $(3, 4)$
18. $(0, 0)$, $(5, 0)$, and $(5, 12)$
19. $(-2, 3)$, $(1, 3)$, and $(1, -1)$
20. $(-3, -1)$, $(5, -1)$, and $(5, 14)$

Find the perimeter of each rectangle having the given points as vertices.

21. $(0, 0)$, $(5, 0)$, $(5, 2)$, and $(0, 2)$
22. $(0, 0)$, $(0, -3)$, $(4, -3)$, and $(4, 0)$
23. $(-1, 4)$, $(-3, 4)$, $(-3, -3)$, and $(-1, -3)$
24. $(-3, -1)$, $(-3, -5)$, $(4, -5)$, and $(4, -1)$
25.–28. *Find the length of the diagonal in each rectangle in Exercises 21–24.*

CHAPTER SUMMARY

11.1 The **nonnegative square root** of $a, a \in R, \ a \geq 0$, designated by $\sqrt{a}$, is the number such that

$$\sqrt{a}\sqrt{a} = a.$$

The negative square root is given by $-\sqrt{a}$.

Since for $a \in R$, $a \geq 0$, $\sqrt{a}$ is a real number, the axioms for the real numbers and the theorems previously proved are also applicable to algebraic expressions containing radicals.

The definition of $\sqrt{a}$ leads to the following consequences. *If $a, b \in R$, then*

$$\sqrt{a}\sqrt{b} = \sqrt{ab} \quad and \quad \sqrt{ab} = \sqrt{a}\sqrt{b} \quad (a, b \geq 0);$$

$$\frac{\sqrt{a}}{\sqrt{b}} = \sqrt{\frac{a}{b}} \quad and \quad \sqrt{\frac{a}{b}} = \frac{\sqrt{a}}{\sqrt{b}} \quad (a \geq 0, b > 0).$$

Certain forms of radicals are arbitrarily designated as the "simplest" form if they meet the following conditions:

a. The radicand contains no polynomial factor raised to a power equal to or greater than 2.
b. No radical expressions are contained in denominators of fractions.
c. The radicand does not contain a fraction.

11.2   The distributive law in the form

$$a\sqrt{c} + b\sqrt{c} = (a + b)\sqrt{c}$$

justifies rewriting sums of differences, in which each term contains the same radical expression, as a single term.

11.3   The process of rewriting a fraction which contains a radical expression in the denominator as an equivalent fraction that does not contain a radical expression in the denominator is called "rationalizing the denominator." Thus,

$$\frac{a}{\sqrt{c}} = \frac{a\sqrt{c}}{\sqrt{c}\sqrt{c}} = \frac{a\sqrt{c}}{c} \quad (c > 0);$$

$$\frac{a}{b + \sqrt{c}} = \frac{a(b - \sqrt{c})}{(b + \sqrt{c})(b - \sqrt{c})} = \frac{a(b - \sqrt{c})}{b^2 - c} \quad (b + \sqrt{c}, b - \sqrt{c} \neq 0).$$

11.4   The distance between two points (x_1, y_1) and (x_2, y_2) is given by

$$d = \sqrt{(x_2 - x_1)^2 + (y_2 - y_1)^2}.$$

CHAPTER REVIEW

Assume that all variables are positive real numbers.

A

11.1 *Simplify each expression.*

1. $\sqrt{49}$ 2. $-\sqrt{144}$ 3. $-\sqrt{\dfrac{9}{64}x^2}$ 4. $\sqrt{\dfrac{y^4}{x^2}}$

5. $\sqrt{64x^2y}$ 6. $\sqrt{48y^2}$ 7. $-\sqrt{8x^2}$ 8. $-\sqrt{28xy^2}$

9. $\dfrac{1}{\sqrt{16}}$ 10. $\dfrac{1}{\sqrt{3}}$ 11. $\dfrac{x}{\sqrt{x}}$ 12. $\dfrac{6y}{\sqrt{2y}}$

13. $\sqrt{\dfrac{3}{25}}$ 14. $\sqrt{\dfrac{2}{3}}$ 15. $3xy\sqrt{\dfrac{1}{x}}$ 16. $\dfrac{2x}{\sqrt{3x}}$

11.2 *Write each expression as an equal expression of one term.*

17. $8\sqrt{y} - 2\sqrt{y}$ 18. $3\sqrt{2} - 4\sqrt{18}$

19. $5\sqrt{3x^2} - \sqrt{12x^2}$ 20. $3x\sqrt{20} - \sqrt{45x^2}$

21. $3\sqrt{x^4y} + \sqrt{4x^4y}$ 22. $\sqrt{2xy^3} - y\sqrt{72xy}$

Reduce each fraction to lowest terms.

23. $\dfrac{4 - \sqrt{12}}{2}$ 24. $\dfrac{15 + 2\sqrt{45}}{6}$

Write each sum or difference as a single fraction.

25. $\dfrac{1}{3} - \dfrac{\sqrt{2}}{3}$ 26. $\dfrac{\sqrt{3}}{2} - \dfrac{1}{4}$ 27. $\dfrac{\sqrt{5}}{3} - \dfrac{3}{4}$ 28. $\dfrac{\sqrt{2}}{5} - 1$

11.3 *Write each of the following as an equal expression without parentheses and in simplest form.*

29. $\sqrt{3}(1 - \sqrt{3})$ 30. $\sqrt{2}(\sqrt{6} - \sqrt{2})$

31. $(3 - \sqrt{2})(3 + \sqrt{2})$ 32. $(2 - \sqrt{7})^2$

33. $(\sqrt{5} - \sqrt{7})(\sqrt{5} + \sqrt{7})$ 34. $(\sqrt{5} - \sqrt{2})^2$

Write each of the following as an equal expression in which the denominator has been rationalized.

35. $\dfrac{1}{\sqrt{5}}$

36. $\dfrac{3}{\sqrt{3}}$

37. $\dfrac{2}{1 - \sqrt{2}}$

38. $\dfrac{5}{\sqrt{5} - \sqrt{2}}$

39. $\dfrac{\sqrt{5} + \sqrt{3}}{2 - \sqrt{3}}$

40. $\dfrac{2 + \sqrt{7}}{\sqrt{5} - \sqrt{3}}$

11.1–11.3 *State the definition, axiom, or theorem that justifies each statement.*

41. $\sqrt{4x^2 y} = \sqrt{4x^2}\sqrt{y}$

42. $\sqrt{4x^2} = 2x$

43. $\dfrac{2}{\sqrt{5}} = \dfrac{2\sqrt{5}}{\sqrt{5}\sqrt{5}}$

44. $\sqrt{3}\sqrt{3} = 3$

45. $\sqrt{\dfrac{3}{5}} = \dfrac{\sqrt{3}}{\sqrt{5}}$

46. $\dfrac{\sqrt{3}}{\sqrt{5}} = \dfrac{\sqrt{3}\sqrt{5}}{\sqrt{5}\sqrt{5}}$

47. $\sqrt{5}\sqrt{5} = 5$

48. $\sqrt{3}\sqrt{5} = \sqrt{15}$

49. $5\sqrt{x} + 3\sqrt{x} = (5 + 3)\sqrt{x}$

50. $7\sqrt{y} - \sqrt{y} = 7\sqrt{y} + (-\sqrt{y})$

51. $\dfrac{3 - 3\sqrt{2}}{6} = \dfrac{3(1 - \sqrt{2})}{6}$

52. $\dfrac{1}{\sqrt{3} - \sqrt{2}} = \dfrac{1(\sqrt{3} + \sqrt{2})}{(\sqrt{3} - \sqrt{2})(\sqrt{3} + \sqrt{2})}$

11.4 *Find the distance between the given pairs of points.*

53. $(5, 0)$, $(7, 8)$

54. $(4, -2)$, $(2, 1)$

55. $(-3, 4)$, $(3, -5)$

56. $(1, -6)$, $(-3, -1)$

B

Simplify each expression.

57. $\dfrac{\sqrt{5xy}\sqrt{18y}}{\sqrt{10x}}$

58. $\dfrac{\sqrt{2x^3}\sqrt{8x}}{\sqrt{3x}}$

59. $\dfrac{\sqrt{42x^2 y}\sqrt{3xy^2}}{\sqrt{27x^3}\sqrt{21y^3}}$

60. $\dfrac{\sqrt{40x^3 y}\sqrt{5xy^2}}{\sqrt{20x^4}\sqrt{14y}}$

12
chapter

Second-Degree Equations

In Chapter 8 you learned to solve first-degree equations in one variable and use such equations as mathematical models for certain word problems. Many problems, particularly those involving products and quotients or such physical quantities as areas, volumes, acceleration, etc., may involve equations of the second degree. We shall now consider such equations which are called **quadratic equations.**

Recall that a real number replacement for the variable in a polynomial which results in the polynomial being equal to 0 is called a zero of the polynomial (see page 149). In this chapter we are concerned with finding the zeros of second-degree polynomials, or equivalently, the solution sets of second-degree equations of the form

$$ax^2 + bx + c = 0, \qquad a, b, c \in R \quad (a \neq 0). \tag{1}$$

A quadratic equation in one variable which is written in *descending* powers of the variable and in which the right member is 0 as in Equation (1) above, is said to be in **standard form.**

12.1 Solution by Factoring

Consider the quadratic equation

$$x^2 - x - 6 = 0.$$

Factoring the left-hand member yields

$$(x + 2)(x - 3) = 0.$$

351

If x is replaced by -2, the factor $x + 2$ equals 0 and the equation,

$$(-2 + 2)(-2 - 3) = 0,$$

is a true statement by Theorem 2.3. Also, if x is replaced by 3, the factor $x - 3$ equals 0 and the equation,

$$(3 + 2)(3 - 3) = 0,$$

is a true statement. The above example suggests the following theorem.

Theorem 12.1.   If $c \in R$ such that c is a zero of $P(x)$ [or $Q(x)$], then c is a solution of $P(x) \cdot Q(x) = 0$.

Theorem 12.1 implies that

$$\{x \mid P(x) \cdot Q(x) = 0\} = \{x \mid P(x) = 0\} \cup \{x \mid Q(x) = 0\}.$$

The method of solving an equation by using Theorem 12.1 is referred to as **solution by factoring.**

Example.   Solve $3x^2 - 7x + 4 = 0$.

Solution.   Write the left-hand member of the equation in factored form.

$$(3x - 4)(x - 1) = 0$$

By Theorem 12.1, $(3x - 4)(x - 1) = 0$ for values of x for which

$$3x - 4 = 0 \qquad \text{or} \qquad x - 1 = 0.$$

If

$$3x - 4 = 0, \qquad \text{then} \quad x = \frac{4}{3},$$

and if

$$x - 1 = 0, \qquad \text{then} \quad x = 1.$$

Hence, the solution set is $\{4/3, 1\}$.

Theorem 12.1 justifies the statement that a replacement for x in an equation of the form $P(x) \cdot Q(x) = 0$ for which either $P(x)$ or $Q(x)$ equals zero is a solution of the equation. In general, we find the replacements for the variable for which each linear factor equals zero by inspection or by solving the linear equations $P(x) = 0$ and $Q(x) = 0$. We shall simply refer to Theorem 12.1 as justification for this process.

We can also reverse the above process to write an equation when the solution set is known.

Example. Write an equation with integral coefficients whose solution set is $\{3, -2/3\}$.

Solution. Since 3 and $-2/3$ are the replacements for a variable, say, x, we may write $x = 3$ and $x = -2/3$. Application of the addition law to each equation gives

$$x - 3 = 0 \quad \text{and} \quad x + \frac{2}{3} = 0.$$

Substituting $x - 3$ and $x + 2/3$ in the left-hand member of the equality $0 \cdot 0 = 0$ gives

$$(x - 3)\left(x + \frac{2}{3}\right) = 0,$$

from which

$$x^2 - \frac{7}{3}x - 2 = 0,$$

$$3x^2 - 7x - 6 = 0.$$

Exercise 12.1

A

Find the solution set of each equation.

Examples. a. $(x - 2)(x - 3) = 0$ b. $2x(x + 1) = 0$

Solutions. The values of the variables for which each factor equals zero can be determined by inspection. Thus, by Theorem 12.1, the solution sets are:

       a. $\{2, 3\}$ b. $\{0, -1\}$

1. $3(x - 1) = 0$ 2. $2(x - 4) = 0$

3. $4(x + 2) = 0$ 4. $5(x + 4) = 0$

5. $(x - 3)(x + 2) = 0$ 6. $(x + 3)(x + 1) = 0$

7. $x(x - 5) = 0$ 8. $5y(y - 1) = 0$

Example. $(3x - 2)(2x + 3) = 0$

Solution. If $3x - 2 = 0$, then $3x = 2$ and $x = 2/3$. If $2x + 3 = 0$, then $2x = -3$ and $x = -3/2$. Hence, by Theorem 12.1, the solution set is

$$\left\{\frac{2}{3}, -\frac{3}{2}\right\}.$$

9. $(3x - 1)(x - 1) = 0$ 10. $(2x + 1)(x - 3) = 0$

11. $(2x + 5)(3x - 7) = 0$ 12. $(2x - 3)(3x + 2) = 0$

13. $3(x - 3)(5x + 4) = 0$ 14. $5(2x + 7)(5x - 2) = 0$

Examples.   a. $x^2 + 2x = 0$ b. $x^2 + x = 30$

Solutions

a. Factor the left-hand member.   b. Rewrite in standard form and factor

$$x(x + 2) = 0$$

   the left-hand member.

By Theorem 12.1, the solution

$$x^2 + x - 30 = 0$$

set is $\{0, -2\}$.

$$(x + 6)(x - 5) = 0$$

The solution set is $\{-6, 5\}$.

15. $x^2 - 5x = 0$ 16. $x^2 - 6x = 0$ 17. $2x^2 + 2x = 0$

18. $x^2 - 4 = 0$ 19. $2x^2 - 18 = 0$ 20. $3x^2 - 3 = 0$

21. $x^2 - 3x - 4 = 0$ 22. $x^2 + 3x + 2 = 0$ 23. $x^2 + 5x = 14$

24. $x^2 - x = 42$ 25. $2x^2 = 5x$ 26. $x^2 = 2x - 1$

Example. $x(2x - 3) = -1$

Solution.   Apply the distributive law and write the equation in standard form.

$$2x^2 - 3x + 1 = 0$$

Factor the left-hand member.

$$(2x - 1)(x - 1) = 0$$

By Theorem 12.1, the solution set is $\left\{\frac{1}{2}, 1\right\}$.

27. $3x(x + 1) = 2x + 2$ 28. $2x(x - 2) = x + 3$

29. $(x - 2)(x + 1) = 4$ 30. $x(3x + 2) = (x + 2)^2$

31. $(6x + 1)(x + 1) = 4$ 32. $(x - 2)(x - 1) = 1 - x$

Example. $\dfrac{x}{4} - \dfrac{3}{4} = \dfrac{1}{x}$

Solution.   Multiply each member by the least common denominator, $4x$.

$$x^2 - 3x = 4$$

Write in standard form and factor the left-hand member.

$$x^2 - 3x - 4 = 0$$

$$(x - 4)(x + 1) = 0$$

By Theorem 12.1, the solution set is $\{4, -1\}$.

33. $x + \dfrac{1}{x} = 2$

34. $1 - \dfrac{2}{x} = \dfrac{15}{x^2}$

35. $\dfrac{1}{2} + \dfrac{1}{2x} = \dfrac{1}{x^2}$

36. $1 - \dfrac{3}{2x} - \dfrac{1}{x^2} = 0$

37. $\dfrac{1}{x} - \dfrac{1}{6} = \dfrac{1}{x + 1}$

38. $x = \dfrac{10}{x - 3}$

Write an equation with integral coefficients in standard form with given solutions.

Example. $\{3, -2\}$

Solution. Set the variable x equal to each solution.

$$x = 3, \qquad\qquad x = -2$$

Then

$$x - 3 = 0 \quad \text{or} \quad x + 2 = 0,$$

and substituting for the zeros in the left-hand member of $0 \cdot 0 = 0$ yields

$$(x - 3)(x + 2) = 0.$$

$$x^2 - x - 6 = 0$$

39. $\{4, 7\}$　　40. $\{5, -2\}$　　41. $\{-1, -6\}$　　42. $\{-3, 8\}$

43. $\left\{\dfrac{1}{2}, 2\right\}$　　44. $\left\{\dfrac{1}{3}, -3\right\}$　　45. $\left\{\dfrac{2}{3}, \dfrac{1}{2}\right\}$　　46. $\left\{-\dfrac{2}{3}, -\dfrac{3}{4}\right\}$

12.2 Solution by Extraction of Roots

A quadratic equation of the form

$$x^2 = c \tag{1}$$

where $c \in R$, $c \geq 0$, can be solved by a method which is called the **extraction of roots.**

First, notice that $\sqrt{c}$ is a solution of $x^2 = c$ because $(\sqrt{c})^2 = c$, and $-\sqrt{c}$ is also a solution of $x^2 = c$ because $(-\sqrt{c})^2 = c$. It can be shown, although we shall not do so, that there are no other solutions. Thus, we have the following.

Theorem 12.2.   If $c \in R$ and $c \geq 0$, then the solution set of $x^2 = c$ is

$$\{x \mid x = \sqrt{c}\} \cup \{x \mid x = -\sqrt{c}\} = \{\sqrt{c}, -\sqrt{c}\}.$$

Examples.   a. The solution set of $x^2 = 9$ is $\{3, -3\}$.

b. The solution set of $x^2 = 4/9$ is $\{2/3, -2/3\}$.

c. The solution set of $x^2 = 10$ is $\{\sqrt{10}, -\sqrt{10}\}$.

Equations of the form

$$(x - a)^2 = c \qquad (c \geq 0)$$

can also be solved by the method of extraction of roots. For example, by Theorem 12.2, the solution set of

$$(x + 3)^2 = 25,$$

is

$$\{x \mid x + 3 = 5\} \cup \{x \mid x + 3 = -5\} = \{2, -8\}.$$

Exercise 12.2

A

Solve each equation by the method of extraction of roots.

Examples.   a. $x^2 = 16$ b. $x^2 = 35$

Solutions.   By Theorem 12.2:

a. The solution set is $\{4, -4\}$.   b. The solution set is $\{\sqrt{35}, -\sqrt{35}\}$.

1. $x^2 = 9$ 2. $x^2 = 81$ 3. $x^2 = 43$

4. $x^2 = 7$ 5. $x^2 = 50$ 6. $x^2 = 75$

Example. $5x^2 - 125 = 0$

Solution. Write an equivalent equation with x^2 as the left-hand member.

$$x^2 = 25$$

By Theorem 12.2, the solution set is $\{5, -5\}$.

7. $3x^2 = 27$ 8. $5x^2 - 80 = 0$ 9. $4x^2 - 24 = 0$

10. $\dfrac{x^2}{3} = 27$ 11. $\dfrac{2x^2}{7} = 8$ 12. $9x^2 - 4 = 0$

Example. $(x + 3)^2 = 5$

Solution. By Theorem 12.2, the solution set is

$$\{x \mid x + 3 = \sqrt{5}\} \cup \{x \mid x + 3 = -\sqrt{5}\} = \{-3 + \sqrt{5}, -3 - \sqrt{5}\}.$$

13. $(x - 1)^2 = 4$ 14. $(x + 3)^2 = 9$ 15. $(x - 2)^2 = 25$

16. $(x + 1)^2 = 36$ 17. $(x - 5)^2 = 1$ 18. $(x + 7)^2 = 4$

19. $(x + 3)^2 = 2$ 20. $(x - 2)^2 = 3$ 21. $(x - 6)^2 = 7$

22. $(x + 1)^2 = 6$ 23. $(x - 2)^2 = c$ 24. $(x - a)^2 = 4$

25. What is the solution set of $x^2 = -16$, for $x \in R$?
26. Do all quadratic equations have real number solutions?

B

Solve by extraction of roots. Write irrational roots in simplest form.

27. $(3x + 4)^2 = 25$ 28. $(2x - 7)^2 = 9$ 29. $\left(\dfrac{x}{3} + \sqrt{2}\right)^2 = 5$

30. $\left(\dfrac{2x}{3} - \sqrt{7}\right)^2 = 12$ 31. $\left(\dfrac{x}{\sqrt{3}} - 4\right)^2 = 6$ 32. $(x\sqrt{2} - \sqrt{5})^2 = 10$

33. $x^2 + 4x + 4 = 9$ 34. $x^2 + 6x + 9 = 16$

2.3 Complex Numbers*

The equation $x^2 = -16$ has no solution in the set of real numbers, because there is no real number which multiplied by itself equals a

* This section can be omitted without losing continuity in the following sections. Exercises in those sections which relate to this discussion are appropriately indicated.

negative number. See Problem 25, Exercise 12.2. However, there are situations in which solutions of such equations are required. Hence, we define a number that has this property.

Definition 12.1.   For all $b \in R$, $b > 0$, $\sqrt{-b}$ is the number such that

$$\sqrt{-b} \cdot \sqrt{-b} = -b.$$

For the special case $b = 1$, we have $\sqrt{-1} \cdot \sqrt{-1} = -1$.

It is also convenient to use other symbols for $\sqrt{-b}$ where $b > 0$. Two such symbols are given by the following:

$$\sqrt{-b} = \sqrt{-1} \cdot \sqrt{b} = \sqrt{b} \cdot \sqrt{-1}.$$

The symbol i is often used to represent $\sqrt{-1}$. Thus,

$$\sqrt{-b} = i\sqrt{b} = \sqrt{b}\, i.$$

Examples.   a. $\sqrt{-9} = \sqrt{9} \cdot \sqrt{-1} = 3\sqrt{-1} = 3i$.

b. $\sqrt{-2} = \sqrt{2} \cdot \sqrt{-1} = \sqrt{2}\, i$.

Numbers such as $3i$, $\sqrt{2}\, i$, and $2\sqrt{2}\, i$, which are square roots of negative numbers, are called **pure imaginary numbers.** Observe from Definition 12.1 and the meaning we have given to i, that

$$i \cdot i = i^2 = -1.$$

Also,

$$i^3 = i^2 \cdot i = -1 \cdot i = -i,$$

$$i^4 = i^2 \cdot i^2 = -1 \cdot (-1) = 1, \text{ etc.}$$

Definition 12.1 and the theorems of Chapter 11 can be used to simplify expressions involving square roots of negative numbers if such numbers are first written in the i form.

Examples

a. $\dfrac{2 + \sqrt{-4}}{2} = \dfrac{2 + \sqrt{4} \cdot \sqrt{-1}}{2} = \dfrac{2 + 2i}{2} = \dfrac{2(1 + i)}{2} = 1 + i.$

b. $\dfrac{\sqrt{3} + \sqrt{-6}}{\sqrt{3}} = \dfrac{\sqrt{3} + \sqrt{6} \cdot \sqrt{-1}}{\sqrt{3}}$

$$= \dfrac{\sqrt{3} + \sqrt{3}\sqrt{2}\, i}{\sqrt{3}} = \dfrac{\sqrt{3}(1 + \sqrt{2}\, i)}{\sqrt{3}} = 1 + \sqrt{2}\, i.$$

Now let us consider the solution of $x^2 = -16$. By Definition 12.1
$$x = \sqrt{-16} = \sqrt{16}\, i = 4i$$
or
$$x = -\sqrt{-16} = -\sqrt{16}\, i = -4i.$$
Hence, the solution set of the equation is $\{4i, -4i\}$.

This example suggests that the solution set of an equation of the form $x^2 = c$, where $c < 0$, is $\{\sqrt{c}\, i, -\sqrt{c}\, i\}$, a result similar to Theorem 12.2 applicable to real numbers. We shall simply cite this theorem for imaginary numbers also.

Theorem 12.2 can also be used to find the solution set of an equation of the form $(x - a)^2 = -b^2$.

Example. Find the solution set of the equation $(x - 3)^2 = -4$.

Solution. From Theorem 12.2,

$$x - 3 = \sqrt{-4} \quad \text{or} \quad x - 3 = -\sqrt{-4}.$$

Then by Definition 12.1

$$x - 3 = \sqrt{4} \cdot \sqrt{-1} = 2i \quad \text{or} \quad x - 3 = -\sqrt{4} \cdot \sqrt{-1} = -2i,$$

from which

$$x = 3 + 2i \quad \text{or} \quad x = 3 - 2i,$$

and the solution set is $\{3 + 2i, 3 - 2i\}$.

Numbers of the form $3 + 2i$, or more generally $a + bi$, where $a, b \in R$, are called **complex numbers**. The set of all such numbers is sometimes designated by C. Hence,

$$C = \{a + bi \mid a, b \in R\}.$$

Note that if $a = 0$, a complex number is of the form bi, a pure imaginary number; if $b = 0$, the number is of the form a, a real number. Hence the set of pure imaginary numbers and the set of real numbers are subsets of the set of complex numbers.

Exercise 12.3

A

Write each pure imaginary number in the form bi, b ∈ R.

Examples.   a. $\sqrt{-144} = \sqrt{144} \cdot \sqrt{-1} = 12\sqrt{-1} = 12i$

b. $\sqrt{-8} = \sqrt{8} \cdot \sqrt{-1} = 2\sqrt{2}\, i.$

1. $\sqrt{-4}$ 2. $\sqrt{-25}$ 3. $\sqrt{-9}$

4. $\sqrt{-36}$ 5. $\sqrt{-20}$ 6. $\sqrt{-45}$

7. $\sqrt{-12}$ 8. $\sqrt{-28}$ 9. $\sqrt{-108}$

10. $\sqrt{-18}$ 11. $\sqrt{-54}$ 12. $\sqrt{-125}$

Write each expression as one of the complex numbers i, −i, 1, or −1.

Examples.   a. $i^4 = (i^2)(i^2) = (-1)(-1) = 1$

b. $i^7 = (i^2)(i^2)(i^2)\, i = (-1)^3 \cdot i = (-1) \cdot i = -i$

13. i^3 14. i^5 15. i^6

16. i^8 17. i^9 18. i^{21}

Write each expression in the form bi or a + bi, where a, b ∈ R.

Example. $\dfrac{6 + \sqrt{-27}}{3} = \dfrac{6 + \sqrt{27}\sqrt{-1}}{3} = \dfrac{6 + \sqrt{9}\sqrt{3}\sqrt{-1}}{3}$

$$= \frac{6 + 3\sqrt{3}\sqrt{-1}}{3} = \frac{3(2 + \sqrt{3}i)}{3} = 2 + \sqrt{3}i.$$

19. $2 - \sqrt{-9}$ 20. $2 + \sqrt{-25}$ 21. $\dfrac{\sqrt{-8}}{2}$

22. $\dfrac{\sqrt{-27}}{3}$ 23. $\dfrac{-4 + \sqrt{-16}}{4}$ 24. $\dfrac{-3 - \sqrt{-63}}{3}$

25. $\dfrac{-3 - \sqrt{-18}}{3}$ 26. $\dfrac{-4 + \sqrt{-32}}{4}$ 27. $\dfrac{\sqrt{2} + \sqrt{-6}}{\sqrt{2}}$

28. $\dfrac{\sqrt{3} - \sqrt{-12}}{\sqrt{3}}$ 29. $\dfrac{\sqrt{5} - \sqrt{-10}}{\sqrt{5}}$ 30. $\dfrac{\sqrt{6} + \sqrt{-18}}{\sqrt{6}}$

Find the solution set of each equation.

Example. $x^2 = -4$

Solution. By Theorem 12.2, the solution set is
$$\{\sqrt{-4}, -\sqrt{-4}\} = \{2i, -2i\}.$$

31. $x^2 = -1$ 32. $x^2 = -9$ 33. $x^2 = -24$

34. $x^2 = -45$ 35. $x^2 = -50$ 36. $x^2 = -60$

Example. $(x - 1)^2 = -9$

Solution. From Theorem 12.2, the solution set is
$$\{x \mid x - 1 = \sqrt{-9}\} \cup \{x \mid x - 1 = -\sqrt{-9}\}.$$

Since
$$x - 1 = \sqrt{-9} \text{ is equivalent to } x = 1 + 3i$$
and
$$x - 1 = -\sqrt{-9} \text{ is equivalent to } x = 1 - 3i,$$
the solution set is $\{1 + 3i, 1 - 3i\}$.

37. $(x + 1)^2 = -1$ 38. $(x + 2)^2 = -4$

39. $(x - 2)^2 = -3$ 40. $(x - 3)^2 = -5$

41. $(x + 3)^2 = -12$ 42. $(x + 4)^2 = -20$

43. $\left(x + \dfrac{1}{2}\right)^2 = -\dfrac{9}{4}$ 44. $\left(x - \dfrac{1}{3}\right)^2 = -\dfrac{4}{9}$

12.4 Solution by Completing the Square

The square of a binomial can always be written as a trinomial. Thus,
$$(x + p)^2 = x^2 + 2px + p^2$$
and
$$\left(x + \dfrac{p}{2}\right)^2 = x^2 + px + \left(\dfrac{p}{2}\right)^2.$$

Notice in each case that the first and third terms of the trinomial are the squares of the terms of the binomial and that the second term is twice the product of the terms of the binomial. Such a trinomial is called a **perfect square trinomial.**

Now consider the polynomial

$$x^2 + px. \tag{1}$$

We can form the perfect square trinomial,

$$x^2 + px + \left(\frac{p}{2}\right)^2, \tag{2}$$

from Expression (1) by adding $(p/2)^2$, the square of one-half the coefficient of x, the first degree term of (1). Expression (2) can now be written as

$$\left(x + \frac{p}{2}\right)^2,$$

the square of a binomial. This process is called **completing the square** and can be applied to finding the solution set of any quadratic equation of the form $ax^2 + bx + c = 0$.

First, we consider as an example the equation

$$x^2 + 4x - 7 = 0 \tag{3}$$

in which the coefficient of x^2 is 1. We write the equivalent equation

$$x^2 + 4x \qquad = 7$$

by adding 7 to each member of Equation (3). The left-hand member can now be written as a perfect square trinomial by adding $(4/2)^2$ or 4, the square of one-half the coefficient of x. Thus, adding 4 to each member we obtain the equivalent equation

$$x^2 + 4x + 4 = 7 + 4.$$

The left-hand member is now the square of the binomial, $x + 2$, and the equation can be written as

$$(x + 2)^2 = 11.$$

By Theorem 12.2 the solution set is

$$\{x \mid x + 2 = \sqrt{11}\} \cup \{x \mid x + 2 = -\sqrt{11}\},$$

which is equal to

$$\{-2 + \sqrt{11}, \ -2 - \sqrt{11}\}.$$

Now consider the equation $4x^2 - 8x - 3 = 0$ where the coefficient of x^2 does not equal 1. Adding 3 to each member and then multiplying each member by 1/4, the reciprocal of 4, yields the equivalent equations

$$4x^2 - 8x \quad = 3,$$

$$x^2 - 2x \quad = \frac{3}{4}.$$

Adding the square of one-half the coefficient of x, $(-2/2)^2$ or 1, to each member gives

$$x^2 - 2x + 1 = \frac{3}{4} + 1,$$

which can be written equivalently as

$$(x - 1)^2 = \frac{7}{4}.$$

By Theorem 12.2, the solution set is

$$\left\{ x \mid x - 1 = \frac{\sqrt{7}}{2} \right\} \cup \left\{ x \mid x - 1 = \frac{-\sqrt{7}}{2} \right\} = \left\{ \frac{2 + \sqrt{7}}{2}, \frac{2 - \sqrt{7}}{2} \right\}.$$

Exercise 12.4

A

What number must be added to each expression to make it a perfect square trinomial? Write the expression in the form $(x + p)^2$.

Examples.   a. $x^2 + 8x$ b. $x^2 - 9x$

Solutions.   a. $\left(\frac{8}{2}\right)^2 = 16;$ b. $\left(\frac{-9}{2}\right)^2 = \frac{81}{4};$

$$x^2 + 8x + 16 = (x + 4)^2 \qquad x^2 - 9x + \frac{81}{4} = \left(x - \frac{9}{2}\right)^2$$

1. $x^2 + 2x$ 2. $x^2 + 4x$ 3. $x^2 - 6x$

4. $x^2 - 8x$ 5. $x^2 - 10x$ 6. $x^2 - 14x$

7. $x^2 - 12x$ 8. $x^2 + 20x$ 9. $x^2 + 3x$

10. $x^2 - 7x$ 11. $x^2 + 11x$ 12. $x^2 + \frac{2}{3}x$

13. $x^2 - \frac{4}{3}x$ 14. $x^2 + \frac{3}{2}x$

Find the solution set of each equation by the method of completing the square.

Example. $x^2 - 3x + 2 = 0$

Solution.   Write the equivalent equation

$$x^2 - 3x \qquad = -2.$$

Add the square of one-half the numerical coefficient of the first-degree term, $(-3/2)^2$, or $9/4$, to both members.

$$x^2 - 3x + \frac{9}{4} = -2 + \frac{9}{4}$$

Write the left-hand member as the square of a binomial and the right-hand member as a basic fraction.

$$\left(x - \frac{3}{2}\right)^2 = \frac{1}{4}$$

By Theorem 12.2, the solution set is

$$\left\{x \mid x - \frac{3}{2} = \frac{1}{2}\right\} \cup \left\{x \mid x - \frac{3}{2} = -\frac{1}{2}\right\} = \{1, 2\}.$$

15. $x^2 + 6x + 8 = 0$ 16. $x^2 + 8x + 15 = 0$

17. $x^2 - 14x + 24 = 0$ 18. $x^2 + 4x - 5 = 0$

19. $x^2 + 7x - 8 = 0$ 20. $x^2 + 5x + 6 = 0$

21. $x^2 - 5x + 3 = 0$ 22. $x^2 - 9x + 18 = 0$

*23. $x^2 + x + 1 = 0$ *24. $x^2 - 2x + 3 = 0$

*25. $x^2 - 3x + 4 = 0$ *26. $x^2 + 4x + 6 = 0$

Example. $2x^2 - 6x - 5 = 0$

Solution.   Write the equivalent equation by adding 5 to both members and then multiplying both members by 1/2.

$$x^2 - 3x \qquad = \frac{5}{2}$$

Add the square of one-half the numerical coefficient of the first-degree term, $(-3/2)^2$, or $9/4$, to both members.

$$x^2 - 3x + \frac{9}{4} = \frac{5}{2} + \frac{9}{4}$$

* Starred problems have imaginary solutions and should be omitted if Section 12.3 has not been studied.

Write the left-hand member as a square of a binomial and the right-hand member as a basic fraction.

$$\left(x - \frac{3}{2}\right)^2 = \frac{19}{4}$$

By Theorem 12.2, the solution set is

$$\left\{x \mid x - \frac{3}{2} = \frac{\sqrt{19}}{2}\right\} \cup \left\{x \mid x - \frac{3}{2} = -\frac{\sqrt{19}}{2}\right\} = \left\{\frac{3 + \sqrt{19}}{2}, \frac{3 - \sqrt{19}}{2}\right\}.$$

27. $4x^2 + 4x - 3 = 0$ 28. $2y^2 - 6y + 3 = 0$

29. $6x^2 - 13x + 6 = 0$ 30. $2x^2 - x - 15 = 0$

31. $6x^2 + x - 1 = 0$ 32. $3x^2 - 5x + 1 = 0$

33. $2x^2 + 7x + 4 = 0$ 34. $5x^2 + 8x + 2 = 0$

*35. $3y^2 - y + 1 = 0$ *36. $4y^2 - 5y + 2 = 0$

B

Solve each quadratic equation for x in terms of a, b, and c.

Example. $x^2 - 2x + c = 0$

Solution. Write the equivalent equation

$$x^2 - 2x \quad = -c.$$

Add the square of one-half the numerical coefficient of the first degree term, $(-2/2)^2$, or 1, to both members.

$$x^2 - 2x + 1 = 1 - c$$

Write the left-hand member as the square of a binomial.

$$(x - 1)^2 = 1 - c$$

By Theorem 12.2, the solution set is

$$\{x \mid x - 1 = \sqrt{1 - c}\} \cup \{x \mid x - 1 = -\sqrt{1 - c}\}$$
$$= \{1 + \sqrt{1 - c}, 1 - \sqrt{1 - c}\}.$$

37. $x^2 + 3x + c = 0$ 38. $x^2 + bx + c = 0$

39. $ax^2 + 2x - 4 = 0$ 40. $ax^2 + 4x + c = 0$

41. $ax^2 + bx + 5 = 0$ 42. $ax^2 + bx + c = 0$

12.5 Solution by the Quadratic Formula

While the technique of completing the square is very valuable for certain purposes, its application as a means of finding the solution set of a quadratic equation is usually quite tedious. To develop a more direct method of solving quadratic equations, we apply the method of completing the square to the general quadratic equation

$$ax^2 + bx + c = 0 \qquad (a \neq 0). \tag{1}$$

Adding $-c$ to each member and multiplying each member by $1/a$, we obtain the equivalent equation

$$x^2 + \frac{b}{a}x \quad = -\frac{c}{a}.$$

The addition of $b^2/4a^2$, the square of one-half of the coefficient of x, to each member yields

$$x^2 + \frac{b}{a}x + \frac{b^2}{4a^2} = -\frac{c}{a} + \frac{b^2}{4a^2},$$

which is equivalent to

$$\left(x + \frac{b}{2a}\right)^2 = \frac{b^2 - 4ac}{4a^2}.$$

By Theorem 12.2, the solution set is

$$\left\{x \mid x + \frac{b}{2a} = \frac{\sqrt{b^2 - 4ac}}{2a}\right\} \cup \left\{x \mid x + \frac{b}{2a} = \frac{-\sqrt{b^2 - 4ac}}{2a}\right\} \tag{2}$$

which is equal to

$$\left\{\frac{-b + \sqrt{b^2 - 4ac}}{2a}, \frac{-b - \sqrt{b^2 - 4ac}}{2a}\right\}.$$

For $a, b, c \in R$ $(a \neq 0)$, we may use the equations in (2) as formulas to find the solution set of any quadratic equation. To be more concise, we usually rewrite and combine the two equations as

$$x = \frac{-b \pm \sqrt{b^2 - 4ac}}{2a} \qquad (a \neq 0).$$

This equation is commonly called the **quadratic formula.** If $b^2 - 4ac \geq 0$, the solutions are real numbers; if $b^2 - 4ac < 0$, the solutions are not real numbers.

Since this formula was derived from the general quadratic equation in standard form, any equation should be written in standard form before the values of a, b, and c are substituted in the formula. Care should be taken to insure that the *sign* of a coefficient is also used in the substitution. It is also helpful to rewrite any equation containing fractions as an equivalent equation without fractions.

Example. $\dfrac{x^2}{2} - \dfrac{5}{6}x - \dfrac{1}{3} = 0$

Solution.   Write an equivalent equation without fractions by multiplying both members of the equation by the least common denominator, 6.

$$6\left(\frac{x^2}{2}\right) - 6\left(\frac{5}{6}x\right) - 6\left(\frac{1}{3}\right) = 6(0)$$

$$3x^2 - 5x - 2 = 0$$

Substitute 3 for a, -5 for b, and -2 for c in the quadratic formula.

$$x = \frac{-(-5) \pm \sqrt{(-5)^2 - 4(3)(-2)}}{2(3)}$$

$$= \frac{5 \pm \sqrt{25 + 24}}{6} = \frac{5 \pm \sqrt{49}}{6}$$

Thus,

$$x = \frac{5 + 7}{6} \qquad \text{or} \qquad x = \frac{5 - 7}{6}.$$

The solution set is $\left\{2, -\dfrac{1}{3}\right\}$.

Exercise 12.5

A

Determine the constants a, b, and c for each quadratic equation when written in standard form $ax^2 + bx + c = 0$.

Examples.   a. $x^2 = 3 - 4x$ b. $5x^2 = 4x$

Solutions on page 368.

Solutions. Rewrite the equations in standard form and compare the constants with a, b, and c as specified by Equation (1), page 366.

a. $x^2 + 4x - 3 = 0$
$\quad a = 1, b = 4, c = -3$

b. $5x^2 - 4x = 0$
$\quad a = 5, b = -4, c = 0$
$\quad$ (there is no constant term)

1. $x^2 + 3x + 2 = 0$
2. $x^2 - 5x + 4 = 0$
3. $r^2 - r - 30 = 0$

4. $s^2 + 2s - 8 = 0$
5. $3x^2 - 7x + 1 = 0$
6. $-2x^2 - 8x + 7 = 0$

7. $x = 2x^2 - 1$
8. $y^2 - 3y = 2$
9. $t^2 = 7t$

10. $3x^2 - 5 = 0$
11. $8r^2 - 10 = 0$
12. $3x^2 = 5x$

Find the solution set of each equation by using the quadratic formula.

Example. $3x^2 = x + 2$

Solution. Write the equation in standard form.

$$3x^2 - x - 2 = 0$$

Substitute 3 for a, -1 for b, and -2 for c in the quadratic formula and simplify.

$$x = \frac{-(-1) \pm \sqrt{(-1)^2 - 4(3)(-2)}}{2(3)}$$

$$= \frac{1 \pm \sqrt{1 + 24}}{6} = \frac{1 \pm \sqrt{25}}{6}$$

Thus,

$$x = \frac{1 + 5}{6} \quad \text{or} \quad x = \frac{1 - 5}{6}.$$

The solution set is $\left\{1, -\dfrac{2}{3}\right\}$.

13. $x^2 - 3x + 2 = 0$
14. $s^2 + 5s + 4 = 0$

15. $t^2 - 4t - 12 = 0$
16. $y^2 - y - 30 = 0$

17. $y^2 + 3y = 1$
18. $x^2 + 5x = -5$

19. $x^2 = 3x + 2$
20. $x^2 = -x + 1$

21. $-6 = 2x^2 - 7x$
22. $-5 = x^2 - 6x$

23. $6x^2 + x = 1$
24. $z = 2z^2 - 3$

*25. $4x^2 + 9 = 0$
*26. $2y^2 = -12$

*27. $x^2 - x + \dfrac{5}{4} = 0$ *28. $2y^2 - \dfrac{7}{3}y + 1 = 0$

*29. $t^2 - \dfrac{5}{6}t + 2 = 0$ *30. $\dfrac{3}{4}x^2 + \dfrac{3}{2}x + 2 = 0$

B

Solve for x in terms of the other variables.

Example. $x^2 - tx - t^2 = 0$

Solution. Substituting 1 for a, $-t$ for b, and $-t^2$ for c in the quadratic formula and simplifying yields

$$x = \frac{-(-t) \pm \sqrt{(-t)^2 - 4(1)(-t^2)}}{2(1)}$$

$$= \frac{t \pm \sqrt{t^2 + 4t^2}}{2} = \frac{t \pm t\sqrt{5}}{2} = \frac{t(1 \pm \sqrt{5})}{2}.$$

Thus,

$$x = \frac{t(1 + \sqrt{5})}{2} \quad \text{or} \quad x = \frac{t(1 - \sqrt{5})}{2}.$$

The solution set is $\left\{ \dfrac{t(1 + \sqrt{5})}{2}, \dfrac{t(1 - \sqrt{5})}{2} \right\}$.

31. $2x^2 - 7x + q = 0$ 32. $tx^2 + 3x - t = 0$

33. $x^2 + 3tx - 4t^2 = 0$ 34. $4x^2 - 4sx + s^2 = 0$

35. $rx^2 + sx + t = 0$ 36. $rx^2 + sx = 0$

12.6 Word Problems

In Chapter 8 you were given several suggestions for finding the solution to a word problem in which you used a first-degree equation in one variable as the mathematical model. In Chapter 10 you solved certain types of word problems in which you used a system of two first-degree equations in two variables as the mathematical model. However, all word problems do not lead to first-degree equations. In this section you will encounter word problems which can be formulated in terms of a model that is a **quadratic equation in one variable.**

Since a quadratic equation often has two solutions it is possible that one or both members of the solution set will not be solutions of the word problem. Any result should be checked carefully with the stated conditions of the problem. For example, consider the following problem:

Find two positive integers whose difference is 2 and whose product is 48.

If x represents a positive integer greater than 2, then $x - 2$ will also represent a positive integer. Thus, the stated condition on the variable x is given by the mathematical model

$$x(x - 2) = 48.$$

Rewriting this equation in standard form and factoring the left-hand member gives

$$x^2 - 2x - 48 = 0$$
$$(x - 8)(x + 6) = 0,$$

and by Theorem 12.1, the solution set is $\{8, -6\}$.

The condition stated in the problem specified *positive* integers. Hence, -6 *cannot* be a solution of the word problem even though it *is* a solution of the quadratic equation. Therefore, the number represented by x can only be 8, and the number represented by $x - 2$ is 6. As a check, we note that $8 \cdot 6 = 48$.

The following example illustrates the use of a quadratic equation as a mathematical model for a problem involving distances and rates.

Example. A man rode his bicycle for a distance of 6 miles and then walked an additional 4 miles. His walking rate was 2 miles per hour less than his riding rate, and the entire trip took 6 hours. What was his rate for each part of the trip?

Solution. Let x represent a rate for riding; then $x - 2$ represents the walking rate. Since $t = d/r$,

$\dfrac{6}{x}$ represents the riding *time* and $\dfrac{4}{x - 2}$ represents the walking *time*.

The total time of the trip is 6 hours. Therefore,

$$\frac{6}{x} + \frac{4}{x - 2} = 6.$$

Solving this equation, we have

$$(x)(x - 2)\left(\frac{6}{x} + \frac{4}{x - 2}\right) = (x)(x - 2)6, \qquad (x \neq 0, 2),$$

$$6x - 12 + 4x = 6x^2 - 12x,$$

$$6x^2 - 22x + 12 = 0,$$

$$2(3x - 2)(x - 3) = 0.$$

The solution set is {2/3, 3}.

Although 2/3 is a replacement for x in the mathematical model, it cannot be the riding rate because the walking rate, which is 2 miles per hour less than the riding rate, would then be negative. The other element in the solution set, 3, meets the conditions stated in the problem. Therefore, the riding rate is 3 miles per hour and the walking rate is 1 mile per hour. With these rates, did the entire trip take 6 hours? That is, does $\frac{6}{3} + \frac{4}{1} = 6$? Yes.

Exercise 12.6

A

Solve. Use one variable in the mathematical model.

Number Problems

Example. The sum of twice the square of an integer and the integer itself is 15. Find the integer.

Solutions. Let $x =$ an integer.
The mathematical model for the word problem is

$$2x^2 + x = 15,$$

from which

$$2x^2 + x - 15 = 0,$$

$$(2x - 5)(x + 3) = 0.$$

The solution set of each equation is {5/2, −3}. Since 5/2 is not an integer, it is not a solution to the word problem. Since $2(-3)^2 + (-3) = 15$, the integer we wished to find is −3.

1. Four times an integer is equal to the square of the same integer. Find all such integers.
2. Three times the square of an integer diminished by four times the integer is equal to 15. Find the integer.
3. Find two consecutive *positive* integers whose product is 42.
4. Find two consecutive *negative* integers whose product is 72.
5. One integer is 3 more than another and their product is 54. Find the integers.
6. Find two consecutive *odd* integers whose product is −1.
7. Two integers differ by 3. The sum of their squares is 89. Find the two integers.
8. The sum of the squares of two consecutive positive even integers is 244. Find the integers.
9. One integer is 8 more than another. The sum of the squares of the two integers is 32. Find the integers.
10. The sum of an integer and its multiplicative inverse is 10/3. Find the integer.
11. The sum of the multiplicative inverses of two consecutive odd integers is 8/15. Find the integers.
12. Find two consecutive even integers such that the sum of their multiplicative inverses is 5/12.
13. One rational number is three more than a second rational number. The sum of the multiplicative inverse of the second and twice the multiplicative inverse of the first is 18/7. Find the rational numbers.
14. The reciprocal of the sum of a certain real number and 2 is equal to the quotient of the sum of the number and 3, divided by 30. Find the number.

Geometry Problems

15. The area of a rectangle is 60 square inches. The length is 2 inches longer than twice the width. Find the dimensions of the rectangle.
16. The sides of a rectangle are in the ratio of 5 to 3. Find the dimensions of the rectangle if the area is 240 square inches.
17. A rectangular lawn whose dimensions are 40 and 60 feet is to have its area increased 438 square feet by a border of uniform width along both ends and one side. Find the width of the border.
18. The base of a triangle is 11 inches longer than its altitude. Find the base and altitude if the area of the triangle is 40 square inches.

Formula Problems

19. The formula for finding the distance (s) in feet, traveled by an object with a starting velocity of v_0 feet per second and an acceleration of a feet per second is $s = v_0 t + \frac{1}{2}at^2$. Find t when $s = 63$, $v_0 = 15$, and $a = 4$.

20. The sum of the consecutive natural numbers $1, 2, 3, \cdots, n$ is given by the formula $s = \frac{1}{2}n(n + 1)$. How many consecutive natural numbers, starting with 1, must be added to give a sum of 561?

21. The number of diagonals, D, of a polygon of n sides is given by $D = \frac{1}{2}n(n - 3)$. Would it be possible for a polygon to have 75 diagonals? If so, how many sides would it have?

22. If an object is projected vertically upward, the height it reaches is given by the formula $h = v_0 t - 16t^2$, and its velocity at any time is given by the formula $v = v_0 - 32t$, where h is the height in feet, v_0 is the initial velocity in feet per second, and t is the time in seconds. How high does an object rise if its initial velocity is 160 feet per second? (*Hint:* $v = 0$ at maximum height.)

B

Uniform Motion Problems

23. An airplane takes $1\frac{1}{2}$ hours less time to fly a distance of 360 miles with a tail wind of 20 mph than it does to fly the same distance against a head wind of the same speed. Find the speed of the airplane if there is no wind.

24. A man made a trip of 360 miles. If his average speed had been 5 miles per hour faster, the trip would have taken 1 hour less. What was his average speed?

12.7 The Quadratic Function

As was the case for a linear equation in two variables (see Section 9.1), the solution set of a quadratic equation in two variables of the form

$$y = ax^2 + bx + c, \qquad a, b, c, x, y \in R \qquad (a \neq 0)$$

is an infinite set of ordered pairs. Since any real number replacement for x produces one and only one value for y, an equation of this form defines a function

$$\{(x, y) \mid y = ax^2 + bx + c\} \quad \text{where} \quad a, b, c, x, y \in R \quad (a \neq 0).$$

This function is called the **quadratic function.**

Consider the equation

$$y = x^2 - 3x - 4.$$

Solutions of the equation can be found by arbitrarily selecting values for x and determining the associated value of y for each replacement of x. For example, if $x = -3$,

$$y = (-3)^2 - 3(-3) - 4$$
$$= 9 + 9 - 4$$
$$= 14.$$

Hence, the ordered pair $(-3, 14)$ is a solution of the equation. Similarly, $(-2, 6)$, $(-1, 0)$, $(0, -4)$, $(1, -6)$, $(2, -6)$, $(3, -4)$, $(4, 0)$, and $(5, 6)$ are also solutions to the equation. (Check these.) By plotting the points corresponding to these ordered pairs, we have the graph in Figure 12.1a.

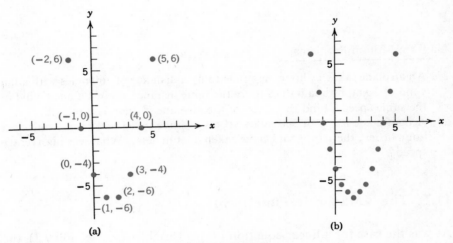

Figure 12.1

You might wonder if any "meaningful" pattern can be formed by graphing the entire solution set

$$\{(x, y) \mid y = x^2 - 3x - 4\}$$

for every $x \in R$. We can get a more definite idea by forming ordered pairs with first components between those already selected. Thus we could obtain

$$\left(\frac{-3}{2}, \frac{11}{4}\right), \quad \left(\frac{-1}{2}, \frac{-9}{4}\right), \quad \left(\frac{1}{2}, \frac{-21}{4}\right), \quad \left(\frac{3}{2}, \frac{-13}{2}\right),$$

$$\left(\frac{5}{2}, \frac{-21}{4}\right), \quad \left(\frac{7}{2}, \frac{-9}{4}\right), \quad \text{and} \quad \left(\frac{9}{2}, \frac{11}{4}\right)$$

as additional members in the solution set. (Check these.) Now, if these ordered pairs are graphed in addition to the ordered pairs in Figure 12.1a, we have the graph in Figure 12.1b.

If we connect these points with a smooth curve as in Figure 12.2, we may assume that this curve is a reasonable approximation to the graph (an

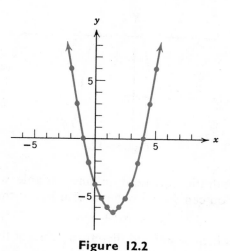

Figure 12.2

infinite set of points) of the solution set of the equation (an infinite set of ordered pairs). A curve such as this, which is obtained by graphing a quadratic function, is called a **parabola.**

In Figure 12.2, notice that as values of x increase in the positive direction, the curve first descends and then begins to rise. The point at which the curve stops descending and begins to rise is called a **low point** or **minimum.**

Now consider the equation

$$y = -x^2 + 3x - 2$$

The graph of the solution set of this equation is shown in Figure 12.3. (Find several ordered pairs to check this graph.) Notice that the curve first rises as values of x increase and then begins to descend. The point at which the curve stops rising and begins to descend is called a **high point** or **maximum.**

Since the graph of a quadratic function is not a straight line, more than two points are necessary to establish its pattern. In fact, a sufficient number of values for x should be chosen to show the *high or low point* and *intercepts* of the curve if they exist.

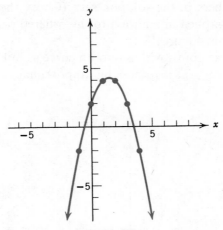

Figure 12.3

Solutions of quadratic equations in one variable were considered in earlier sections. You can see now that a solution of the equation

$$ax^2 + bx + c = 0$$

is a value of x for which $y = 0$ and therefore a zero of the function defined by the equation in two variables

$$y = ax^2 + bx + c. \tag{1}$$

Since the value of y is 0 at any point where the graph of Equation (1) crosses the x-axis, the intercept (the value of x at such a point) is a solution of the equation $ax^2 + bx + c = 0$.

Exercise 12.7

A

Specify the function by listing the members defined by each equation over the domain $\{x \mid -3 \leq x \leq 3, x \in J\}$.

Example. $y = x^2 - 3$

Solution.   Replacing x with $-3, -2, -1, 0, 1, 2,$ and 3 in turn yields

$$\{(-3, 6), (-2, 1), (-1, -2), (0, -3), (1, -2), (2, 1), (3, 6)\}.$$

1. $y = x^2$ 2. $y = -x^2$

3. $y = x^2 + 2$ 4. $y = -x^2 + 7$

5. $f(x) = -x^2 + 5$

6. $f(x) = -x^2 + 12$

7. $y = x^2 - 4x + 3$

8. $y = x^2 - x - 2$

9. $g(x) = 2x^2 + 8$

10. $h(x) = -2x^2 + 8$

11. $y = 3x^2 + 6x - 9$

12. $y = -x^2 + x - 4$

13. $y = -\frac{1}{2}x^2 + 2x + 1$

14. $y = \frac{1}{2}x^2 - \frac{1}{2}x + 3$

Graph each equation in Problems 15 to 26 over a domain sufficient to estimate the intercepts and the coordinates of the high or low point.

Example. $y = x^2 - 3$

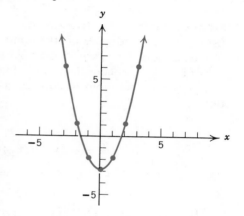

Solution. Plotting the points corresponding to the ordered pairs found in the example above and connecting these points with a smooth curve gives the graph shown in the figure. The x-intercepts are estimated as $\frac{3}{2}$ and $\frac{-3}{2}$ to the nearest $\frac{1}{2}$ unit. The y-intercept is -3. The coordinates of the low point are the components of $(0, -3)$.

15. Problem 3, above

16. Problem 4

17. Problem 5

18. Problem 6

19. Problem 7

20. Problem 8

21. Problem 9

22. Problem 10

23. Problem 11

24. Problem 12

25. Problem 13

26. Problem 14

B

27. Compare the graphs of Problems 21 and 22. What do you notice? How do the equations differ? What conjecture would you make about the graph of $y = ax^2 + bx + c$ for $a > 0$ and for $a < 0$.

28. Graph the equations $y = \frac{1}{2}x^2$, $y = 2x^2$, $y = 3x^2$, and $y = 4x^2$ on the same set of axes. Make a conjecture about the graphs of equations of the form $y = ax^2$ for different values of a.

29. Graph the equations $y = x^2 - 4$, $y = x^2 - 2$, $y = x^2$, $y = x^2 + 2$, and $y = x^2 + 4$ on the same set of axes. Make a conjecture about the graphs of equations of the form $y = x^2 + c$ for different values of c.

30. Graph the equations $y = x^2$, $y = x^2 + 9$, and $y = x^2 - 9$. Solve the respective quadratic equations in one variable in which $y = 0$. Interpret the solutions of the equations in one variable in terms of the graphs.

By graphical methods, estimate the components of the ordered pairs in each of the following sets.

Example
$\{(x, y) \mid y = x + 2\} \cap \{(x, y) \mid y = x^2\}$

Solution. Graph the two equations. Estimate the coordinates of the points of intersection. Since the ordered pairs consisting of the coordinates of the two points are solutions to both equations,

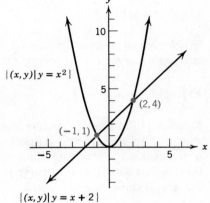

$$\{(x, y) \mid y = x + 2\} \cap \{(x, y) \mid y = x^2\} = \{(-1, 1), (2, 4)\}.$$

31. $\{(x, y) \mid x + y = 3\} \cap \{(x, y) \mid y = x^2 - 2x + 1\}$
32. $\{(x, y) \mid x + y = 5\} \cap \{(x, y) \mid y = -x^2 + 4\}$
33. $\{(x, y) \mid y = 1 - x\} \cap \{(x, y) \mid y = x^2 - 3x + 2\}$
34. $\{(x, y) \mid y - x = 3\} \cap \{(x, y) \mid y = x^2 + 2x + 1\}$
35. $\{(x, y) \mid y = x^2 - 4\} \cap \{(x, y) \mid y = -x^2 + 4\}$
36. $\{(x, y) \mid y = x^2 + 2x + 1\} \cap \{(x, y) \mid y = -x^2 + 2x - 1\}$

CHAPTER SUMMARY

An equation of the form

$$ax^2 + bx + c = 0, \qquad a, b, c \in R \quad (a \neq 0),$$

*is a **quadratic equation in standard form.***

12.1 The following theorem can be used to determine the solution set of a quadratic equation of the form $ax^2 + bx + c = 0$ if the left-hand member can be factored.

If $c \in R$ such that c is a zero of $P(x)$ [or $Q(x)$], then c is a solution of $P(x) \cdot Q(x) = 0$.

12.2 The following theorem can be used to determine the solution set of a quadratic equation of the form $x^2 = c$.

If $c \in R$ and $c \geq 0$, *then the solution set of $x^2 = c$ is*

$$\{x \mid x = \sqrt{c}\} \cup \{x \mid x = -\sqrt{c}\}.$$

12.3 If $b \in R$, $b > 0$, then $\sqrt{-b}$ is a number (called a **pure imaginary number**) such that

$$\sqrt{-b} \cdot \sqrt{-b} = -b.$$

Other symbols for $\sqrt{-b}$, $b > 0$ are:

$$\sqrt{-1} \cdot \sqrt{b}, \qquad i\sqrt{b}, \qquad \text{or} \qquad \sqrt{b}\,i,$$

where $i = \sqrt{-1}$. As a special case of $\sqrt{-b}$ where $b = 1$,

$$\sqrt{-1}\sqrt{-1} = i \cdot i = i^2 = -1.$$

The solution set of $x^2 = c$, where $c < 0$, is $\{\sqrt{c}\,i, -\sqrt{c}\,i\}$. A number of the form $a + bi$, $a, b \in R$, is a **complex number.**

12.4 The solution set of any quadratic equation $ax^2 + bx + c = 0$ can be found by first writing the equation equivalently in a form in which the left member is a perfect square. This method of solving an equation is referred to as **completing the square.**

12.5 The solution set of any quadratic equation $ax^2 + bx + c = 0$ can be found by using the **quadratic formula**

$$x = \frac{-b \pm \sqrt{b^2 - 4ac}}{2a} \qquad (a \neq 0).$$

If $b^2 - 4ac \geq 0$, the solutions are real numbers; if $b^2 - 4ac < 0$, the solutions are not real numbers.

12.6 A quadratic equation is sometimes a useful mathematical model for a word problem. However, every solution of a satisfactory model is not always a solution of the word problem.

12.7 The graph of a quadratic function defined by $y = ax^2 + bx + c$ can be found by graphing a number of ordered pairs that are solutions of the equation and connecting the graphs (points) with a smooth curve. The curve is called a **parabola.**

CHAPTER REVIEW

A

12.1 *Write each equation in standard form.*

1. $7 = 2x - 5x^2$
2. $bx = c - 3ax^2$

Find the solution set of each equation by factoring.

3. $x^2 + 2x - 3 = 0$
4. $2x^2 = 9 - 7x$

12.2, 12.3 *Find the solution set of each equation by the extraction of roots.*

5. $x^2 = 25$
6. $x^2 = 7$
7. $(x - 2)^2 = 4$

8. $(x + 3)^2 = 5$
*9. $x^2 = -3$
*10. $x^2 = -11$

*11. $(x + 4)^2 = -1$
*12. $(x - 1)^2 = -6$

12.4 *What term should be added to each expression to make the expression a perfect square trinomial?*

13. $x^2 + 4x$
14. $x^2 - 9x$
15. $x^2 + bx$
16. $x^2 + \dfrac{b}{a}x$

Find the solution set of each equation by completing the square.

17. $x^2 - 7x + 6 = 0$
18. $4x^2 - 8x - 21 = 0$

12.5 *Find the solution set of each equation by use of the quadratic formula.*

19. $x^2 - 5x + 6 = 0$
20. $2x^2 - 3x = 5$

21. $5x^2 = -7x + 10$
22. $3x^2 - 2 = 4x$

*23. $x^2 - 3x = -6$
*24. $4x^2 + 6x + 3 = 0$

12.7 25. Graph $y = x^2 - 7x + 10$. Estimate the x-intercepts and the coordinates of the low point.

B

26. By graphical means, estimate the solution set of

$$\{(x, y) \mid y = x^2 + 3x + 2\} \cap \{(x, y) \mid y = x^2\}.$$

chapters 11 & 12

Cumulative Review

CHAPTER II

1. If $a \in R$ and $a \geq 0$, then $\sqrt{a}$ is the non-negative number such that $\sqrt{a} \cdot \sqrt{a} = $ _____.

2. $\sqrt{a}$ is called the positive _____ _____ of a if $a > 0$.

3. If $a < 0$, there is no real square root of a because the square of a real number is always _____ or zero.

4. Since $\sqrt{a}$ is defined to be a non-negative real number for $a \geq 0$, $-\sqrt{a}$ is a _____ real number or zero.

5. If $a, b \in R$, $a, b \geq 0$, then $\sqrt{a}\sqrt{b} = $ _____.

6. If $a, b \in R$, $a, b > 0$, then $\dfrac{\sqrt{a}}{\sqrt{b}} = $ _____.

7. If $a, b, c \in R$, $c > 0$, then $a\sqrt{c} + b\sqrt{c} = (a + b)$ _____.

8. The process by which $\dfrac{1}{\sqrt{3}}$ is written as $\dfrac{\sqrt{3}}{3}$ is called _____ _____.

9. A two-term radical expression can be rationalized by multiplying it by its _____.

10. $b - \sqrt{c}$ is the conjugate of _____.

11. $(b - \sqrt{c})(b + \sqrt{c}) = $ _____.

12. The fraction $\dfrac{3}{2 - \sqrt{3}}$ can be rewritten as an equivalent fraction in which the denominator has been rationalized by multiplying both numerator and denominator by _____.

In Problems 13 to 20, justify each statement with a definition or theorem.

13. $\sqrt{6}\sqrt{12} = \sqrt{6 \cdot 12}$

14. $\sqrt{6 \cdot 6 \cdot 2} = \sqrt{6 \cdot 6}\sqrt{2}$

15. $\sqrt{6 \cdot 6\sqrt{2}} = 6\sqrt{2}$

16. $\sqrt{\dfrac{2}{7}} = \dfrac{\sqrt{2}}{\sqrt{7}}$

17. $\dfrac{\sqrt{2}}{\sqrt{7}} = \dfrac{\sqrt{2}\sqrt{7}}{\sqrt{7}\sqrt{7}}$

18. $\dfrac{\sqrt{2}\sqrt{7}}{\sqrt{7}\sqrt{7}} = \dfrac{\sqrt{2}\sqrt{7}}{7}$

19. $\dfrac{\sqrt{2}\sqrt{7}}{7} = \dfrac{\sqrt{2 \cdot 7}}{7}$

20. $\dfrac{\sqrt{12}}{\sqrt{3}} = \sqrt{\dfrac{12}{3}}$

21. The distance between two points (x_1, y_1) and (x_2, y_2) is given by $d =$ _____.

CHAPTER 12

22. An equation of the second degree is also called a _____ equation.

23. When a second-degree equation in one variable is written in descending powers of the variable and the right-hand member is 0, for example $ax^2 + bx + c = 0$, it is said to be in _____ form.

24. If $c \in R$ and $P(c) = 0$, then c is a _____ of $P(x)$.

25. If $x \in R$ and if $c \in R$ such that c is a zero of $P(x)$ [or $Q(x)$], then c is a _____ of $P(x) \cdot Q(x) = 0$.

26. If $P(c) = 0$, then $P(c) \cdot Q(x) = 0$ by the _____ _____law.

27. The process of solving $ax^2 + bx + c = 0$ by first writing it in the form $P(x) \cdot Q(x) = 0$ is called solution by _____.

28. If $x, c \in R$, $c \geq 0$, and if $x^2 = c$, then $x = \sqrt{c}$ or $x =$ _____.

29. The method of solving a quadratic equation by using the theorem stated in Problem 28 is called solution by _____ _____ _____.

30. $x^2 = -9$ does not have real solutions because the square of a real number is (always/sometimes/never) positive or zero.

31. If $b \in R$, $b > 0$, then $\sqrt{-b}$ is a number such that $\sqrt{-b} \cdot \sqrt{-b} =$ _____.

32. A number of the form $a + bi$, where $a, b \in R$, is called a _____ number.

33. A polynomial of the form $x^2 + bx$ can be changed to a perfect square trinomial by adding the square of _____.

34. The process used in Problem 33 is called _____ _____ _____.

35. To find the solution of a quadratic equation after completing the square in one member, the method of _____ _____ _____ can be used.

36. If $ax^2 + bx + c = 0$ $(a \neq 0)$, then $x = \dfrac{-b \pm \sqrt{b^2 - 4ac}}{2a}$. The latter equation is called the _____ formula.

37. In the quadratic formula, if $b^2 - 4ac \geq 0$, the solutions are _____ numbers.

38. In the quadratic formula, if $b^2 - 4ac$ _____, the solutions are not real numbers.

39. The solution set of the quadratic equation which is the mathematical model for a word problem (always/sometimes/never) contains the solution to the problem.

40. The solution set of $y = ax^2 + bx + c$ is an infinite set of _____ _____.

41. A function defined by $y = ax^2 + bx + c$ is called a _____ function.

42. The graph of a quadratic function is a curve called a _____.

43. The x-coordinate of a point where the graph of a function crosses the x-axis is called the _____.

44. The x-intercept of the graph of $y = ax^2 + bx + c$ is also a _____ of the equation $ax^2 + bx + c = 0$.

Appendices

Appendices

appendix 1
Axioms, Definitions, and Theorems

The variables represent elements in the set of real numbers, R, unless otherwise stated. The reference following each axiom, definition, or theorem indicates the page that the statement first appears in the text and where the variables represent elements of a subset of R.

Equality Axioms

$a = a$.	Reflexive law	(p. 22)
If $a = b$, then $b = a$.	Symmetric law	(p. 22)
If $a = b$ and $b = c$, then $a = c$.	Transitive law	(p. 22)
If $a = b$, then a may be replaced by b or b may be replaced by a in any collection of symbols without affecting the truth or falsity of the statement.	Substitution law	(p. 22)

Order Axioms

Exactly one of the following is true: $a < b$, $a = b$, or $a > b$.	Trichotomy law	(p. 23)
If $a < b$ and $b < c$, then $a < c$.	Transitive law	(p. 23)

Axioms for Operations

$a + b \in R$.	Closure law for addition	(p. 29)
$a + b = b + a$.	Commutative law of addition	(p. 29)
$(a + b) + c = a + (b + c)$.	Associative law of addition	(p. 29)
$a + 0 = a$; $0 + a = a$.	Identity law of addition	(p. 30)
$a \cdot b \in R$.	Closure law for multiplication	(p. 34)
$a \cdot b = b \cdot a$.	Commutative law of multiplication	(p. 34)
$(a \cdot b) \cdot c = a \cdot (b \cdot c)$.	Associative law of multiplication	(p. 35)
$1 \cdot a = a$; $a \cdot 1 = a$.	Identity law of multiplication	(p. 35)
$a \cdot (b + c) = a \cdot b + a \cdot c$.	Distributive law	(p. 38)
$a + (-a) = 0$; $-a + a = 0$.	Additive inverse law	(p. 65)
$a \cdot \dfrac{1}{a} = 1$; $\dfrac{1}{a} \cdot a = 1$	Multiplicative inverse law	(p. 90)

387

Definitions

1.1. The union of two sets, A and B, is the set of all elements that belong either to A or B, or to both. (p. 6)

1.2. The intersection of two sets, A and B, is the set of all elements that belong to both A and B; that is, those elements common to both sets. (p. 6)

1.3. The complement of a set A in a universal set is the set of all elements of the universal set that do not belong to A. (p. 7)

1.4. The Cartesian product of two sets, A and B, denoted by $A \times B$, is the set of all possible ordered pairs such that the first component is an element of A and the second component is an element of B. (p. 12)

2.1. The sum of two whole numbers a and b is that number, $a + b$, which is the cardinality of the set formed by the union of two disjoint sets whose cardinalities are respectively a and b. (p. 28)

2.2. The product of two whole numbers a and b is that number $a \cdot b$ which is the cardinality of $A \times B$, the Cartesian product of A and B, where A has cardinality a and B has cardinality b. (p. 33)

2.3. $b - a$ is the unique number d, such that $a + d = b$. The number d, or $b - a$, is called the difference of b and a. (p. 46)

2.4. $\dfrac{a}{b}$ is the unique number q, such that $b \cdot q = a$. The number q, or $\dfrac{a}{b}$, is called the quotient of a divided by b. (p. 47)

2.5. A solution of an equation or inequality is a replacement for the variable that forms a true statement. The set of all solutions is called the solution set. (p. 55)

3.1. If a is positive or zero, then $|a| = a$; if a is negative, then $|a| = -a$. (p. 68)

3.2. $a < b$ if there is a positive number c such that $a + c = b$. (p. 73)

5.1. $\sqrt{a}$ is the non-negative number such that $\sqrt{a} \cdot \sqrt{a} = a$. (p. 124)

6.1. $a^m = a \cdot a \cdot a \cdots a$ (m factors); $(m \in N)$. (p. 144)

6.2. $a^0 = 1$; $(a \neq 0)$. (p. 163)

6.3. $a^{-m} = \dfrac{1}{a^m}$; $(a \neq 0, m \in J)$. (p. 163)

9.1. A relation is an association between the elements of two sets. (p. 266)

9.2. A function is a relation in which each element in the domain is associated with only one element in the range. (p. 270)

9.3. The slope of the line containing two points $P_1(x_1, y_1)$ and $P_2(x_2, y_2)$ is

$$m = \frac{y_2 - y_1}{x_2 - x_1} ; \qquad (x_2 \neq x_1). \text{ (p. 287)}$$

12.1. $\sqrt{-b}$, $b > 0$, is the number such that $\sqrt{-b} \cdot \sqrt{-b} = -b$. (p. 358)

Theorems

2.1. If $a = b$, then $a + c = b + c$. (Addition law of equality) (p. 42)

2.2. If $a = b$, then $a \cdot c = b \cdot c$. (Multiplication law of equality) (p. 43)

2.3. $a \cdot 0 = 0$ and $0 \cdot a = 0$. (Zero factor law) (p. 44)

3.1. $-(-a) = a$. (Double negative law) (p. 65)

3.2. $-a + (-b) = -(a + b)$. (p. 70)

3.3. If $a > b > 0$, then $a + (-b) = +(a - b)$;
   if $b > a > 0$, then $a + (-b) = -(b - a)$. (p. 71)

3.4. $a - b = a + (-b)$. (p. 73)

3.5. $a \cdot (-b) = -(a \cdot b)$. (p. 76)

3.6. $(-a) \cdot (-b) = a \cdot b$. (p. 77)

4.1. $\dfrac{a}{b} = a \cdot \dfrac{1}{b}$; $(b \neq 0)$. (p. 91)

4.2. If $\dfrac{a}{b} = \dfrac{c}{d}$, then $a \cdot d = b \cdot c$;

   if $a \cdot d = b \cdot c$, then $\dfrac{a}{b} = \dfrac{c}{d}$; $(b, d \neq 0)$. (p. 97)

4.3. $\dfrac{a \cdot c}{b \cdot c} = \dfrac{a}{b}$; $(b, c \neq 0)$. (Fundamental principle of fractions). (p. 97)

4.4. $\dfrac{a}{c} + \dfrac{b}{c} = \dfrac{a + b}{c}$; $(c \neq 0)$. (p. 102)

4.5. $\dfrac{a}{b} = -\dfrac{-a}{b} = -\dfrac{a}{-b} = \dfrac{-a}{-b}$;

$-\dfrac{a}{b} = \dfrac{-a}{b} = \dfrac{a}{-b} = -\dfrac{-a}{-b}$; $(b \neq 0)$. (p. 106)

4.6. $\dfrac{a}{c} - \dfrac{b}{c} = \dfrac{a-b}{c}$; $(c \neq 0)$. (p. 107)

4.7. $\dfrac{a}{b} \cdot \dfrac{c}{d} = \dfrac{a \cdot c}{b \cdot d}$; $(b, d \neq 0)$. (p. 110)

4.8. $\dfrac{\frac{1}{a}}{\frac{a}{b}} = \dfrac{b}{a}$; $(a, b \neq 0)$. (p. 111)

4.9. $\dfrac{a}{b} \div \dfrac{c}{d} = \dfrac{a}{b} \cdot \dfrac{d}{c}$; $(b, c, d \neq 0)$. (p. 112)

6.1. $a^m \cdot a^n = a^{m+n}$; $(m, n \in N)$. (p. 158)

6.2. $\dfrac{a^m}{a^n} = a^{m-n}$; $(a \neq 0; \ m, n \in N, \ \text{and} \ m > n)$. (p. 159)

In the following theorems $P(x)$, $Q(x)$, $R(x)$, and $S(x)$ are expressions that represent real numbers.

7.1. $\dfrac{P(x)}{Q(x)} = P(x) \cdot \dfrac{1}{Q(x)}$; $[Q(x) \neq 0]$. (p. 192)

7.2. $\dfrac{P(x)}{Q(x)} = \dfrac{R(x)}{S(x)}$, if $P(x) \cdot S(x) = Q(x) \cdot R(x)$;

$P(x) \cdot S(x) = Q(x) \cdot R(x)$, if $\dfrac{P(x)}{Q(x)} = \dfrac{R(x)}{S(x)}$; $[Q(x), S(x) \neq 0]$.

(p. 192)

7.3. $\dfrac{P(x) \cdot R(x)}{Q(x) \cdot R(x)} = \dfrac{P(x)}{Q(x)}$ and $\dfrac{P(x)}{Q(x)} = \dfrac{P(x) \cdot R(x)}{Q(x) \cdot R(x)}$; $[Q(x), R(x) \neq 0]$.

(p. 192)

7.4. $\dfrac{P(x)}{R(x)} + \dfrac{Q(x)}{R(x)} = \dfrac{P(x) + Q(x)}{R(x)}$; $[R(x) \neq 0]$. (p. 198)

7.5. $\dfrac{P(x)}{Q(x)} = -\dfrac{-P(x)}{Q(x)} = -\dfrac{P(x)}{-Q(x)} = \dfrac{-P(x)}{-Q(x)}$;

$-\dfrac{P(x)}{Q(x)} = \dfrac{-P(x)}{Q(x)} = \dfrac{P(x)}{-Q(x)} = -\dfrac{-P(x)}{-Q(x)}$; $[Q(x) \neq 0]$. (p. 203)

7.6. $\dfrac{P(x)}{R(x)} - \dfrac{Q(x)}{R(x)} = \dfrac{P(x) - Q(x)}{R(x)}$; $[R(x) \neq 0]$. (p. 204)

7.7. $\dfrac{P(x)}{Q(x)} \cdot \dfrac{R(x)}{S(x)} = \dfrac{P(x) \cdot R(x)}{Q(x) \cdot S(x)}$; $[Q(x), S(x) \neq 0]$. (p. 208)

7.8. $\dfrac{1}{P(x)/Q(x)} = \dfrac{Q(x)}{P(x)}$; $[P(x), Q(x) \neq 0]$ (p. 208)

7.9. $\dfrac{P(x)}{Q(x)} \div \dfrac{R(x)}{S(x)} = \dfrac{P(x)}{Q(x)} \cdot \dfrac{S(x)}{R(x)}$; $[Q(x), R(x), S(x) \neq 0]$. (p. 208)

8.1. The equations
$$P(x) = Q(x)$$
and
$$P(x) + R(x) = Q(x) + R(x)$$
are equivalent. (p. 221)

8.2. The equations
$$P(x) = Q(x)$$
and
$$P(x) \cdot R(x) = Q(x) \cdot R(x), \quad [R(x) \neq 0]$$
are equivalent. (p. 222)

8.3. The inequalities
$$P(x) < Q(x)$$
and
$$P(x) + R(x) < Q(x) + R(x)$$
are equivalent. (p. 234)

8.4. The inequalities
$$P(x) < Q(x)$$
and
$$P(x) \cdot R(x) < Q(x) \cdot R(x)$$

are equivalent if $R(x) > 0$; the inequalities

$$P(x) < Q(x)$$

and

$$P(x) \cdot R(x) > Q(x) \cdot R(x)$$

are equivalent if $R(x) < 0$. (p. 235)

9.1. The graph of an equation of the form

$$y = mx + b$$

has a slope m and a y-intercept b. (p. 290)

10.1. The equations in the system

$$a_1 x + b_1 y = c_1$$

$$a_2 x + b_2 y = c_2$$

are inconsistent if

$$\frac{a_1}{a_2} = \frac{b_1}{b_2} \neq \frac{c_1}{c_2},$$

and are dependent if

$$\frac{a_1}{a_2} = \frac{b_1}{b_2} = \frac{c_1}{c_2}. \qquad \text{(p. 305)}$$

10.2. Any ordered pair (x, y) that satisfies the equations

$$a_1 x + b_1 y + c_1 = 0$$
$$a_2 x + b_2 y + c_2 = 0$$

also satisfies the equation

$$s \cdot (a_1 x + b_1 y + c_1) + t \cdot (a_2 x + b_2 y + c_2) = 0$$

for any real numbers s and t. (p. 308)

11.1. $\sqrt{a}\sqrt{b} = \sqrt{ab}$ and $\sqrt{ab} = \sqrt{a}\sqrt{b}$; $(a, b \geq 0)$. (p. 331)

11.2. $\dfrac{\sqrt{a}}{\sqrt{b}} = \sqrt{\dfrac{a}{b}}$ and $\sqrt{\dfrac{a}{b}} = \dfrac{\sqrt{a}}{\sqrt{b}}$; $(a \geq 0 \text{ and } b > 0)$. (p. 332)

12.1. If c is a zero of $P(x)$ [or $Q(x)$], then c is a solution of $P(x) \cdot Q(x) = 0$. (p. 352)

12.2. The solution set of $x^2 = c$ is

$$\{x \mid x = \sqrt{c}\} \cup \{x \mid x = -\sqrt{c}\} = \{\sqrt{c}, -\sqrt{c}\}; \qquad (c \geq 0). \quad \text{(p. 356)}$$

(If $c < 0$, the solutions are not real numbers.)

appendix II

Formulas from Geometry

1. Square

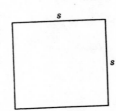

Perimeter:
  $P = 4s$
Area:
  $A = s^2$

2. Rectangle

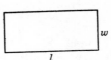

Perimeter:
  $P = l + l + w + w$
  $P = 2l + 2w$
Area:
  $A = lw$

3. Triangle

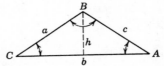

Perimeter:
  $P = a + b + c$
Area:
  $A = \dfrac{1}{2} bh$

4. Right triangle

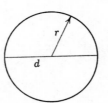

$m\angle A + m\angle B + m\angle C = 180°$
$c^2 = a^2 + b^2$

5. Circle

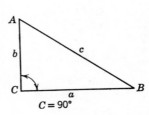

Diameter:
  $d = 2r$
Circumference:
  $C = 2\pi r = \pi d$
Area:
  $A = \pi r^2$

Answers

Exercise 1.1 (Page 3)

1. {January, February, March, April} 3. {r, e, p, s, n, t}
5. {5, 7, 9, 11, 13} 7. {0, 1, 2, 3, 4} 9. $t \in U$
11. $a \in U$ 13. $c \in \{b, c\}$ 15. $f \notin \{a, b, c\}$
17. $a \in L$ 19. $c \in K$ 21. $P \subset K$ 23. $\varnothing \subset L$
25. {a, b, c, d} 27. {a, b}, {a, c}, {a, d}, {b, c}, {b, d}, {c, d}
29. $\varnothing$ 31. $K = L$ 33. $N \neq P$
35. $K \neq 3$ 37. False 39. True
41. False 43. False 45. True
47. True 49. No 51. $A = B$
53. $A = C$

Exercise 1.2 (Page 7)

1. {a, b, c, d, e, f, g, h} 3. {a, b, c, d, e, f, g, h}
5. {a, c, e, g} 7. {a, c} 9. {a, b, c, d, e, f, g, h}
11. $\varnothing$ 13. {m, n, p} 15. {j, k, n, p} 17. {m}
19. {j} 21. {f, g, h, i, j} 23. {a, c, e}
25. {a, b, c, d, e, g, i} 27. {a, b, c, d, e, g, i}
29. A and C 31. R and S; S and T 33. No
35. Yes 37. S and T are disjoint
39. $S \subset T$ 41. $T \subset S$

Exercise 1.3 (Page 10)

(Different Venn diagrams are possible for each answer).

1. 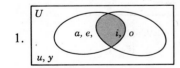 $A \cap B = \{i\}$

3.  $A' = \{o, u, y\}$

5. 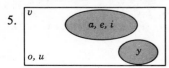 $A \cup E = \{a, e, i, y\}$

7. $(A \cup B)' = \{u, y\}$

9. $(C \cup D)' = \{a, e, i\}$

11. $(A \cap E)' = \{a, e, i, o, u, y\}$

13. $(A \cup B) \cup C = \{a, e, i, o, u, y\}$

15. $(B \cup C) \cap A = \{i\}$

17.   19.

21.   23.

25. $\{r, s, t\}$ 27. $\{s, t, u\}$ 29. $\{t\}$
31. $\{s, t, u, w\}$ 33. $\{r, s, w, x, y, z\}$ 35. $\{r, s, t, u, v\}$

Exercise 1.4 (Page 13)

1. $\{(a, r), (a, s), (a, t), (e, r), (e, s), (e, t), (i, r), (i, s), (i, t), (o, r), (o, s), (o, t)\}$
3. $\{(a, a), (a, e), (a, t), (e, a), (e, e), (e, t), (i, a), (i, e), (i, t), (o, a), (o, e),$
   $(o, t)\}$

5. $\{(i, r), (i, s), (i, t), (o, r), (o, s), (o, t), (s, r), (s, s), (s, t)\}$
7. $\{(a, r), (a, s), (a, t), (e, r), (e, s), (e, t), (t, r), (t, s), (t, t)\}$
9. $\{(a, a), (a, e), (a, i), (a, o), (e, a), (e, e), (e, i), (e, o), (i, a), (i, e), (i, i),$
   $(i, o), (o, a), (o, e), (o, i), (o, o)\}$
11. $\{(a, a), (a, e), (a, t), (e, a), (e, e), (e, t), (t, a), (t, e), (t, t)\}$
13. $\varnothing$ 15. 28   17. 8   19. 14   21. $\{r\}$
23. If A or B equals $\varnothing$ or both equal $\varnothing$ 25. $\{(a, r), (b, r)\}$
27. $\{(a, r), (a, s), (b, r), (b, s), (c, r), (c, s)\}$

Chapter I Review (Page 14)

1. $A \subset U$ 2. $f \in C$ 3. $A \subset A$ 4. $D \subset A$
5. $b \in U$ 6. $E \subset U$ 7. $\{e, f\}, \{e\}, \{f\}, \varnothing$; 4
8. $\{a, c, e\}, \{a, c\}, \{a, e\}, \{c, e\}, \{a\}, \{c\}, \{e\}, \varnothing$; 8
9. $C \cup D$ 10. $C \cap D$ 11. $C \times D$
12. $E \times D$ 13. $B \cap C$ 14. $A \cup B$
15. $\{a, c, e, f\}$ 16. $\{a, c\}$ 17. $\{b, f\}$
18. $\{(a, e), (a, f), (c, e), (c, f), (e, e), (e, f)\}$
19. $\{(e, c), (e, d), (e, e), (f, c)(f, d), (f, e)\}$
20. $\{(a, a), (b, a), (f, a)\}$
21. $\{a, c, e\}$ 22. $\varnothing$ 23. $S \subset T$ 24. $T \subset S$
25. S and T are disjoint   26. $S = T$ 27. Yes
28. Yes   29. Yes   30. No

Exercise 2.1 (Page 18)

1. 7   3. 6   5. 2   7. 6
9. 0   11. 1   13. 0   15. 1
17. 12   19. $6 < 16$ 21. $4 \neq 9$ 23. $7 > 3$
25. $2 + 3 = 5$ 27. $5 > 2$ 29. $7 < 10$ 31. $12 = 12$
33. Cardinal   35. Ordinal; cardinal   37. Ordinal; cardinal
39. Cardinal; ordinal   41. 0   43. 0
45. 5   47. 6
49. $\varnothing$ represents an empty set; $\{0\}$ represents a set with one member, namely 0.

Exercise 2.2 (Page 23)

1. a. The set of all x such that 2 is less than x, x is less than 6, and x is a natural number.   b. $\{3, 4, 5\}$
3. a. The set of all x such that x is less than 7, and x is a whole number.
   b. $\{0, 1, 2, 3, 4, 5, 6\}$

5. a. The set of all z such that 0 is less than z, z is less than 8, and z is a whole number.   b. {1, 2, 3, 4, 5, 6, 7}

7. a. The set of all x such that x is greater than 3, and x is a natural number.
   b. {4, 5, 6, 7, $\cdots$}

9. a. The set of all y such that y is greater than 0, and y is a whole number.
   b. {1, 2, 3, 4, $\cdots$}

11. $\{x \mid 0 < x < 3, x \in W\}$

13. $\{x \mid 4 < x < 8, x \in W\}$

15. $\{x \mid 0 \le x < 4, x \in W\}$

17. $\{x \mid 3 < x < 9, x \in W\}$

19. $\{x \mid x \in W\}$

21. $\{x \mid x > 4, x \in W\}$

23. Symmetric law

25. Transitive law of equality or substitution law

27. Reflexive law

29. Transitive law of inequality

31. Trichotomy law

33. Substitution law of equality

35. Trichotomy law

37. Symmetric law

39. Substitution law

41. {0, 1, 2, 3, 4}

43. {1, 3, 5, 7}

45. {0, 1, 2, 3, $\cdots$}

Exercise 2.3 (Page 26)

21. $b > c$

23. $a < d$

25. $0 < b$

27. $a < b$

29. $c = c$

31. a. {3, 4, 5, 6}
    b. $\{x \mid 3 \le x \le 6, x \in N\}$

33. a. {8, 9, 10, $\cdots$}
    b. $\{x \mid x \ge 8, x \in N\}$

35. (number line)

37. (number line)

Exercise 2.4 (Page 31)

1. Closure law for addition
3. Commutative law of addition
5. Identity law of addition
7. Commutative law of addition
9. Associative law of addition
11. Closure law for addition
13. Associative law of addition
15. Substitution law
17. Commutative law of addition
19. Associative law of addition
21. Closure law for addition
23. Commutative law of addition
25. Substitution law
27. $7 + (5 + 95)$; 107
29. $(997 + 3) + 36$; 1036
31. $(87 + 13) + (3 + 7)$; 110
33. $(22 + 78) + (14 + 86)$; 200
35. $(4 + 96) + (65 + 35) + 20$; 220
37. Not closed; $6 + 8$ is not an element in the set
39. Closed   41. Closed   43. Closed
45. a. For all $a, b \in W$, then $a + b \in W$
  b. For all $a, b \in W$, then $a + b = b + a$.
  c. For all $a, b, c \in W$, then $(a + b) + c = a + (b + c)$
  d. For all $a \in W$, $0 + a = a$ and $a + 0 = a$.

Exercise 2.5 (Page 36)

1. Closure law for multiplication
3. Associative law of multiplication
5. Identity law of multiplication
7. Commutative law of multiplication
9. Associative law of multiplication
11. Closure law for multiplication
13. Substitution law
15. Commutative law of multiplication
17. Closure law for multiplication
19. Commutative law of multiplication
21. Substitution law
23. $(5 \times 2) \times 16$; 160
25. $(2 \times 50) \times 8$; 800
27. $13 \times (5 \times 20)$; 1,300
29. $(4 \times 25) \times (5 \times 3)$; 1,500
31. $(5 \times 2) \times (8 \times 3)$; 240
33. Not closed; 4×5 is not an element of the set
35. Closed   37. Closed

39. ● (1, 10) ● (2, 10) ● (3, 10)
    ● (1, 5) ● (2, 5) ● (3, 5)

41.
● (1, 8) ● (3, 8) ● (5, 8)
● (1, 6) ● (3, 6) ● (5, 6)
● (1, 4) ● (3, 4) ● (5, 4)
● (1, 2) ● (3, 2) ● (5, 2)

43 a. For all $a, b \in W, a \cdot b \in W$
 b. For all $a, b \in W, a \cdot b = b \cdot a$
 c. For all $a, b, c \in W$, then $(a \cdot b) \cdot c = a \cdot (b \cdot c)$
 d. For all $a \in W, 1 \cdot a = a$ and $a \cdot 1 = a$

Exercise 2.6 (Page 40)

1. $4 \cdot 2 + 4 \cdot 8$
3. $6 \cdot 7 + 6 \cdot 9$
5. $1 \cdot 2 + 5 \cdot 2$
7. $6 \cdot 9 + 5 \cdot 9$
9. $5 \cdot 6 + 5 \cdot x$
11. $y \cdot 8 + y \cdot x$
13. $a \cdot c + b \cdot c$
15. $5 \cdot 7 + 1 \cdot 7 + 5 \cdot 9 + 1 \cdot 9$
17. $x \cdot y + 3 \cdot y + x \cdot 5 + 3 \cdot 5$
19. $6 \cdot 8 + y \cdot 8 + 6 \cdot x + y \cdot x$
21. $x \cdot a + y \cdot a + x \cdot b + y \cdot b$
23. $3 \cdot 10 + 3 \cdot 7 = 51$
25. $10 \cdot 10 + 3 \cdot 10 + 10 \cdot 7 + 3 \cdot 7 = 221$
27. $10 \cdot 10 + 2 \cdot 10 + 10 \cdot 8 + 2 \cdot 8 = 216$
29. $10 \cdot 10 + 5 \cdot 10 + 10 \cdot 9 + 5 \cdot 9 = 285$
31. $2 \cdot 4 + 2 \cdot 5 + 2 \cdot 6 + 3 \cdot 4 + 3 \cdot 5 + 3 \cdot 6$
33. $x \cdot 5 + x \cdot y + x \cdot z + 2 \cdot 5 + 2 \cdot y + 2 \cdot z$
35. $x \cdot w + x \cdot z + x \cdot 6 + y \cdot w + y \cdot z + y \cdot 6 + 8 \cdot w + 8 \cdot z + 8 \cdot 6$
37. For all $a, b, c \in W, a \cdot (b + c) = a \cdot b + a \cdot c$

Exercise 2.7 (Page 44)

1. Addition law of equality
3. Zero factor law
5. Multiplication law of equality
7. Zero factor law
9. Addition law of equality
11. Addition law of equality
13. Identity law of multiplication
15. Distributive law
17. Closure law for multiplication
19. Associative law of addition
21. Zero factor law
23. Commutative law of addition
25. Commutative law of addition
27. Distributive law
29. Identity law of multiplication
31 b. Closure law for multiplication
 e. Substitution law

Exercise 2.8 (Page 48)

1. 1
3. 2
5. Does not exist
7. 1
9. $x - 2, x \geq 2$
11. $x - y, x \geq y$
13. $x - 0, x \geq 0$
15. $(5 + x) - x$
17. 2
19. 5
21. Does not exist
23. 0
25. Does not exist
27. $4/x; \{1, 2, 4\}$

29. $12/y$; $\{1, 2, 3, 4, 6, 12\}$
33. $(8 - y)/2$; $\{0, 2, 4, 6, 8\}$

31. $2/(y - 1)$; $\{2, 3\}$
35. 3

Exercise 2.9 (Page 50)

1. $\{7, 11, 13\}$
5. $\{2, 3, 5, 7, 11, 13, 17, 19, 23\}$
9. $\{37, 41, 43, 47\}$
13. $3 \cdot 7$
19. $3 \cdot 3 \cdot 5 \cdot 5$
25. $5 \cdot 5$
31. $1 \cdot 4$
37. $3 \cdot (x + 2)$

3. $\{8, 9, 10, 12\}$
7. $\{14, 15, 16, 18, 20, 21, 22\}$
11. $\{101, 103, 107, 109, 113\}$
15. $2 \cdot 3 \cdot 7$
21. $2 \cdot 2 \cdot 2 \cdot 2 \cdot 31$
27. $2 \cdot 24$
33. $12 \cdot 12$
39. $2 \cdot (y + 5)$

17. $2 \cdot 2 \cdot 2 \cdot 2 \cdot 2 \cdot 2 \cdot 2$
23. $2 \cdot 2 \cdot 3 \cdot 103$
29. $4 \cdot 12$
35. 18.8
41. $6 \cdot (x - 2)$ 43. 89

45. The factors of 14, 2, and 7 are contained in the factors of $6 \cdot 21$, which are 2, 3, and 7.

Exercise 2.10 (Page 53)

1. 9
9. 14
17. 7
25. 1
33. 1
41. 10

3. 10
11. 18
19. 3
27. 6
35. 17

5. 6
13. 2
21. 4
29. 7
37. 9

7. 15
15. 8
23. 3
31. 5
39. 5

Exercise 2.11 (Page 56)

1. $\{5\}$
9. $\{1\}$
17. $\{3\}$
25. $\{4\}$
33. $\{0, 1, 2, 3, 4\}$
37. $\{0, 1\}$

3. $\{1\}$
11. $\{9\}$
19. $\{6\}$
27. $\{0\}$

39. $\{1\}$

5. $\{4\}$
13. $\{3\}$
21. $\varnothing$
29. $\{4\}$
35. $\{0, 1, 2, 3, 4\}$
41. $\{2\}$

7. $\varnothing$
15. $\{3\}$
23. $\varnothing$
31. $\varnothing$

43. $\{6, 7, 8\}$

Chapter 2 Review (Page 59)

1. 2
5. $6 > 3$

2. 0
6. $2 < 9$

3. 4
7. Cardinal

4. 4
8. Ordinal

9. $\{1, 2, 3, 4\}$

10. $\{2, 3, 4, 5, 6, 7\}$

11. $\{x \mid x < 3, x \in W\}$

12. $\{x \mid x \in N\}$

13. Transitive law of inequality

14. Symmetric law

15.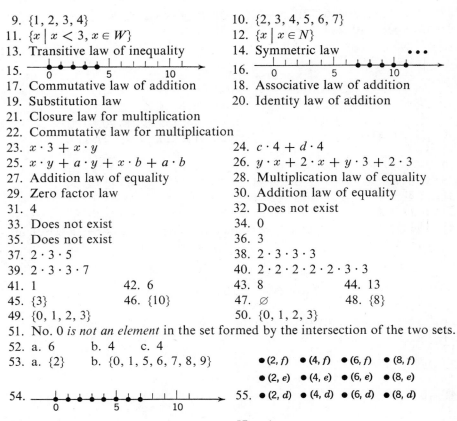

16.

17. Commutative law of addition

18. Associative law of addition

19. Substitution law

20. Identity law of addition

21. Closure law for multiplication

22. Commutative law for multiplication

23. $x \cdot 3 + x \cdot y$

24. $c \cdot 4 + d \cdot 4$

25. $x \cdot y + a \cdot y + x \cdot b + a \cdot b$

26. $y \cdot x + 2 \cdot x + y \cdot 3 + 2 \cdot 3$

27. Addition law of equality

28. Multiplication law of equality

29. Zero factor law

30. Addition law of equality

31. 4

32. Does not exist

33. Does not exist

34. 0

35. Does not exist

36. 3

37. $2 \cdot 3 \cdot 5$

38. $2 \cdot 3 \cdot 3 \cdot 3$

39. $2 \cdot 3 \cdot 3 \cdot 7$

40. $2 \cdot 2 \cdot 2 \cdot 2 \cdot 2 \cdot 3 \cdot 3$

41. 1   42. 6

43. 8   44. 13

45. $\{3\}$ 46. $\{10\}$

47. $\varnothing$ 48. $\{8\}$

49. $\{0, 1, 2, 3\}$

50. $\{0, 1, 2, 3\}$

51. No. 0 *is not an element* in the set formed by the intersection of the two sets.

52. a. 6   b. 4   c. 4

53. a. $\{2\}$ b. $\{0, 1, 5, 6, 7, 8, 9\}$

$\bullet (2, f)$ $\bullet (4, f)$ $\bullet (6, f)$ $\bullet (8, f)$

$\bullet (2, e)$ $\bullet (4, e)$ $\bullet (6, e)$ $\bullet (8, e)$

54.

55. $\bullet (2, d)$ $\bullet (4, d)$ $\bullet (6, d)$ $\bullet (8, d)$

56. $y = 0$

57. $y \geq x$

58. The factors of 15, 3, and 5 are contained in the factors of $35 \cdot 36$ which are 5, 7, 2, and 3.

59. $\{3\}$

60. $\{6, 7, 8, 9, 10\}$

Exercise 3.1 (Page 66)

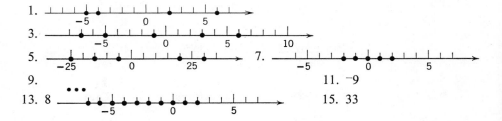

1.

3.

5.   7.

9.   11. $^-9$

13. 8   15. 33

17. ^-x

19. Negative

21. Additive inverse law

23. Double negative law

25. Additive inverse law

27. a. For every $a \in J$, $a + {}^-a = 0$ and $^-a + a = 0$.

 b. If $a \in J$, then $^-({}^-a) = a$.

Exercise 3.2 (Page 68)

1. $e > a$

3. $d > c$

5. $f > b$

7. $e > f$

9. $a < d$

11. $1 < 7$

13. $^-5 > {}^-9$

15. $^-4 < 4$

17. $^-3 < 7$

19. $^-21 < 0$

21. $12 = {}^-({}^-12)$

23. $^-3 + 3 = 0$

25. $6 + {}^-6 < 3$

27. 25

29. 7

31. 5

33. 6

35. $|{}^-3| = |3|$

37. $|{}^-5| > |2|$

39. $|{}^-10| > |{}^-8|$

41. $|{}^-1| > |0|$

43. $|{}^-2 + 2| = |0|$

45. $|{}^-y| = |y|$

Exercise 3.3 (Page 74)

1. Closure law for addition

3. Associative law of addition

5. Addition law of equality

7. $+6$

9. -3

11. -8

13. -15

15. 12

17. -14

19. -17

21. 4

23. -3

25. -10

27. -8

29. 7

31. -4

33. 13

35. -2

37. 4

39. Positive

41. 0

43. Negative

45. $x > y$

47. $x = -y$ or both x and y equal 0; x is the additive inverse of y

49. b. Additive inverse law   c. Substitution law

Exercise 3.4 (Page 79)

1. Closure law for multiplication

3. Associative law of multiplication

5. Identity law of multiplication

7. Multiplication law of equality

9. 45

11. -30

13. 54

15. 0

17. -8

19. 40

21. 0

23. -42

25. 288

27. -2

29. -5

31. 7

33. Does not exist

35. 0

37. -6

39. $(-2)(-2)$

41. $(4)(-3)$

43. $(8)(-1)$

45. $(6)(-3)$

47. Negative

49. Positive

51. Positive

53. Negative

55. $\{2, 4, 6\}$

57. {4, 8, 12, 16, 20, 24, 28}
61. {2, 5, 8, 11}

59. {3, 5, 7, 9}
63. a. Additive inverse law
    c. Distributive law
    d. Zero factor law

Exercise 3.5 (Page 82)

1. 19	3. -31	5. -4	7. 2
9. -22	11. 3	13. -5	15. 9
17. 4	19. -2	21. -7	23. 2
25. -3	27. 8	29. -12	31. 27
33. -5	35. 1	37. 24	

Exercise 3.6 (Page 85)

1. {-4}	3. {-9}	5. {5}	7. {-3}
9. {5}	11. {-2}	13. {15}	15. {7}
17. {4}	19. $\emptyset$	21. {4}	23. {0}
25. {12}	27. {-2}	29. {-6}	31. {16}
33. $\emptyset$		35. {3}	

37. {0, 1, 2, 3}   39. {1, 2, 3}

41. {$-2, -1, 0, 1, 2$}

Chapter 3 Review (Page 87)

1. -17
2. 14
3. y
4. $-x$
5. Distributive law
6. Multiplication law of equality
7. Associative law of addition
8. Zero factor law
9. Additive inverse law
10. Associative law of multiplication
11. Closure law for multiplication
12. Commutative law of addition
13. Identity law of multiplication
14. Identity law of addition
15. Commutative law of multiplication
16. Closure law for addition
17. $-7 < 3$
18. $-2 > -5$
19. $-(-3) > -3$
20. $0 < -(-5)$
21. $|5| = |-5|$
22. $|-8| > |3|$
23. $|-3| > |0|$
24. $-|-4| < 4$
25. 15
26. -6
27. -3
28. -16
29. 4
30. 0
31. 1
32. -4

33. -35 34. 18   35. -15 36. 20

37. -2 38. 4   39. 0   40. Does not exist

41. 29   42. -7 43. -10 44. -15

45. $\{-13\}$ 46. $\{8\}$ 47. $\varnothing$ 48. $\{-6\}$

49. $\{-2, -1, 0\}$ 50. $\{-5\}$ 51. $x = y$ 52. $x > y$

53. $x < 0$ and $y > 0$; $x > 0$ and $y < 0$

54. Both x and $y > 0$; both x and $y < 0$

55. $\{4, 8, 12, 16\}$ 56. $\{7, 14, 21, 28\}$

57. $\{-6, -4, -2, 0\}$ 58. $\{4, 7, 10, 13\}$

Exercise 4.1 (Page 92)

1. $\dfrac{1}{2}$ 3. $\dfrac{1}{-7}$ 5. Does not exist

7. $\dfrac{1}{\frac{4}{3}}$ 9. $\dfrac{1}{y}$ 11. $\dfrac{1}{y-7}$

13. $\dfrac{1}{a+b}$ 15. $4 \cdot \dfrac{1}{7}$ 17. $9 \cdot \dfrac{1}{5}$

19. $x \cdot \dfrac{1}{6}$ 21. $(x - y) \cdot \dfrac{1}{10}$ 23. $x \cdot \dfrac{1}{y}$

25. Reflexive law

27. Transitive law of equality or substitution law

29. Closure law for multiplication

31. Commutative law of multiplication

33. Associative law of addition

35. Identity law of addition

37. Additive inverse law

39. Zero factor law

41. Multiplication law of equality

43. $\{0, 1\}$

45. $\left\{\dfrac{1}{2}, 1, \dfrac{7}{4}\right\}$

47. $\{6, 4\}$

49. $\left\{\dfrac{1}{3}, \dfrac{3}{2}, 6, \dfrac{1}{8}, 0, 4\right\}$

51. $\dfrac{1}{0} \notin Q$

53. $x = 0$

55. For every nonzero rational number, $\dfrac{a}{b}$, $\dfrac{a}{b} \cdot \dfrac{1}{\frac{a}{b}} = 1$ and $\dfrac{1}{\frac{a}{b}} \cdot \dfrac{a}{b} = 1$.

Exercise 4.2 (Page 99)

1. $\dfrac{7}{9} = \dfrac{14}{18}$ 3. $\dfrac{7}{8} \neq \dfrac{14}{17}$ 5. $\dfrac{0}{6} = \dfrac{0}{7}$

7. $\dfrac{1}{3} \neq \dfrac{0}{2}$ 9. $\dfrac{2 \cdot 5}{3} = \dfrac{4 \cdot 5}{6}$ 11. $\dfrac{-9}{2} = \dfrac{18}{-4}$

13. $\dfrac{3}{4}$ 15. $\dfrac{6}{5}$ 17. $\dfrac{3}{5}$

19. $\dfrac{2}{3}$ 21. $\dfrac{4}{5}$ 23. $\dfrac{4 \cdot x}{9 \cdot y}$

25. $\dfrac{8}{10}$ 27. $\dfrac{0}{36}$ 29. $\dfrac{9 \cdot x}{21 \cdot y}$

31. $\dfrac{3 \cdot y}{4 \cdot y}$ 33. c., d., and e.

35. b. Multiplicative inverse law
    c. Closure law for multiplication
    g. Identity law of multiplication
    h. Theorem 4.1

37. a. $b \cdot \dfrac{a}{b} = a$ Definition 2.4

    b. $\dfrac{1}{b} \cdot \left(b \cdot \dfrac{a}{b} \right) = \dfrac{1}{b} \cdot a$ Theorem 2.2

    c. $\left(\dfrac{1}{b} \cdot b \right) \cdot \dfrac{a}{b} = \dfrac{1}{b} \cdot a$ Associative law of multiplication

    d. $1 \cdot \dfrac{a}{b} = \dfrac{1}{b} \cdot a$ Multiplicative inverse law

    e. $\dfrac{a}{b} = \dfrac{1}{b} \cdot a$ Identity law of multiplication

    f. $\dfrac{a}{b} = a \cdot \dfrac{1}{b}$ Commutative law of multiplication

Exercise 4.3 (Page 104)

1. 45   3. 105   5. 294   7. 210

9. $6 \cdot x \cdot y$ 11. $18 \cdot x \cdot y$ 13. $\dfrac{4}{5}$ 15. 1

17. 3   19. $\dfrac{9}{5}$ 21. $\dfrac{5}{14}$ 23. $\dfrac{19}{15}$

25. $\dfrac{17}{15}$ 27. $\dfrac{35}{24}$ 29. $\dfrac{209}{210}$ 31. $<$

33. $>$ 34. $<$ 37. $>$

39. a. Reflexive law of equality
    d. Distributive law
    e. Theorem 4.1

Exercise 4.4 (Page 107)

1. $\dfrac{3}{7}$

3. 1

5. $\dfrac{3}{8}$

7. $-\dfrac{1}{4}$

9. $-\dfrac{22}{21}$

11. $\dfrac{43}{32}$

13. $-\dfrac{1}{30}$

15. $\dfrac{23}{18}$

17. $-\dfrac{392}{165}$

19. $\neq$

21. $\neq$

23. $\neq$

25. Theorem 4.3

27. Theorem 4.1, Theorem 3.5

29. Theorem 4.1, Theorem 3.5, Theorem 3.6

31. c. Theorem 4.4

    d. Theorem 3.4

Exercise 4.5 (Page 112)

1. $\dfrac{4}{21}$

3. $\dfrac{18}{35}$

5. $\dfrac{2x}{15}$

7. $\dfrac{x \cdot y}{36}$

9. $-\dfrac{35}{117}$

11. $\dfrac{6}{5}$

13. $-\dfrac{3 \cdot x}{20}$

15. $\dfrac{x}{28}.$

17. $\dfrac{1}{8}$

19. $-\dfrac{4}{125}$

21. $\dfrac{1}{4}$

23. $\dfrac{5}{-3}$

25. $\dfrac{4}{3}$

27. $\dfrac{5}{y}$ $(y \neq 0)$

29. $\dfrac{y}{x}$ $(x \neq 0)$

31. $\dfrac{1}{x+2}$ $(x \neq -2)$

33. $\dfrac{15}{8}$

35. $\dfrac{10}{9}$

37. $\dfrac{3 \cdot x}{10}$

39. $\dfrac{4 \cdot x}{3 \cdot y}$ $(y \neq 0)$

41. $\dfrac{63}{65 \cdot y}$ $(y \neq 0)$

43. $\dfrac{4 \cdot x}{11 \cdot y}$ $(y \neq 0)$

45. $\dfrac{2}{3} \cdot (x + y)$

47. $\dfrac{1}{8} \cdot (5 \cdot x - y)$

49. $\dfrac{1}{12} \cdot (9 \cdot x + 5 \cdot y)$

51. Elements in the set cannot be divided by zero.

Exercise 4.6 (Page 115)

1. $\{1\}$ 3. $\left\{\dfrac{2}{7}\right\}$ 5. $\left\{\dfrac{-3}{4}\right\}$ 7. $\left\{\dfrac{-4}{7}\right\}$

9. $\left\{\dfrac{4}{3}\right\}$ 11. $\left\{\dfrac{5}{4}\right\}$ 13. $\left\{\dfrac{5}{2}\right\}$ 15. $\left\{\dfrac{3}{2}\right\}$

17. $\left\{\dfrac{-3}{2}\right\}$ 19. $\left\{\dfrac{5}{8}\right\}$ 21. $\{-1\}$ 23. $\left\{\dfrac{1}{4}\right\}$

25. $\left\{\dfrac{6}{5}\right\}$ 27. $\left\{\dfrac{1}{6}\right\}$ 29. $\left\{\dfrac{5}{6}\right\}$ 31. $\left\{\dfrac{5}{6}\right\}$

Exercise 4.7 (Page 118)

1. 0.25 3. 0.75 5. 0.625 7. 2.625
9. $0.4\overline{4}$ 11. $4.3\overline{3}$ 13. $4.72\overline{72}$ 15. 4.44

17. $\dfrac{7}{10}$ 19. $\dfrac{1}{5}$ 21. $\dfrac{1}{8}$ 23. $\dfrac{1}{16}$

Chapter 4 Review (Page 120)

1. $\dfrac{1}{5}$ 2. 8 3. $\dfrac{1}{x+y}$ $(x \neq -y)$

4. $x + y$ 5. $\dfrac{4}{7} = \dfrac{20}{35}$ 6. $\dfrac{3 \cdot 5}{8} = \dfrac{6 \cdot 5}{2 \cdot 8}$

7. $\dfrac{2}{5}$ 8. $\dfrac{5 \cdot x}{6 \cdot y}$ 9. $\dfrac{18}{42}$

10. $\dfrac{8 \cdot x}{12 \cdot y}$ 11. 120 12. $60 \cdot x \cdot y$

13. $\dfrac{6}{7}$ 14. $\dfrac{3 + x}{11}$ 15. $\dfrac{17}{60}$

16. $\dfrac{21 + 6 \cdot x - 2 \cdot y}{42}$ 17. $\dfrac{12}{35}$ 18. $\dfrac{-4 \cdot y}{7 \cdot x}$

19. 2 20. $\dfrac{-7 \cdot x}{5 \cdot y}$ 21. $\left\{\dfrac{3}{2}\right\}$ 22. $\left\{\dfrac{-4}{3}\right\}$

23. $\left\{\frac{1}{4}\right\}$ 24. $\left\{\frac{4}{5}\right\}$ 25. $\left\{\frac{9}{2}\right\}$ 26. $\left\{\frac{-4}{7}\right\}$

27. 0.625 28. 7.11$\overline{1}$ 29. $\frac{17}{20}$ 30. $\frac{423}{1,000}$

31. a., d., e., and f. 32. $\frac{3}{2}, \frac{6}{4}, \frac{12}{8}, \frac{18}{12}, \frac{24}{16}$

33. $\left\{-\frac{3}{10}\right\}$ 34. $\left\{\frac{43}{60}\right\}$

Exercise 5.1 (Page 125)

1. 3 3. 7 5. -2 7. -9

9. $\frac{2}{3}$ 11. $-\frac{7}{9}$ 13. $\sqrt{81}$ 15. $\sqrt{196}$

17. $-\sqrt{25}$ 19. $-\sqrt{100}$ 21. $\sqrt{\frac{9}{64}}$ 23. $-\sqrt{\frac{25}{81}}$

25. 8 27. -2 29. $\frac{11}{6}$ 31. $\frac{1}{2}$

33. $\frac{35}{12}$ 35. $\frac{35}{6}$ 37. $x \geq 5$

39. $x \geq -3$ 41. (Did you?)

Exercise 5.2 (Page 128)

1. $\left\{1, \frac{2}{3}, \sqrt{36}, -\sqrt{9}, 0, \sqrt{81}, -\sqrt{16}\right\}$

3. $\{\sqrt{37}, \sqrt{31}\}$ 5. $\{-\sqrt{9}, -\sqrt{16}\}$

7. $\{-\sqrt{16}\}$ 9. $\varnothing$

11. $\left\{-\sqrt{2}, 1, \frac{2}{3}, \sqrt{36}, \sqrt{37}, -\sqrt{9}, \sqrt{31}, 0, -\sqrt{17}, \sqrt{81}, -\sqrt{16}\right\}$

13. 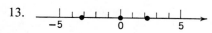 15.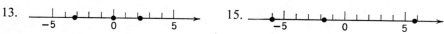

17.  19.

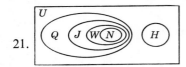

21.

23. False

25. True   27. False   29. True   31. 6.95
33. 28.16   35. 2.52

Exercise 5.3 (Page 132)

1. True   3. False   5. True   7. False
9. True   11. True

13.    15.

17.   19.

21.   23.

25. Closure law for multiplication   27. Additive inverse law
29. Distributive law   31. Commutative law of addition
33. Identity law of addition   35. Identity law of multiplication
37. Reflexive law   39. Multiplicative inverse law
41. Distributive law   43. Associative law of multiplication
45. Zero factor law   47. Addition law of equality
49. Multiplication law of equality   51. Multiplication law of equality
53. Fundamental principle of fractions
55. $x \geq 6$ 57. $x \geq -8$ 59. $\{4\}$ 61. $\varnothing$
63. $\{9\}$ 65. $\{0\}$ 67. $\{1\}$ 69. $\varnothing$
71. $\{25\}$ 73. $\{5\}$ 75. $\{\sqrt{2}, -\sqrt{2}\}$ 77. $\{\sqrt{3}, -\sqrt{3}\}$

79.   81.

Chapter 5 Review (Page 135)

1. -13 2. $\dfrac{2}{9}$ 3. $\sqrt{49}$

4. $-\sqrt{64}$ 5. $\sqrt{\dfrac{9}{25}}$ 6. $-\sqrt{\dfrac{4}{9}}$

7. $\{-3\}$ 8. $\left\{7, \dfrac{3}{4}, \sqrt{9}, 0, 2\right\}$ 9. $\{-\sqrt{5}\}$

10. $\{\sqrt{10}\}$

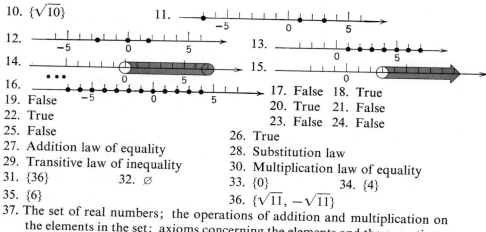

11.

12.

13.

14.

15.

16.

17. False   18. True

19. False

20. True   21. False

22. True

23. False   24. False

25. False

26. True

27. Addition law of equality

28. Substitution law

29. Transitive law of inequality

30. Multiplication law of equality

31. $\{36\}$ 32. $\varnothing$

33. $\{0\}$ 34. $\{4\}$

35. $\{6\}$

36. $\{\sqrt{11}, -\sqrt{11}\}$

37. The set of real numbers; the operations of addition and multiplication on the elements in the set; axioms concerning the elements and the operations; theorems pertaining to the elements and the operations

38.

39. $\varnothing$

40. Infinite number of points; but there is no solid line because there are an infinite number of holes for the points whose coordinates are irrational numbers.

Cumulative Review, Chapters I—5 (Page 138)

1. rule

2. union

3. complement

4. Venn

5. Cartesian

6. $\{1, 2, 3\}$

7. $\{2\}$

8. $\{0, 3\}$

9. $\{(1, 2), (1, 3), (2, 2), (2, 3)\}$

10. cardinal

11. natural

12. variable

13. axiom

14. coordinate

15. sum

16. closed

17. numerals

18. product

19. lattice

20. distributive

21. theorems

22. never

23. fraction bar

24. order

25. true statement

26. solution set

27. integers

28. additive inverses

29. sometimes

30. $-8 < -4 < 6$

31. always

32. absolute

33. always

34. sometimes

35. always

36. always

37. 3

38. never

39. always

40. -30

41. 0

42. undefined

43. 19

44. -8

45. -19

46. 24

47. rational

48. multiplicative inverses

49. $\dfrac{1}{7}$; 5

50. $a \cdot \dfrac{1}{b}$

51. fundamental principle

52. 360

53. $\dfrac{31}{20}$

54. $\dfrac{a}{b}$

55. $-\dfrac{a}{b}$

56. $\dfrac{47}{45}$

57. $\dfrac{4}{7}$

58. $\dfrac{7}{9}$

59. always

60. $\dfrac{97}{100}$

61. $\dfrac{775}{1,000}$

62. rational

63. real

64. $\left\{-\sqrt{48}, -\sqrt{5}, \dfrac{3}{2}, 0\right\}$

65. $\left\{-\sqrt{64}, 0, \dfrac{3}{2}, 2, \sqrt{49}\right\}$

66. $\{-\sqrt{64}, 0, 2, \sqrt{49}\}$

67. always

68.

69.

70. $N \subset J$

71. $Q \cup H = R$

72. $-5 \in J$

73. $Q \subset R$

74. $Q \cap H = \varnothing$

75. $4 \in Q$

76. $\sqrt{5} \in R$

77. $R \cup J = R$

78. {prime numbers} $\cap$ {composite numbers} $= \varnothing$

79. 3

80. $\{-5\}$

81. $\left\{\dfrac{10}{3}\right\}$

82. $\left\{\dfrac{-2}{5}\right\}$

83. $\{10\}$

84. $\left\{\dfrac{-5}{6}\right\}$

85. $\{100\}$

86. $\{\sqrt{12}, -\sqrt{12}\}$

Exercise 6.1 (Page 145)

1. 64

3. 250

5. -64

7. 540

9. 36

11. -72

13. $2 \cdot 2 \cdot 5$

15. $2 \cdot 2 \cdot 2 \cdot 2 \cdot 3 \cdot 3$

17. xxx

19. $xxyy$

21. $2 \cdot 2 \cdot 2 \cdot 3xxyzz$

23. $3 \cdot 3 \cdot 5 \cdot 5xxxy$

25. 2^3

27. x^3

29. $2 \cdot 3^2 y^2 z$

31. $5rs^2 t$

33. $2r^2 s$

35. y^5

37. $2 \cdot 6$ 39. pq 41. $4 \cdot \dfrac{1}{x}$ or $\dfrac{4}{x}$

43. $.5(3 + x)$
45. Associative law of multiplication
47. Identity law of multiplication
49. Closure law for multiplication
51. Transitive law of equality or substitution law

Exercise 6.2 (Page 150)

1. monomial; degree 1
3. binomial; degree 3
5. binomial; degree 2
7. trinomial; degree 2
9. trinomial; degree 3
11. polynomial of four terms; degree 3
13. 2; 3
15. 1st term: 4; 4   2nd term: -1; 2
17. 1st term: 4; 2   2nd term: 1; 3   3rd term: -2; 4
19. 14   21. -20 23. 2   25. 1
27. 32   29. 10   31. 6   33. 4
35. 0   37. 0   39. $\dfrac{15}{4}$ 41 -2.
43. 1   45. 3   47. 1   49. 0
51. -4 53. 0 and 1   55. 1 and -2
57. Closure law for multiplication   59. Distributive law
61. Multiplication law of equality   63. 28

Exercise 6.3 (Page 154)

1. a. Distributive law
   b. Basic numeral
3. a. Distributive law
   b. Theorem 3.4
   c. Basic numeral
5. a. Commutative law of addition
   b. Associative law of addition
   c. Distributive law
   d. Basic numeral
7. $6x^2 - 8x$ 9. $9p^2 - q$ 11. $2x^2 + xy + 2y^2$
13. $8rs + 5u$ 15. $7x^2 + 2x + 2$ 17. $2y^2 + y + 2$
19. $2x^3 + 5x^2 - 7x - 2$ 21. $3x^2y + 2xy + xy^2$ 23. $2x^2 - 4x + 2$
25. $16t^2 + t - 9$ 27. $-2x^2y + 12xy - 5xy^2$
29. $z^3 - 2z^2 - 4z$ 31. $-3p$ 33. $3x^2 + 2$
35. $-3x^2y + 4xy + xy^2$ 37. $10x^2 - 6x + 8$ 39. $14t^2 + 2t - 3$
41. $-12x^2y + 12xy^2$ 43. $2x - 8$ 45. -1
47. -7 49. $x + 12$

Exercise 6.4 (Page 159)

1. a. Theorem 6.1   b. Basic numeral
3. a. Theorem 6.2   b. Basic numeral
5. a. Associative law of multiplication
   b. Theorem 6.1
   c. Basic numeral
7. a. Theorem 4.7   b. Theorem 6.2
   c. Basic numeral   d. Definition 2.4

9. y^6 11. x^9 13. $-x^5$ 15. $6x^2y^3$
17. $-35p^2q^4$ 19. $8x^3y^5$ 21. $24x^3$ 23. $-24x^6$
25. $3x^4y^4$ 27. r 29. $-2x^2$ 31. $-2r^2s^2$
33. $2x^3yz$ 35. $-5x^3yz^3$ 37. $-2xyz$ 39. x^8y
41. $\frac{1}{2}xyz$ 43. $11t^4$ 45. $-2p^2q^2$ 47. $3xy^2 + 2x^2y$
49. $x + 2x^2y$ 51. $x^3 - x^5$ 53. $35x^6$
55. $a^m \cdot a^n = a^{m+n}; \ (m, n \in N)$ 57. x^6 59. $4y^2$

61. $\dfrac{4}{y^2}$ 63. $\dfrac{x^3y^3}{r^3s^3}$ 65. x^{mn} 67. $\dfrac{x^m}{y^m}$

Exercise 6.5 (Page 165)

1. Definitions 6.3 and 6.1   3. Definition 6.2
5. a. Definition 6.3
   b. Definition 6.1
   c. Basic numeral
7. a. Definition 6.3   9. a. Definition 6.2
   b. Theorem 4.7 or 4.8   b. Definition 6.3
   c. Definition 2.4   c. Theorem 4.8
   d. Identity law of multiplication   d. Identity law of multiplication
      (or basic numeral)   and Definition 2.4

11. $\dfrac{2}{9}$ 13. $\dfrac{1}{4}$ 15. $\dfrac{9}{64}$ 17. 9

19. 1   21. $\dfrac{5}{3}$ 23. x 25. $\dfrac{1}{x^7}$

27. $\dfrac{y}{x}$ 29. $\dfrac{y^4}{x^2}$ 31. 1.4×10^1 33. 7.145×10^3
35. 6.2×10^4 37. 5.0×10^{-1} 39. 2.6×10^{-3} 41. 5.16×10^{-6}
43. 1,020   45. -247 47. 207,000,000   49. 0.0202
51. 0.0000545   53. -0.00022 55. 10^{-4} 57. 1.2×10^2

Exercise 6.6 (Page 170)

1. Distributive law

3. a. Distributive law
   b. Definition 6.1

5. a. Distributive law
   b. Theorem 3.6
   c. Theorem 3.5

7. $12x^2 - 3xy^2$

9. $x^2y - xy^2$

11. $x^3 + x^2 - 2x$

13. $6y^4 + 9y^3 - 3y^2$

15. $x^3 - 3x^2 - 2x$

17. a. Distributive law   b. Distributive law
    c. Definition 6.1   d. Commutative law of multiplication
    e. Distributive law   f. Basic numeral

19. $t^2 + 6t + 9$

21. $z^2 - 25$

23. $y^2 - 25$

25. $y^2 + 14y + 49$

27. $10x^2 + 17x + 3$

29. $6y^2 + y - 1$

31. $6x^2 + 13x + 6$

33. $9x^2 - 4y^2$

35. $4x^2 + 4xy + y^2$

37. $2x^2 + 10x + 12$

39. $6x^2 - 16x - 6$

41. $7x^2 - 343$

43. $y^3 + y^2 - 2y$

45. $24s^3 - 24s^2 + 6s$

47. $12r^3s + 26r^2s^2 + 12rs^3$

49. $x^3 + 3x^2 + 3x + 2$

51. $x^4 - 13x^2 + 36$

53. $x^3 - y^3$

55. $x^2 + 2xy + y^2$; $a^2x^2 + 2abxy + b^2y^2$

57. $x^2 - y^2$; $a^2x^2 - b^2y^2$

Exercise 6.7 (Page 175)

1. $3(x + 2)$

3. $2x(3x - 1)$

5. $xy(3y + 2x)$

7. $3y(3xy + 2)$

9. $3(r + 2s - 3t)$

11. $p(x + y - z)$

13. $x(2 + y + x)$

15. $9(2xy - 3x + 5y)$

17. $2rs(1 - 2rs - 4r^2s^2)$

19. $-x(1 + x)$

21. $-xy(x^2 - y)$

23. $-3(x^2 + 2x - 1)$

25. $-x(1 - x + x^2)$

27. $-pq(p - 2 + q)$

29. $-1(2 - x)$

31. $-1(2x - x^2)$

33. $-1(m^2 + m + 1)$

35. $x(x + 2)$; 8

37. $xy(3x - 5)$; 24

Exercise 6.8 (Page 178)

1. $(x + 2)(x + 3)$

3. $(x + 3)(x + 4)$

5. $(x + 2)(x - 5)$

7. $(x + 2)(x - 4)$

9. $(x + 3)(x - 2)$

11. $(x + 4)(x - 3)$

13. Not factorable

15. Not factorable

17. $(x - 2)(x - 4)$

19. $(x - 2)(x - 5)$

21. $(x + 2)(x + 2)$

23. $(x + 4)(x + 4)$

25. $(x + 1)(x + 6)$

27. Not factorable

29. $(x + 5)(x - 4)$

31. $(x + 6)(x - 13)$

33. $(x + 4)(x - 4)$

35. $(x + 6)(x - 6)$

37. $5(x + 1)(x + 3)$

39. $5(x + 4)(x - 2)$

41. $6(x^2 + 6x + 4)$

43. $7(x + 1)(x + 1)$

45. $7(x - 5)(x - 5)$

47. $12(x + 1)(x + 6)$

Exercise 6.9 (Page 182)

1. $(3x + 4)(x + 1)$
3. $(2x + 5)(2x + 1)$
5. Not factorable
7. $(5x + 4)(x - 1)$
9. $(3x - 2)(2x + 3)$
11. $(3x - 2)(x - 2)$
13. $(2x + 1)(2x + 1)$
15. $(2x + 3)(2x + 3)$
17. Not factorable
19. $(5x - 7)(x + 1)$
21. $(6x + 7)(x - 1)$
23. $(2x + 3)(2x - 3)$
25. Not factorable
27. $(4s + t)(s + t)$
29. $(3x - y)(x - 2y)$
31. $(4y + 5z)(4y - 5z)$
33. $2x(4y + 1)(y - 1)$
35. $9(x + 1)(2x - 3)$
37. $2xy(5y + x)(5y - x)$
39. $3xy(3y + x)(3y - x)$
41. $2x(2x + y)(x - 3y)$
43. $(7x + 8)(3x + 2)$
45. $(7x + 11)(3x - 5)$
47. $(17x + 19y)(17x - 19y)$
49. $13(3x + 1)(x - 2)$

Exercise 6.10 (Page 185)

1. $4y - 1$ $(y \neq 0)$
3. $6y - 2 + x$ $(x, y \neq 0)$
5. $-3xy^2 + y - 1$ $(x, y \neq 0)$
7. $2x^2 + x - 3$
9. $2x^2 + x + 1$ $(x \neq 0)$
11. $-x^2y + x - 1$ $(x, y \neq 0)$
13. $x + 6$
15. $r - 1$
17. $p + 3$
19. $x - 4$
21. $2x + 3$
23. $2x - 5$
25. $2x + 4$
27. $2x + 2$
29. $2x - 2$
31. $x^2 + 2x + 5$ $(x \neq -1)$
33. $x^2 + x + 1$ $(x \neq 1)$
35. $x^2 - x + 2$ $(x \neq -2)$
37. $x^3 + x^2 + x + 1$ $(x \neq 1)$

Chapter 6 Review (Page 188)

1. $2 \cdot 2 \cdot 2 \cdot 3 \cdot 3xxyyy$
2. $2^2 r^2 s^2 t^3$
3. Trinomial; fourth degree
4. First term: 7; 4. Second term: 2; 13
5. 5
6. -2
7. 14
8. -5
9. 1, -2
10. $2x^2 + 2y^2$
11. $12x^3y^4$
12. $9r^2/s^2$ or $9r^2 s^{-2}$
13. y^3/x^2
14. 3.72×10^4
15. 2.34×10^{-4}
16. 0.00421
17. 5,760,000
18. $6x^2 - 5x - 21$
19. $3xy(2x + 1 - 6y)$
20. $(3x + 2)(2x + 3)$
21. $(7xy + 11z)(7xy - 11z)$
22. Not factorable
23. $2x^2 - x + 4$
24. $3x + 5$
25. 0
26. 6
27. $\dfrac{3y^5}{x^4}$
28. 100
29. $\dfrac{1}{250}$
30. $4(3x + 7)(5x - 6)$

31. $x^4 + x^3 + x^2 + x + 1$ $\quad (x \neq 1)$
32. $x^5 - x^4 + x^3 - x^2 + x - 1$ $\quad (x \neq -1)$

Exercise 7.1 (Page 193)

1. Theorem 4.1
3. a. Distributive law
   b. Theorem 7.3

5. a. Distributive law
   b. Theorem 7.3

7. $3x^2 - 2x + 1$ $\quad (x \neq 0)$

9. $x - 1$ $\quad (x \neq 0)$

11. $3x^2 + 2x - 1$ $\quad (x \neq 0)$

13. $x + 7$ $\quad (x \neq 2)$

15. $\dfrac{x - 2y}{x + 2y}$ $\quad (x \neq 2y, -2y)$

17. $\dfrac{x - 2}{x - 3}$ $\quad (x \neq 3, -3)$

19. $\dfrac{p + 3}{p - 1}$ $\quad (p \neq 1, -3)$

21. $3x + 2 + \dfrac{1}{x}$

23. $2x + 1 - \dfrac{1}{4x}$

25. $x - 2 + \dfrac{1}{x + 5}$

27. $x + 1 - \dfrac{3}{x - 7}$

29. $2x - 3 + \dfrac{1}{x + 1}$

31. $3x + 1 + \dfrac{3}{2x + 1}$

33. $2x^2 - x + 1 - \dfrac{1}{3x + 2}$

35. $2x + 2 + \dfrac{x - 5}{3x^2 + x + 1}$

37. True

39. True

41. True

43. No; the denominator may not be a factor of the numerator.

Exercise 7.2 (Page 196)

1. a. Theorem 7.3
   b. Basic numerals for $2 \cdot 5$ and $3 \cdot 5$
3. a. Theorem 7.3
   b. Distributive law (4 times)

5. $\dfrac{x + y}{3x + 3y}$ $\quad (x \neq -y)$

7. $\dfrac{3x^2 - 3}{9x + 9}$ $\quad (x \neq -1)$

9. $\dfrac{5x - 5y}{x^2 - y^2}$ $\quad (x \neq y, -y)$

11. $\dfrac{4x^2 - y^2}{10x - 5y}$ $\quad \left(x \neq \dfrac{y}{2}\right)$

13. $\dfrac{s^2 + s - 6}{5rs - 10r}$ $(r \neq 0, s \neq 2)$

15. $\dfrac{7xy + 7x^2}{y^2 - x^2}$ $(y \neq x, -x)$

17. $\dfrac{x^2 + 2x - 8}{x^2 + 5x + 4}$ $(x \neq -1, -4)$

19. $\dfrac{4x^2 + 4x - 3}{2x^2 - x - 6}$ $\left(x \neq -\dfrac{3}{2}, 2\right)$

21. $\dfrac{x^2 - 4y^2}{x^2 - 4xy + 4y^2}$ $(x \neq 2y)$

23. $\dfrac{-1}{1 - x}$ $(x \neq 1)$

25. $\dfrac{3 - 6x}{-3x^2}$ $(x \neq 0)$

27. $\dfrac{36x^2 - 3x - 105}{12x^3 - 21x^2}$ $\left(x \neq \dfrac{21}{12}, 0\right)$

29. $\dfrac{25x^2 - 49}{25x^2 + 70x + 49}$ $\left(x \neq \dfrac{-7}{5}\right)$

31. $\dfrac{x^3 + 1}{x^3 - 2x^2 + 2x - 1}$ $(x \neq 1, x^2 - x + 1 \neq 0)$

33. $\dfrac{2x^3 + x^2 - 4x - 2}{x^3 + 2x^2 - 2x - 4}$ $(x \neq -2, x^2 - 2 \neq 0)$

35. $\dfrac{1 - x^4}{-2x^3 + 3x^2 + 2x - 3}$ $\left(x \neq \dfrac{3}{2}, 1 - x^2 \neq 0\right)$

Exercise 7.3 (Page 201)

1. a. Theorem 7.4
   b. Distributive law
   c. Basic numeral

3. a. Theorem 7.4
   b. Distributive law
   c. Basic numeral

5. $\dfrac{2x + 1}{2y}$

7. $\dfrac{3x - 2y}{x}$

9. $\dfrac{5x - 2}{4y}$

11. $\dfrac{5}{x + 2y}$

13. $\dfrac{6 - 2x}{x^2 - 2x + 1}$

15. $2 \cdot 3xy^2$

17. $2 \cdot 2 \cdot 2 \cdot 3x^2y^2$

19. $(x - y)(x + y)$

21. $(x + 2)(x - 2)(x - 3)$

23. $2 \cdot 2(y + 1)(y - 1)(y - 1)$

25. $\dfrac{4y + 3x}{xy}$

27. $\dfrac{2y + 3x}{x^2y}$

29. $\dfrac{y}{2}$

31. $\dfrac{8x - y}{6}$

33. $\dfrac{11x + 43}{(x + 2)(x + 5)}$

35. $\dfrac{3x - 3}{(x + 3)(x - 3)}$

37. $\dfrac{3x - y}{(x + y)(x - y)}$

39. $\dfrac{2x}{(x-1)(x+1)(x+1)}$

41. $\dfrac{2y-7}{(y+4)(y-4)}$

43. $\dfrac{3x^2+x}{(x+2)(x-2)(x-3)}$

45. $\dfrac{x^2-2xy+y^2+2x+2y}{2(x+2y)(x-y)}$

Exercise 7.4 (Page 204)

1. a. Theorem 7.6
   b. Distributive law
   c. Basic numeral for $3-2$

3. a. Theorem 7.6
   b. Theorem 3.4
   c. Distributive law
   d. Basic numeral for $4+(-2)$

5. $\dfrac{2}{3x}$

7. $\dfrac{2x-3}{x}$

9. $\dfrac{y-3}{4x}$

11. -2

13. $\dfrac{-5}{x+2y}$

15. $\dfrac{x-4}{x^2-3x+2}$

17. $\dfrac{3y-2x}{xy}$

19. $\dfrac{3x-y}{xy^2}$

21. $\dfrac{-x-7}{6}$

23. $\dfrac{7x-5y}{6}$

25. $\dfrac{x+11y}{6}$

27. $\dfrac{w}{(w+3)(w+2)}$

29. $-\dfrac{-(2x-y)}{3}=-\dfrac{2x-y}{-3}=-\dfrac{(2x-y)}{-3}=\dfrac{-2x+y}{-3}=\dfrac{y-2x}{-3}$

31. $\dfrac{-(3y-x)}{5}=\dfrac{3y-x}{-5}=-\dfrac{-(3y-x)}{-5}=\dfrac{-3y+x}{-5}=\dfrac{x-3y}{-5}$

33. $-\dfrac{-3}{2y-x}=-\dfrac{3}{-(2y-x)}=\dfrac{-3}{-(2y-x)}=\dfrac{-3}{-2y+x}=\dfrac{-3}{x-2y}$

35. $\dfrac{4}{3-x}$

37. $\dfrac{3}{x-4}$

39. $\dfrac{6}{x-y}$

41. $\dfrac{-2}{(x+1)(x+1)(x-1)}$

43. $\dfrac{-2x+12}{(x-1)(x+1)(x+4)}$

45. $\dfrac{x^2-4x}{(x-2)(x+2)(x-3)}$

47. $\dfrac{-x^2-x}{(x+2)(x-1)(x+3)}$

Exercise 7.5 (Page 209)

1. Theorem 7.7

3. a. Theorem 7.7
   b. Commutative law of multiplication
   c. Theorem 7.3

5. 5

7. $\dfrac{y}{8x}$

9. $2y$

11. $\dfrac{1}{2}$

13. 5

15. $\dfrac{x-3}{x+7}$

17. $\dfrac{x+y}{x+3y}$

19. $\dfrac{4x}{5}$

21. 1

23. $\dfrac{x-3y}{4x+y}$

25. $\dfrac{s-1}{s+7}$

27. $\dfrac{(x-4)(x+3)}{(x-1)(2x+1)}$

29. $\dfrac{x}{x+1}$

31. $\dfrac{(x+2)(x+3)}{(x+6)(x+1)}$

33. $\dfrac{2-t}{2t(t+1)}$

35. -4

37. $\dfrac{-(x+3)}{2x^2}$

39. $\dfrac{x^2(x-1)}{2(x+3)}$

Exercise 7.6 (Page 212)

1. $\dfrac{8}{27}$

3. $\dfrac{2}{3}$

5. $\dfrac{1}{2}$

7. $\dfrac{16}{7}$

9. $\dfrac{x^2-1}{x^2+1}$

11. $\dfrac{xy+x}{y-1}$

13. $\dfrac{2x-3y}{8-3y}$

15. $\dfrac{7}{10x+2}$

17. $\dfrac{1}{x-1}$

19. $\dfrac{3}{y+1}$

21. $\dfrac{5}{4}$

23. $\dfrac{x-2}{x}$

25. $\dfrac{x^2+5x+2}{x+1}$

Chapter 7 Review (Page 215)

1. $\dfrac{x+5}{x+1}$ $(x \neq 5, -1)$

2. $\dfrac{x-7}{x+1}$ $(x \neq -1)$

3. $\dfrac{x^2-x-2}{x^2-4}$

4. $\dfrac{9x+27}{3x^2-27}$

5. $3x(x + 6)(x + 1)(x - 1)$

6. $(x + 2)(x + 2)(x + 1)(x + 1)$

7. $\dfrac{43x - 25}{15}$

8. $\dfrac{13x}{6}$

9. $\dfrac{5x - 5}{(x + 2)(x - 2)}$

10. $\dfrac{4x + 11}{(x - 3)(x + 2)}$

11. $\dfrac{7x - 9}{(x + 3)(x - 3)(x + 1)}$

12. $\dfrac{-5}{(y + 4)(y - 4)(y - 1)}$

13. $\dfrac{5}{2}$

14. $\dfrac{1}{8x}$

15. 5

16. $\dfrac{4}{3}$

17. $\dfrac{-5}{3x + 2}$

18. $-s$

19. $2x$

20. $\dfrac{2 - xy}{1 + xy}$

21. $x - 1$

22. $\dfrac{x + y}{x - y}$

23. $\dfrac{-2x^2 + 10x - 4}{(x - 1)(x - 1)(x + 1)}$

24. $\dfrac{2x^2 + 13x + 6}{(x + 1)(x + 2)(x + 2)}$

25. $\dfrac{x + 3}{x - 5}$

26. $\dfrac{s(2s - 1)}{s + 4}$

27. $\dfrac{y - 5}{y - 2}$

28. $\dfrac{(t + 2)(t + 4)}{(t + 3)(t + 1)}$

29. $\dfrac{x}{1 - x}$

30. $\dfrac{2y^2 - 3y - 3}{y - 3}$

Exercise 8.1 (Page 218)

1. Conditional; $\{7\}$ 3. Conditional; $\{3\}$ 5. Conditional; $\{13\}$
7. Conditional; $\{8\}$ 9. Conditional; $\{-5\}$ 11. Identity
13. Conditional; $\{5\}$ 15. Identity   17. Conditional; $\{20\}$
19. Conditional; $\{-24\}$ 21. Identity   23. Identity
25. Identity   27. $y \in R, y \neq 0$ 29. $x \in R, x \neq -4$
31. $x \in R, x \neq -2$ 33. $y \in R, y \neq 4, -5$

Exercise 8.2 (Page 224)

1. Theorem 8.1: If $P(x) = Q(x)$, then $P(x) + R(x) = Q(x) + R(x)$
3. Distributive law

5. Theorem 8.1: If $P(x) = Q(x)$, then $P(x) + R(x) = Q(x) + R(x)$
7. Theorem 8.1: If $P(x) = Q(x)$, then $P(x) + R(x) = Q(x) + R(x)$

9. Distributive law   11. $\{9\}$ 13. $\left\{\dfrac{-2}{5}\right\}$

15. $\left\{\dfrac{10}{3}\right\}$ 17. $\{0\}$ 19. $\left\{\dfrac{-5}{7}\right\}$

21. $\{10\}$ 23. $\{18\}$ 25. $\{-20\}$
27. $\{0\}$ 29. $\{3\}$ 31. $\{3\}$

33. $\{-3\}$ 35. $\{2\}$ 37. $\left\{\dfrac{3}{4}\right\}$ 39. $\{4\}$

Exercise 8.3 (Page 227)

1. $\{1\}$ 3. $\{1\}$ 5. $\{6\}$ 7. $\{4\}$
9. $\{4\}$ 11. $\{3\}$ 13. $\{-5\}$ 15. $\{2\}$

17. $\left\{\dfrac{1}{5}\right\}$ 19. $\{60\}$ 21. $\{-1\}$ 23. $\left\{\dfrac{22}{3}\right\}$

25. $\{2\}$ 27. $\{5\}$ 29. $\varnothing$
31. $\{4\}$ 33. $\{7\}$ 35. $\{-2\}$

37. $\{3\}$ 39. $\{3\}$ 41. $\left\{\dfrac{7}{2}\right\}$

43. For any replacement value for x, the denominator is always 1 unit larger than the numerator. Thus, the fraction can never equal 1.

Exercise 8.4 (Page 231)

1. $b - a + c = y$; Theorem 8.1 (add $-a + c$)

3. $8y = a^2b$; Theorem 8.1 (add $4y$); $y = \dfrac{a^2b}{8}$; Theorem 8.2 $\left(\text{multiply by } \dfrac{1}{8}\right)$

5. $\dfrac{3x}{4} = b + c$; Theorem 8.1 (add c);

   $x = \dfrac{4(b + c)}{3}$; Theorem 8.2 $\left(\text{multiply by } \dfrac{4}{3}\right)$

7. $y = 7 - x$ 9. $b = p - a - c$

11. $r = \dfrac{I}{pt}$ $(p, t \neq 0)$

13. $m = \dfrac{E}{c^2}$ $(c \neq 0)$

15. $w = \dfrac{p - 2l}{2}$

17. $t = \dfrac{v - a}{g}$ $(g \neq 0)$

19. $x = \dfrac{k}{y}$ $(x, y \neq 0)$

21. $y = \dfrac{-2}{4 + x}$ $(x \neq -4)$

23. $x = \dfrac{4}{y + 12}$ $(y \neq -12)$

25. $y = \dfrac{x + 3}{x}$ $(x \neq 0)$

27. $r = \dfrac{s - a}{s}$ $(s \neq 0)$

29. $b = \dfrac{2A - hc}{h}$ $(h \neq 0)$

Exercise 8.5 (Page 235)

1. $x < 5$; Theorem 8.3 (add 3); $\{x \mid x < 5, x \in J\}$

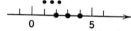

3. $x < 2$; Theorem 8.4 $\left(\text{multiply by } \dfrac{1}{4}\right)$; $\{x \mid x < 2, x \in J\}$

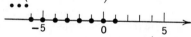

5. $x \geq 7$; Theorem 8.3 (add 6), Theorem 8.4 $\left(\text{multiply by } \dfrac{1}{2}\right)$;

$\{x \mid x \geq 7, x \in R\}$

7. $\{x \mid x > -6, x \in R\}$

9. $\{x \mid x > 7, x \in J\}$

11. $\{x \mid x > 1, x \in J\}$

13. $\{x \mid x < 1, x \in R\}$

15. $\{x \mid x > 3, x \in R\}$

17. $\{x \mid x \leq -1, x \in R\}$

19. $\{x \mid x < -6, x \in R\}$

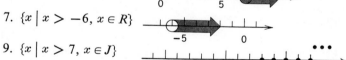
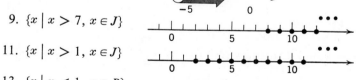
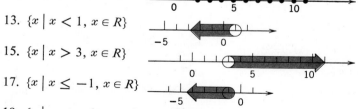
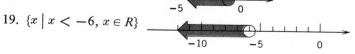

21. $\{x \mid x \geq -7, x \in R\}$
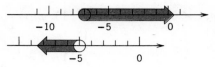

23. $\{x \mid x < -5, x \in R\}$

Exercise 8.6 (Page 237)

1. $14 + n$ 3. $3n - n$
5. a. $23 - s$ 7. a. 105 9. a. $12x$
   b. $6s$ b. $10(y - 3)$ b. $3(x + 15)$
   c. $\dfrac{1}{2}(23 - s)$ c. $50(3t + 4)$

                             d. $25\left(\dfrac{x}{2} - 4\right)$

11. {whole numbers}   13. {whole numbers}
15. {positive real numbers}   17. {integers}
19. {real numbers}

Exercise 8.7 (Page 242)

(There are several possible models for each exercise.)

1. a. $x + x - 72 = 112$ 3. a. $4x + x = 135$
   b. $\{92\}$ b. $\{27\}$
   c. 92 and 20   c. 108 and 27

5. a. $x + \dfrac{2}{5}x = 105$

   b. $\{75\}$
   c. 75 and 30
7. a. $x + x + 1 + x + 2 + x + 3 = 206$
   b. $\{50\}$
   c. 50, 51, 52, and 53
9. a. $x + x + 2 + x + 4 = 237$ 11. a. $(x + 2)^2 - x^2 = 52$
   b. $\{77\}$ b. $\{12\}$
   c. 77, 79, and 81   c. 12 and 14
13. a. $x + 5 = 3x - 9$ 15. a. $x + 4(x - 3) = 123$
    b. $\{7\}$ b. $\{27\}$
    c. 7 and 21   c. 27 and 24

17. a. $2x = 5(25 - x) + 1$
   b. $\{18\}$
   c. 18 and 7

19. a. $x(x + 2) = (x + 2)^2 - 22$
   b. $\{9\}$
   c. No such even integers.

21. a. $x^2 + (x + 1)^2 + (x + 2)^2 = 3x^2 + 59$
   b. $\{9\}$
   c. 9, 10, and 11

23. a. $(9x + 7x) - (9x - 7x) = 56$ b. $\{4\}$ c. 36 and 28

25. a. $\dfrac{3}{2}(x + x + 2) = (x + 4) - 9$ b. $\{-4\}$ c. -4, -2, and 0

Exercise 8.8 (Page 245)

(There are several possible models for each exercise.)

1. a. $x + x + 10 = 36$
   b. $\{13\}$
   c. 13 ft. and 23 ft.

3. a. $210x + 75(166 - x) = 29{,}325$
   b. $\{125\}$
   c. 125 adults, 41 children

5. a. $x + x - 6{,}900 = 72{,}100$
   b. $\{39{,}500\}$
   c. 32,600 votes cast by 18–21 age group.

7. a. $x + x + x - 8 + x - 8 = 76$
   b. $\{23\}$
   c. Length, 23 feet; width, 15 feet.

9. a. $x + x + 2 + x + 3 = 50$
   b. $\{15\}$
   c. 15, 17, and 18 inches

11. a. $10(x + 3) + 25x = 240$
   b. $\{6\}$
   c. 9 dimes, 6 quarters

13. a. $10(2x) + x + 5(102 - 3x) = 648$
   b. $\{23\}$
   c. 23 pennies, 46 dimes, 33 nickels

15. a. $0.20x + 0.50(30) = 0.40(30 + x)$
   b. $\{15\}$
   c. 15 quarts

17. a. $0.50x + 0.70(30 - x) = 0.55(30)$
   b. $\{22.5\}$
   c. 22.5 ounces

19. a. $0.04x + 0.05(6{,}200 + x) = 472$
   b. $\{1{,}800\}$
   c. \$1,800 at 4%; \$8,000 at 5%

21. a. $0.06x = 0.07(26{,}000 - x)$
   b. $\{14{,}000\}$
   c. \$840 interest

23. a. $w = 5.4(65) - 220$
   b. $\{131\}$
   c. 131 lbs.

25. a. $I.Q. = \dfrac{18(100)}{15.5}$

    b. $\{116.1\}$

    c. 116.1

27. a. $\dfrac{3,600}{x} = \dfrac{12,600}{x + 2,700}$

    b. $\{1,080\}$

    c. Plane: 1,080 m.p.h.;
       rocket: 3,780 m.p.h.

29. a. $2(3x) - 2(x) = 100$

    b. $\{25\}$

    c. Slower, 25 m.p.h.;
       faster, 75 m.p.h.

31. a. $80 \leq \dfrac{76 + 92 + 95 + 60 + x}{5} < 90$

    b. $\{x \mid 77 \leq x < 127\}$

    c. 77% or higher

33. a. $-20 < \dfrac{9}{5} C + 32 < 10$

    b. $\{C \mid -28.9 < C < -12.2\}$

    c. Between $-28.9°C.$ and $-12.2°C.$

Chapter 8 Review (Page 254)

1. $\{17\}$
2. $\{22\}$
3. Identity
4. $\{48\}$
5. Distributive law
6. Theorem 8.1
7. Theorem 8.1
8. Theorem 8.2
9. $\{0\}$
10. $\{3\}$
11. $\{-8\}$
12. $\{0\}$

13. $\{3\}$

14. $\left\{\dfrac{5}{7}\right\}$

15. $x = -4a$

16. $c = \dfrac{2y}{3(a^2 - 1)} \quad (a \neq 1, -1)$

17. $y = \dfrac{3}{2 - x} \quad (x \neq 2)$

18. $h = \dfrac{2A}{b + c} \quad (b + c \neq 0)$

19. $\{x \mid x < 5, x \in J\}$

20. $\{x \mid x > -6, x \in R\}$

21. $\{x \mid x \geq 4, x \in R\}$

22. $\{x \mid x \leq 0, x \in J\}$

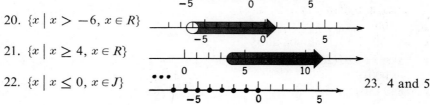

23. 4 and 5

24. 13 and 70
25. Width: 9 ft; length: 14 feet
26. 11 in., 14 in., 21 in.
27. 8 quarters, 17 dimes
28. 9 dimes, 12 quarters, 6 halves
29. 40 ounces
30. 750 gallons
31. $8,000 at 4%, $9,600 at 5%
32. $4,000 at 6%, $2,000 at 7%, $6,000 at 5.5%
33. 75 lbs. per sq. ft.
34. 15°C.
35. 5 m.p.h. riding, 2 m.p.h. walking
36. 5,000 m.p.h. and 2,500 m.p.h.
37. 88% or greater
38. Between 816 and 1,156 m.p.h.

Cumulative Review, Chapters 6–8 (Page 257)

1. factored
2. completely (or prime)
3. trinomial
4. coefficient
5. greatest
6. 3
7. real numbers
8. a^{m-n}

9. 1
10. $\dfrac{1}{a^m}$
11. scientific

12. distributive
13. b
14. $Q(x)$
15. $P(x)$
16. $P(x) + Q(x)$
17. $-P(x)$
18. $-P(x)$
19. $P(x) - R(x)$
20. $P(x) \cdot R(x)$

21. $\dfrac{S(x)}{R(x)}$
22. $\dfrac{x^2 + 2}{x - 1}$
23. solution

24. solution set
25. identities
26. equivalent
27. equivalent
28. multiplying
29. same
30. same
31. opposite
32. mathematical model

Exercise 9.1 (Page 263)

1. $(1, 2), (0, 1), (-2, -1)$
3. $(-5, -11), (0, -1), (3, 5)$
5. $(0, 5), (2, -1), (4, -7)$
7. $(-2, 1), (0, 5), (2, 9)$
9. $(-2, -2), (0, 2), \left(5, \dfrac{1}{3}\right)$
11. $(0, -4), (2, 2), \left(\dfrac{4}{3}, 0\right)$
13. $(2, 1), (0, 3), (-2, -3)$

15. $(-2, 5), (-1, 4), (0, 3), (1, 2), (2, 1)$

17. $\left(\frac{1}{2}, -3\right), \left(\frac{1}{4}, -2\right), \left(\frac{1}{8}, -\frac{3}{2}\right), \left(\frac{1}{16}, -\frac{5}{4}\right)$

19. $(1, 8), (2, 4), \left(3, \frac{8}{3}\right), (4, 2)$ 21. $y = -x + 6$

23. $y = 3x - 12$ 25. $y = \frac{2x + 3}{2}$ 27. $y = 7x - 3$

29. $y = -12x + 24$ 31. $y = \frac{3}{x}$ 33. $y = \frac{3}{1 - x^2}$

35. $y = \frac{1}{1 - 3x}$ 37. Theorem 8.1

39. Theorem 8.2   41. Theorem 8.2, Theorem 4.1
43. Theorem 8.2, Theorem 4.1, Theorem 8.1
45. One; 12

Exercise 9.2 (Page 268)

1. Domain: $\{1\}$; range: $\{4, 5\}$ 3. Domain: $\{0, 2\}$; range: $\{1, 2\}$
5. Domain: $\{1, 2, 3\}$; range: $\{0, 2\}$
7. Domain: $\{-5\}$; range: $\{4, 5, 6, 7, 8, 9\}$
9. a. $\{-2, -1, 0, 1, 2\}$ b. $y = 2x - 3$ c. $\{-7, -5, -3, -1, 1\}$
   d. $\{(-2, -7), (-1, -5), (0, -3), (1, -1), (2, 1)\}$
11. a. $\{0, 1, 4, 9\}$ b. $2x - y = -4$ c. $\{4, 6, 12, 22\}$
   d. $\{(0, 4), (1, 6), (4, 12), (9, 22)\}$
13. a. $\{-3, -1, 1, 3\}$ b. $x + 2y = 1$ c. $\{2, 1, 0, -1\}$
   d. $\{(-3, 2), (-1, 1), (1, 0), (3, -1)\}$
15. a. $\{-2, -1, 0, 1\}$; b. $y = x + 3$; c. $\{1, 2, 3, 4\}$;
   d. $\{(-2, 1), (-1, 2), (0, 3), (1, 4)\}$
17. a. $\{2, 4, 6, 8\}$; b. $y = 2x - 1$; c. $\{3, 7, 11, 15\}$;
   d. $\{(2, 3), (4, 7), (6, 11), (8, 15)\}$
19. a. $\{-3, -1, 1, 3\}$;   b. $y = \frac{-2}{x}$;   c. $\left\{\frac{2}{3}, 2, -2, -\frac{2}{3}\right\}$;

   d. $\left\{\left(-3, \frac{2}{3}\right), (-1, 2), (1, -2), \left(3, -\frac{2}{3}\right)\right\}$

21. a. $\{2, 3, 4, 5\}$;   b. $y = \frac{2}{x - 1}$; c. $\left\{2, 1, \frac{2}{3}, \frac{1}{2}\right\}$;

   d. $\left\{(2, 2), (3, 1), \left(4, \frac{2}{3}\right), \left(5, \frac{1}{2}\right)\right\}$

23. a. $\{0, 1, 2, 3\}$; b. $y = \sqrt{x + 4}$; c. $\{2, \sqrt{5}, \sqrt{6}, \sqrt{7}\}$;
    d. $\{(0, 2), (1, \sqrt{5}), (2, \sqrt{6}), (3, \sqrt{7})\}$

25. a. $\{3, 4, 5, 6\}$; b. $y = \sqrt{x - 2}$; c. $\{1, \sqrt{2}, \sqrt{3}, 2\}$;
    d. $\{(3, 1), (4, \sqrt{2}), (5, \sqrt{3}), (6, 2)\}$

27. a. $\{-4, -2, 2, 4\}$; b. $y = \sqrt{x^2 - 4}$; c. $\{0, \sqrt{12}\}$;
    d. $\{(-4, \sqrt{12}), (-2, 0), (2, 0), (4, \sqrt{12})\}$

29. a. $\{0, 1, 2, 3\}$; b. $y = \sqrt{9 - x^2}$; c. $\{3, \sqrt{8}, \sqrt{5}, 0\}$;
    d. $\{(0, 3), (1, \sqrt{8}), (2, \sqrt{5}), (3, 0)\}$

31. a. $\{-2, -1, 0, 1, 2\}$; b. $y = |x|$; c. $\{0, 1, 2\}$;
    d. $\{(-2, 2), (-1, 1), (0, 0), (1, 1), (2, 2)\}$

33. $x = 3$ 35. $x = \dfrac{3}{2}$ 37. $x > 9$

Exercise 9.3 (Page 271)

1. 4   3. 16   5. -1 7. -8
9. -2 11. 11   13. 2   15. -5

17. $\dfrac{1}{2}$ 19. 2   21. 0   23. $\dfrac{2}{3}$

25. $\dfrac{7}{9}$ 27. 1   29. $-\dfrac{2}{5}$ 31. $-\dfrac{3}{4}$

33. -1 35. $\{x \mid x \in R, x \neq 0\}$ 37. $\{x \mid x \in R, x \neq -3\}$

39. $\left\{x \mid x \in R, x \neq \dfrac{1}{2}\right\}$ 41. $\{x \mid x \in R, x \geq 0\}$

43. $\{x \mid x \in R, x \geq 6\}$ 45. $\{x \mid x \in R, x \leq 3\}$

Exercise 9.4 (Page 274)

1. -3 3. -6 5. $\dfrac{-5}{2}$

7. -3 9. -2 11. $a^2 + 2a - 2$
13. Range: $\{-3, -1, 3, 5\}$; $f = \{(-1, -3), (0, -1), (2, 3), (3, 5)\}$
15. Range: $\{7, 3, 4\}$; $g = \{(-2, 7), (0, 3), (1, 4), (2, 7)\}$

17. Range: $\left\{4, 2, \dfrac{4}{3}, 1\right\}$; $G = \left\{\left(\dfrac{1}{4}, 4\right), \left(\dfrac{1}{2}, 2\right), \left(\dfrac{3}{4}, \dfrac{4}{3}\right), (1, 1)\right\}$

19. Range: $\{1, \sqrt{3}, \sqrt{5}, \sqrt{7}\}$; $P = \{(5, 1), (7, \sqrt{3}), (9, \sqrt{5}), (11, \sqrt{7})\}$

21. Range: $\{3, \sqrt{8}, \sqrt{5}, 0\}$; $f = \{(0, 3), (1, \sqrt{8}), (2, \sqrt{5}), (3, 0)\}$

23. a. $x^2 + 4x + 3$ 25. $2x - 3 + h$

    b. $x^2 - 1$

    c. $4x + 4$

Exercise 9.5 (Page 279)

1. I   3. II   5. IV   7. III

9. II   11. I   13. x-axis   15. y-axis

17.

19.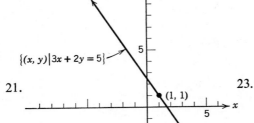

21.

23.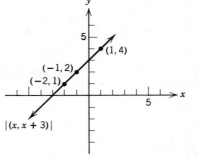

25.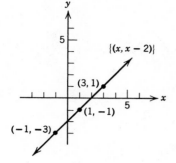

27.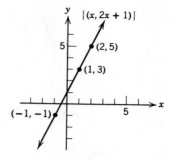

29.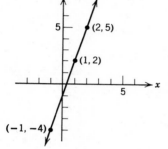

Exercise 9.6 (Page 282)

1.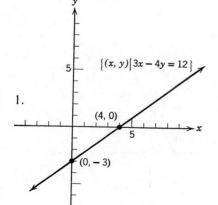

3.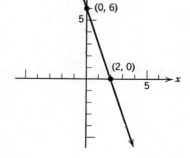

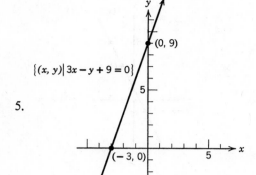

5.

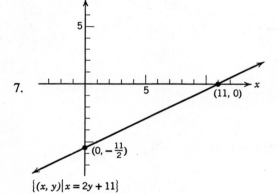

7.

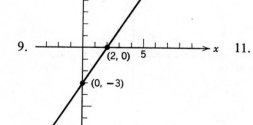

9.

11.

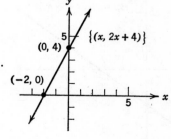

13.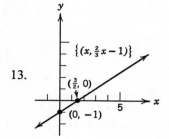

15.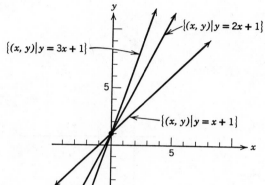

The greater the coefficient of x, the "steeper" the slope.

Exercise 9.7 (Page 284)

1.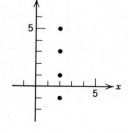

3.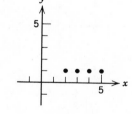

5.

7.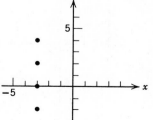

9.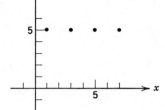

11.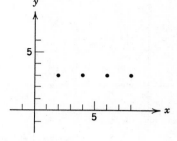

13.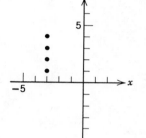

15.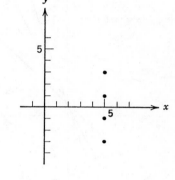

17.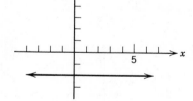

19.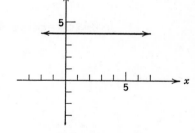

21.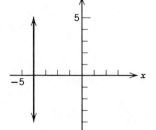

23.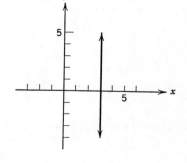

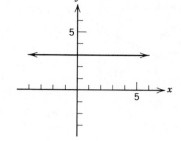

25.

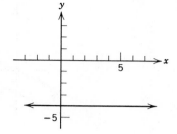

27.

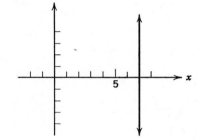

29.

Exercise 9.8 (Page 291)

1. $\dfrac{3}{2}$ 3. $\dfrac{2}{3}$ 5. $\dfrac{3}{2}$

7. $\dfrac{1}{2}$ 9. $\dfrac{3}{16}$ 11. $\dfrac{1}{3}$

13. $y = -x + 3$; slope is -1 15. $y = -2x + 4$; slope is -2

17. $y = \dfrac{-2}{3}x + \dfrac{5}{3}$; slope is $\dfrac{-2}{3}$ 19. $y = 3x - 2$; slope is 3

21. $y = \dfrac{5}{4}x$; slope is $\dfrac{5}{4}$ 23. $y = -\dfrac{1}{5}x + \dfrac{3}{5}$; slope is $\dfrac{-1}{5}$

25. Slope between $(1, 6)$ and $(-3, -2)$ is 2; slope between $(-3, -2)$ and $(4, 12)$ is also 2. The three points lie on the same line.

27. $x - 2y = 0$

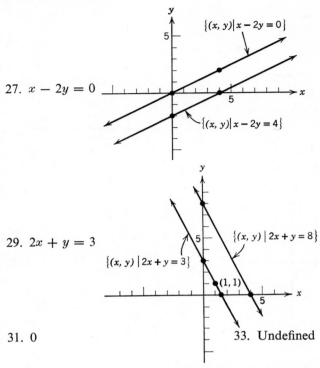

29. $2x + y = 3$

31. 0   33. Undefined

Exercise 9.9 (Page 295)

1. $(1, 0)$ is a member of the solution set.
3. $(2, 2)$ is a member of the solution set.

5. $(-2, 0)$ is not a member of the solution set.
7. $(1, 7)$ is a member of the solution set.
9. $(0, 4)$ is a member of the solution set.
11. $(0, 0)$ is a member of the solution set.

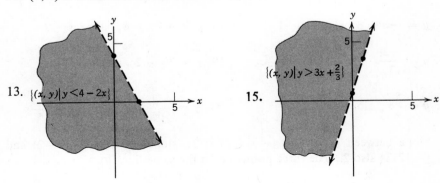

13. $\{(x, y)\,|\,y < 4 - 2x\}$ 15.

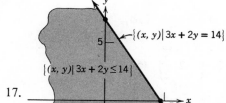

17.

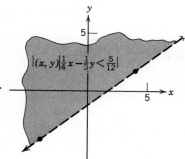

19.

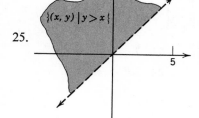

21.

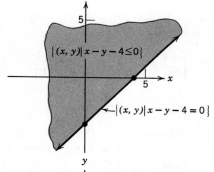

23.

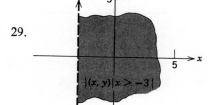

25.

27.

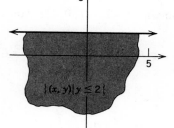

29.

Chapter 9 Review (Page 298)

1. $(-4, -14), (0, -2), (4, 10)$ 2. $\left(-2, \frac{5}{2}\right), (5, 13), (7, 16)$

3. $(-\sqrt{2}, 3), (0, 3), (5, 3)$ 4. $(-3, -12), (0, -6), (4, 2)$

5. $(0, -3), \left(\frac{5}{4}, 2\right), \left(\frac{3}{4}, 0\right)$ 6. $(9, 1), \left(\frac{13}{2}, 0\right), (4, -1)$

7. $(-4, 0), (-4, -3), (-4, \sqrt{6})$ 8. $(4, 0), (6, -1), (0, 2)$

9. $(-1, -8), (0, -7), (1, -6)$ 10. $\left(0, \frac{2}{3}\right), (1, 1), \left(2, \frac{4}{3}\right)$

11. $\left(\frac{-1}{3}, -11\right), (0, -10), \left(\frac{1}{3}, -9\right)$ 12. $(-1, 8), (-2, 10), (-3, 12)$

13. $y = -8x + 12$ 14. $y = \frac{1}{2}(-2x + 3)$

15. Domain: $\{-1, 1\}$; range: $\{2, 3\}$
16. Domain: $\{3, 4, 5, 6\}$; range: $\{0\}$
17. Domain: $\{5, 7, 9, 11\}$; range: $\{16, 22, 28, 34\}$

18. Domain: $\{-2, -1, 0, 1\}$; range: $\left\{\frac{1}{3}, \frac{2}{5}, \frac{1}{2}, \frac{2}{3}\right\}$

19. $\{x \mid x \in R, x \neq 4\}$ 20. $\{x \mid x \in R, x \leq 3\}$ 21. 7
22. 4   23. 19   24. $2a^2 + 3a + 5$

25. $\{(x, y) \mid 4x - y = 5\}$

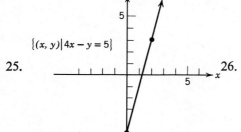

26. $\{(x, y) \mid 3x - 2y = 7\}$

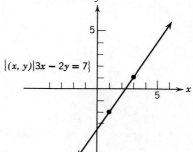

27. $\{(x, 5x + 1)\}$

28. $\{(x, 2x + 5)\}$

29.

30.

31.

32.

33. $\{(x, y) \mid y = 1\}$

34. $\{(x, y) \mid x = 2\}$

35. $-\dfrac{4}{5}$

36. $-\dfrac{2}{7}$

37. $y = \dfrac{2}{3}x - \dfrac{1}{12}$; slope is $\dfrac{2}{3}$

38. $y = \dfrac{-3}{4}x - \dfrac{1}{4}$; slope is $\dfrac{-3}{4}$

39. $(4, -6)$ is a member of the solution set.

40. $(1, -1)$ is not a member of the solution set.

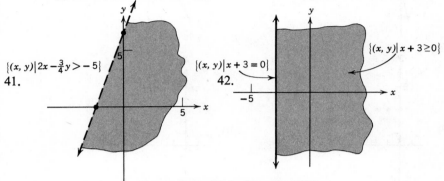

41. $\{(x, y) \mid 2x - \frac{3}{4}y > -5\}$

42. $\{(x, y) \mid x + 3 = 0\}$ $\{(x, y) \mid x + 3 \geq 0\}$

Exercise 10.1 (Page 305)

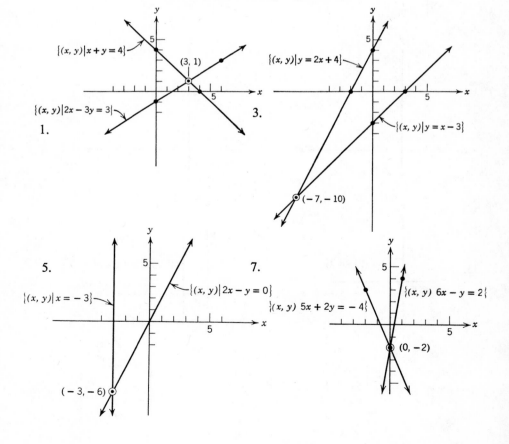

1. $\{(x, y) \mid x + y = 4\}$ $(3, 1)$ $\{(x, y) \mid 2x - 3y = 3\}$

3. $\{(x, y) \mid y = 2x + 4\}$ $\{(x, y) \mid y = x - 3\}$ $(-7, -10)$

5. $\{(x, y) \mid x = -3\}$ $\{(x, y) \mid 2x - y = 0\}$ $(-3, -6)$

7. $\{(x, y) \; 5x + 2y = -4\}$ $\{(x, y) \; 6x - y = 2\}$ $(0, -2)$

9. Inconsistent   11. Dependent   13. Inconsistent   15. Dependent

17. $\{(4, -1)\}$ 19. $\{(1, 2)\}$ 21. Dependent   23. $\{(1, 1)\}$

25. Inconsistent   27. $\{(3, -2)\}$ 29. $(2, 2)$

31. If $\dfrac{a_1}{a_2} = \dfrac{b_1}{b_2} \neq \dfrac{c_1}{c_2}$

Exercise 10.2 (Page 309)

1. $x - 2y = 6$
   $-3x + y = 4$

3. $2x - y = 4$
   $x - y = 2$

5. $\dfrac{1}{2}x - y = 2$

   $x - \dfrac{2}{3}y = 3$

7. $\{(4, 1)\}$ 9. $\{(1, 0)\}$ 11. $\{(15, 5)\}$

13. $\{(0, 1)\}$ 15. $\{(5, -1)\}$ 17. $\{(1, 3)\}$

19. $\{(2, 1)\}$ 21. $\{(-10, -4)\}$ 23. $\{(6, 1)\}$

25. $\{(-6, -5)\}$ 27. $\{(6, -1)\}$ 29. $\{(100, 200)\}$

31. $\left\{ \left(\dfrac{a + b}{2}, \dfrac{a - b}{2} \right) \right\}$ 33. $a = 3, b = -2$

35. The equations are inconsistent.

Exercise 10.3 (Page 314)

1. 8, 10   3. Numerator, 21; denominator, 36

5. 17 feet; 37 feet   7. 21 nickels; 13 quarters

9. 401 adults; 53 children

11. 15 oz. of the 12% solution and 30 oz. of the 30% solution

13. $52\frac{1}{2}$ lb. of 50% alloy and 35 lb. of 75% alloy.

15. $1,100 at 5%; $1,900 at 6%

17. $1,500 at 7%; $3,000 at $6\frac{1}{2}$%   19. 1,325 miles from A

21. Freight, 30 mph; passenger, 45 mph

23. 50 lbs. and 20 lbs.

25. Fulcrum should be 2 feet from the 20-lb. weight.

Exercise 10.4 (Page 323)

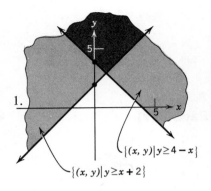

1.

$\{(x, y)\,|\,y \geq 4 - x\}$

$\{(x, y)\,|\,y \geq x + 2\}$

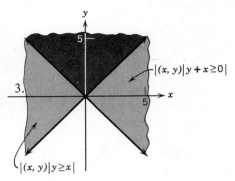

3.

$\{(x, y)\,|\,y + x \geq 0\}$

$\{(x, y)\,|\,y \geq x\}$

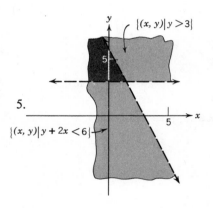

5.

$\{(x, y)\,|\,y > 3\}$

$\{(x, y)\,|\,y + 2x < 6\}$

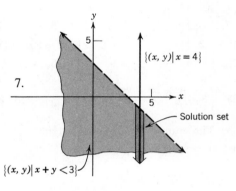

7.

$\{(x, y)\,|\,x = 4\}$

Solution set

$\{(x, y)\,|\,x + y < 3\}$

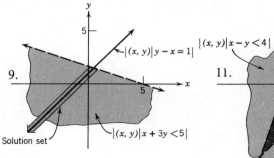

9.

$\{(x, y)\,|\,y - x = 1\}$

$\{(x, y)\,|\,x + 3y < 5\}$

Solution set

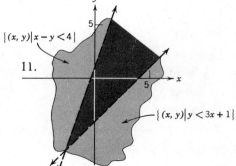

11.

$\{(x, y)\,|\,x - y < 4\}$

$\{(x, y)\,|\,y < 3x + 1\}$

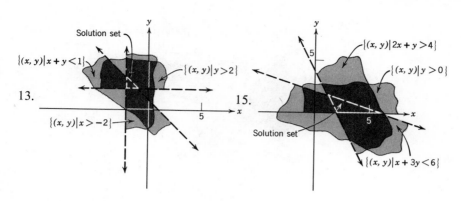

13.

15.

Chapter 10 Review (Page 325)

3. $\{(4, 3)\}$

4. $\{(-3, 7)\}$

5. Inconsistent

6. $\{(-1, -5)\}$

7. Inconsistent

8. Dependent

9. $\left\{ \left(\dfrac{-1}{3}, 6 \right) \right\}$

10. $\left\{ \left(2, \dfrac{1}{4} \right) \right\}$

11. $\{(5, 3)\}$

12. $\{(-1, 3)\}$

13. $\{(-1, -5)\}$

14. $\{(8, 6)\}$

15. $\dfrac{4}{3}, \dfrac{17}{3}$

16. 19 dimes, 22 quarters

17. Boat, 21 mph; freight, 45 mph

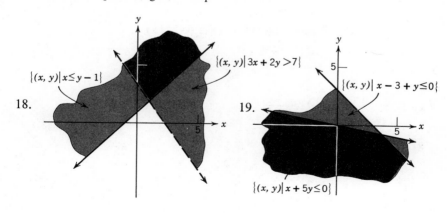

18.

19.

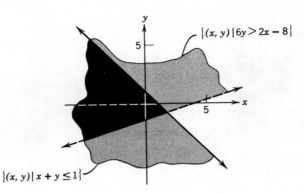

20.

Cumulative Review, Chapters 9–10 (Page 327)

1. second component
2. is not
3. solution
4. solution set
5. infinite
6. correspondence
7. ordered pairs
8. domain
9. only one
10. dependent
11. $\{x \mid x \geq 9\}$
12. range
13. 4
14. ordered pair
15. linear

16. straight lines
17. intercept
.18. $(0, -15)$; $\left(\dfrac{15}{2}, 0\right)$

19. slope
20. slope
21. $\dfrac{-4}{5}$

22. do not
23. half-plane
24. system
25. intersection
26. inconsistent
27. dependent
28. inconsistent
29. dependent
30. linear combination
31. 2
32. standard form
33. two
34. common

Exercise 11.1 (Page 333)

1. 3
3. 9
5. Simplest form
7. $\dfrac{2}{3}$

9. $-xy$
11. $\dfrac{1}{x^2y^3}$
13. $\dfrac{y}{x^2}$

15. $\sqrt{81}$
17. $\sqrt{36}$
19. $\sqrt{\dfrac{4}{25}}$

21. $\sqrt{4x^2}$
23. $\sqrt{\dfrac{1}{x^2}}$
25. $\sqrt{\dfrac{1}{36y^6}}$

27. Definition 5.1

29. a. Theorem 11.1
    b. Definition 5.1

31. a. Theorem 11.1
    b. Definition 5.1
    c. Commutative law of multiplication

33. 4

35. $3\sqrt{x}$

37. $2\sqrt{5}$

39. $2\sqrt{2x}$

41. $3x\sqrt{2}$

43. $4xy\sqrt{2y}$

45. a. Fundamental principle of fractions
    b. Definition 5.1

47. a. Fundamental principle of fractions
    b. Definition 5.1
    c. Fundamental principle of fractions

49. $\dfrac{1}{2}$

51. $\dfrac{\sqrt{2}}{2}$

53. $2\sqrt{3}$

55. $\dfrac{4\sqrt{2}}{3}$

57. $-\sqrt{3}$

59. $x\sqrt{y}$

61. a. Theorem 11.2
    b. Definition 5.1

63. a. Theorem 11.2
    b. Fundamental principle of fractions
    c. Definition 5.1
    d. Theorem 11.1

65. $\dfrac{\sqrt{2}}{3}$

67. $\sqrt{3}$

69. $2\sqrt{6x}$

71. $\dfrac{3\sqrt{xy}}{4}$

73. 1

75. $\sqrt{15}$

77. y

Exercise 11.2 (Page 337)

1. a. Distributive law
   b. Basic numeral

3. a. Theorem 11.1
   b. Definition 5.1
   c. Distributive law
   d. Basic numeral

5. a. Theorem 11.1
   b. Definition 5.1
   c. Distributive law
   d. Basic numeral

7. $4\sqrt{5}$

9. $-4\sqrt{x}$

11. $\sqrt{y}$

13. $7\sqrt{2}$

15. $17\sqrt{x}$

17. $5\sqrt{2xy}$

19. a. Theorem 11.1
    b. Definition 5.1
    c. Fundamental principle of fractions

21. a. Theorem 11.1
    b. Definition 5.1
    c. Distributive law
    d. Fundamental principle of fractions

23. $3 + 2\sqrt{3}$ 25. $-1 + \sqrt{2}$ 27. $\dfrac{6 - \sqrt{2}}{2}$

29. $\dfrac{3 + \sqrt{2}}{5}$ 31. $\dfrac{2\sqrt{5} - 3}{x}$ 33. $\dfrac{4 - \sqrt{3}}{8}$

35. $\dfrac{6 + 5\sqrt{3}}{10}$ 37. $\dfrac{9\sqrt{3} - 2\sqrt{2}}{6}$ 39. $\dfrac{\sqrt{2} - 5}{5}$

41. $\dfrac{2\sqrt{3} + 3\sqrt{7}}{3x}$ 43. $\dfrac{2\sqrt{3} - y}{y}$ 45. $\dfrac{\sqrt{3} - 4y}{2y}$

47. $2\sqrt{6}$ 49. $-5\sqrt{3}$ 51. $43 - 8\sqrt{6}$

53. $4\sqrt{2}$ 55. $4\sqrt{3x} - 12\sqrt{2x}$ 57. $3y\sqrt{xy}$

59. $3 + \sqrt{3}$ 61. 1 63. $5x + 1$

Exercise 11.3 (Page 342)

1. a. Distributive law b. Theorem 11.1 3. $2\sqrt{3} + 2$
5. $4\sqrt{2} - 12$ 7. $2\sqrt{3} + \sqrt{15}$ 9. $3\sqrt{2} - 2$
11. $2\sqrt{3} - 4\sqrt{2}$ 13. 5
15. a. Distributive law
    b. Distributive law
    c. Identity element for multiplication
    d. Definition 5.1
    e. Commutative law of addition; basic numeral

17. $1 + \sqrt{5}$ 19. $x + 2\sqrt{x} - 8$ 21. $y - 16$

23. $y + 8\sqrt{y} + 16$ 25. $-10 + 3\sqrt{6}$ 27. 1

29. $x - 2\sqrt{5x} + 5$ 31. $\sqrt{2} + 1$ 33. $\dfrac{3 + 3\sqrt{x}}{1 - x}$

35. $\dfrac{-7 + 5\sqrt{3}}{2}$ 37. $\dfrac{3 + \sqrt{21}}{4}$ 39. $\dfrac{1}{\sqrt{3}}$

41. $\dfrac{-1}{3 - 3\sqrt{2}}$ 43. $\dfrac{1}{2 + \sqrt{6}}$

Exercise 11.4 (Page 346)

1. 8   3. 6   5. $\sqrt{5}$ 7. $2\sqrt{2}$

9. $\sqrt{157}$ 11. $3\sqrt{5}$ 13. $\sqrt{178}$ 15. $\sqrt{41}$

17. 12   19. 12   21. 14   23. 18

25. $\sqrt{29}$ 27. $\sqrt{53}$

Chapter 11 Review (Page 349)

1. 7   2. -12 3. $\dfrac{-3}{8}x$

4. $\dfrac{y^2}{x}$ 5. $8x\sqrt{y}$ 6. $4y\sqrt{3}$

7. $-2x\sqrt{2}$ 8. $-2y\sqrt{7x}$ 9. $\dfrac{1}{4}$

10. $\dfrac{\sqrt{3}}{3}$ 11. $\sqrt{x}$ 12. $3\sqrt{2y}$

13. $\dfrac{\sqrt{3}}{5}$ 14. $\dfrac{\sqrt{6}}{3}$ 15. $3y\sqrt{x}$

16. $\dfrac{2\sqrt{3x}}{3}$ 17. $6\sqrt{y}$ 18. $-9\sqrt{2}$

19. $3x\sqrt{3}$ 20. $3x\sqrt{5}$ 21. $5x^2\sqrt{y}$

22. $-5y\sqrt{2xy}$ 23. $2-\sqrt{3}$ 24. $\dfrac{5+2\sqrt{5}}{2}$

25. $\dfrac{1-\sqrt{2}}{3}$ 26. $\dfrac{2\sqrt{3}-1}{4}$ 27. $\dfrac{4\sqrt{5}-9}{12}$

28. $\dfrac{\sqrt{2}-5}{5}$ 29. $\sqrt{3}-3$ 30. $2\sqrt{3}-2$

31. 7   32. $11-4\sqrt{7}$ 33. -2

34. $7-2\sqrt{10}$ 35. $\dfrac{\sqrt{5}}{5}$ 36. $\sqrt{3}$

37. $-2-2\sqrt{2}$ 38. $\dfrac{5\sqrt{5}+5\sqrt{2}}{3}$

39. $3 + 2\sqrt{5} + 2\sqrt{3} + \sqrt{15}$

40. $\dfrac{2\sqrt{5} + \sqrt{35} + 2\sqrt{3} + \sqrt{21}}{2}$

41. Theorem 11.1   42. Definition 5.1
43. Fundamental principle of fractions
44. Definition 5.1   45. Theorem 11.2
46. Fundamental principle of fractions   47. Definition 5.1
48. Theorem 11.1   49. Distributive law   50. Theorem 3.4
51. Distributive law   52. Fundamental principle of fractions

53. $2\sqrt{17}$ 54. $\sqrt{13}$ 55. $\sqrt{117}$ 56. $\sqrt{41}$

57. $3y$ 58. $\dfrac{4x\sqrt{3x}}{3}$ 59. $\dfrac{\sqrt{2}}{3}$

60. $\dfrac{\sqrt{35y}}{7}$

Exercise 12.1 (Page 353)

1. $\{1\}$ 3. $\{-2\}$ 5. $\{3, -2\}$

7. $\{0, 5\}$ 9. $\left\{\dfrac{1}{3}, 1\right\}$ 11. $\left\{\dfrac{-5}{2}, \dfrac{7}{3}\right\}$

13. $\left\{3, \dfrac{-4}{5}\right\}$ 15. $\{0, 5\}$ 17. $\{0, -1\}$

19. $\{3, -3\}$ 21. $\{4, -1\}$ 23. $\{2, -7\}$

25. $\left\{0, \dfrac{5}{2}\right\}$ 27. $\left\{\dfrac{2}{3}, -1\right\}$ 29. $\{3, -2\}$

31. $\left\{\dfrac{1}{3}, \dfrac{-3}{2}\right\}$ 33. $\{1\}$ 35. $\{1, -2\}$

37. $\{2, -3\}$ 39. $x^2 - 11x + 28 = 0$ 41. $x^2 + 7x + 6 = 0$
43. $2x^2 - 5x + 2 = 0$ 45. $6x^2 - 7x + 2 = 0$

Exercise 12.2 (Page 356)

1. $\{3, -3\}$ 3. $\{\sqrt{43}, -\sqrt{43}\}$ 5. $\{5\sqrt{2}, -5\sqrt{2}\}$

7. $\{3, -3\}$ 9. $\{\sqrt{6}, -\sqrt{6}\}$ 11. $\{2\sqrt{7}, -2\sqrt{7}\}$
13. $\{3, -1\}$ 15. $\{7, -3\}$ 17. $\{6, 4\}$

19. $\{-3 + \sqrt{2}, -3 - \sqrt{2}\}$ 21. $\{6 + \sqrt{7}, 6 - \sqrt{7}\}$

23. $\{2 + \sqrt{c}, 2 - \sqrt{c}\}$ 25. $\varnothing$ 27. $\left\{\dfrac{1}{3}, -3\right\}$

29. $\{-3\sqrt{2} + 3\sqrt{5}, -3\sqrt{2} - 3\sqrt{5}\}$

31. $\{4\sqrt{3} + 3\sqrt{2}, 4\sqrt{3} - 3\sqrt{2}\}$ 33. $\{1, -5\}$

Exercise 12.3 (Page 360)

1. $2i$ 3. $3i$ 5. $2\sqrt{5}\,i$ 7. $2\sqrt{3}\,i$

9. $6\sqrt{3}\,i$ 11. $3\sqrt{6}\,i$ 13. $-i$ 15. -1

17. i 19. $2 - 3i$ 21. $\sqrt{2}\,i$ 23. $-1 + i$

25. $-1 - \sqrt{2}\,i$ 27. $1 + \sqrt{3}\,i$ 29. $1 - \sqrt{2}\,i$

31. $\{i, -i\}$ 33. $\{2\sqrt{6}\,i, -2\sqrt{6}\,i\}$

35. $\{5\sqrt{2}\,i, -5\sqrt{2}\,i\}$ 37. $\{-1 + i, -1 - i\}$

39. $\{2 + \sqrt{3}\,i, 2 - \sqrt{3}\,i\}$ 41. $\{-3 + 2\sqrt{3}\,i, -3 - 2\sqrt{3}\,i\}$

43. $\left\{\dfrac{-1}{2} + \dfrac{3}{2}\,i, \dfrac{-1}{2} - \dfrac{3}{2}\,i\right\}$

Exercise 12.4 (Page 363)

1. $1; \ (x + 1)^2$ 3. $9; \ (x - 3)^2$ 5. $25; \ (x - 5)^2$

7. $36; \ (x - 6)^2$ 9. $\dfrac{9}{4}; \ \left(x + \dfrac{3}{2}\right)^2$ 11. $\dfrac{121}{4}; \ \left(x + \dfrac{11}{2}\right)^2$

13. $\dfrac{4}{9}; \ \left(x - \dfrac{2}{3}\right)^2$ 15. $\{-2, -4\}$ 17. $\{2, 12\}$ 19. $\{1, -8\}$

21. $\left\{\dfrac{5 + \sqrt{13}}{2}, \dfrac{5 - \sqrt{13}}{2}\right\}$ 23. $\left\{\dfrac{-1 + \sqrt{3}\,i}{2}, \dfrac{-1 - \sqrt{3}\,i}{2}\right\}$

25. $\left\{\dfrac{3 + \sqrt{7}\,i}{2}, \dfrac{3 - \sqrt{7}\,i}{2}\right\}$ 27. $\left\{\dfrac{1}{2}, -\dfrac{3}{2}\right\}$

29. $\left\{\dfrac{3}{2}, \dfrac{2}{3}\right\}$ 31. $\left\{\dfrac{1}{3}, -\dfrac{1}{2}\right\}$

33. $\left\{\dfrac{-7 + \sqrt{17}}{4}, \dfrac{-7 - \sqrt{17}}{4}\right\}$ 35. $\left\{\dfrac{1 + \sqrt{11}\,i}{6}, \dfrac{1 - \sqrt{11}\,i}{6}\right\}$

37. $\left\{\dfrac{-3 + \sqrt{9 - 4c}}{2}, \dfrac{-3 - \sqrt{9 - 4c}}{2}\right\}$

39. $\left\{\dfrac{-1 + \sqrt{1 + 4a}}{a}, \dfrac{-1 - \sqrt{1 + 4a}}{a}\right\}$

41. $\left\{\dfrac{-b + \sqrt{b^2 - 20a}}{2a}, \dfrac{-b - \sqrt{b^2 - 20a}}{2a}\right\}$

Exercise 12.5 (Page 367)

1. $a = 1, b = 3, c = 2$ 3. $a = 1, b = -1, c = -30$
5. $a = 3, b = -7, c = 1$ 7. $a = 2, b = -1, c = -1$
9. $a = 1, b = -7, c = 0$ 11. $a = 8, b = 0, c = -10$
                             15. $\{6, -2\}$

13. $\{1, 2\}$

17. $\left\{\dfrac{-3 + \sqrt{13}}{2}, \dfrac{-3 - \sqrt{13}}{2}\right\}$ 19. $\left\{\dfrac{3 + \sqrt{17}}{2}, \dfrac{3 - \sqrt{17}}{2}\right\}$

21. $\left\{2, \dfrac{3}{2}\right\}$ 23. $\left\{\dfrac{1}{3}, \dfrac{-1}{2}\right\}$

25. $\left\{\dfrac{3}{2} i, -\dfrac{3}{2} i\right\}$ 27. $\left\{\dfrac{1 + 2i}{2}, \dfrac{1 - 2i}{2}\right\}$

29. $\left\{\dfrac{5 + \sqrt{263}\,i}{12}, \dfrac{5 - \sqrt{263}\,i}{12}\right\}$ 31. $\left\{\dfrac{7 + \sqrt{49 - 8q}}{4}, \dfrac{7 - \sqrt{49 - 8q}}{4}\right\}$

33. $\{t, -4t\}$

35. $\left\{\dfrac{-s + \sqrt{s^2 - 4rt}}{2r}, \dfrac{-s - \sqrt{s^2 - 4rt}}{2r}\right\}$

Exercise 12.6 (Page 371)

1. 0 or 4   3. 6, 7   5. 6, 9 or −6, −9
7. 5, 8 or −5, −8   9. −4, 4   11. 3, 5
13. $\dfrac{7}{2}, \dfrac{1}{2}$ or $\dfrac{2}{3}, -\dfrac{7}{3}$ 15. Width, 5 in.; length 12 in.   17. 3 ft.

19. 3 or $-\dfrac{21}{2}$ secs.   21. Not possible   23. 100 m.p.h.

Exercise 12.7 (Page 376)

1. $\{(-3, 9), (-2, 4), (-1, 1), (0, 0), (1, 1), (2, 4), (3, 9)\}$
3. $\{(-3, 11), (-2, 6), (-1, 3), (0, 2), (1, 3), (2, 6), (3, 11)\}$
5. $\{(-3, -4), (-2, 1), (-1, 4), (0, 5), (1, 4), (2, 1), (3, -4)\}$
7. $\{(-3, 24), (-2, 15), (-1, 8), (0, 3), (1, 0), (2, -1), (3, 0)\}$
9. $\{(-3, 26), (-2, 16), (-1, 10), (0, 8), (1, 10), (2, 16), (3, 26)\}$
11. $\{(-3, 0), (-2, -9), (-1, -12), (0, -9), (1, 0), (2, 15), (3, 36)\}$
13. $\{(-3, -19/2), (-2, -5), (-1, -3/2), (0, 1), (1, 5/2), (2, 3), (3, 5/2)\}$

The components of the ordered pairs are shown to the nearest $\frac{1}{2}$ unit.

15. 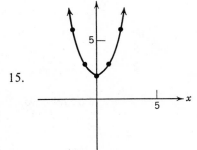 y-intercept: 2; minimum: $(0, 2)$

17. 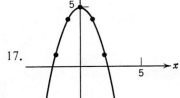 y-intercept: 5; x-intercepts: 2 and -2; maximum: $(0, 5)$

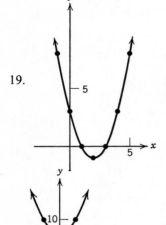

19.   *y*-intercept: 3; *x*-intercepts: 1 and 3; minimum: (2, −1)

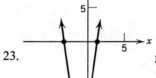

21.   *y*-intercept: 8; minimum: (0, 8)

23.   *y*-intercept: −9; *x*-intercepts: −3 and 1; minimum: (−1, −12)

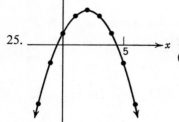

25.   intercept: 1; *x*-intercepts $-\dfrac{1}{2}$ and $\dfrac{9}{2}$; maximum: (2, 3)

27. If $a > 0$, parabola opens upward: if $a < 0$, parabola opens downward.

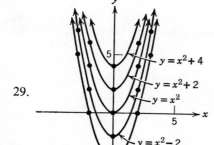

29.   Parabola "moves vertically" as c varies.

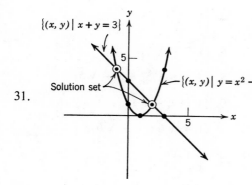

31.   Estimated   solution   set   is
$\{(-1, 4), (2, 1)\}$

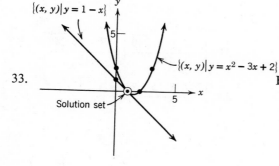

33.   Estimated solution set is $\{(1, 0)\}$

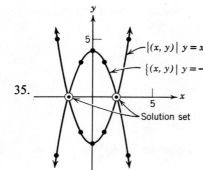

35. $\{(x, y) \mid y = x^2 - 4\}$
$\{(x, y) \mid y = -x^2 + 4\}$

Estimated solution set is $\{(2, 0), (-2, 0)\}$

Chapter 12 Review (Page 380)

1. $5x^2 - 2x + 7 = 0$

2. $3ax^2 + bx - c = 0$

3. $\{1, -3\}$

4. $\left(1, \dfrac{-9}{2}\right)$

5. $\{5, -5\}$

6. $\{\sqrt{7}, -\sqrt{7}\}$

7. $\{0, 4\}$

8. $\{-3 + \sqrt{5}, -3 - \sqrt{5}\}$

9. $\{\sqrt{3}\,i, -\sqrt{3}\,i\}$

10. $\{\sqrt{11}\,i, -\sqrt{11}\,i\}$

11. $\{-4 + i, -4 - i\}$

12. $\{1 + \sqrt{6}\,i, 1 - \sqrt{6}\,i\}$

13. 4

14. $\dfrac{81}{4}$

15. $\dfrac{b^2}{4}$

16. $\dfrac{b^2}{4a^2}$

17. $\{1, 6\}$

18. $\left(\dfrac{7}{2}, \dfrac{-3}{2}\right)$

19. $\{2, 3\}$

20. $\left(\dfrac{5}{2}, -1\right)$

21. $\left(\dfrac{-7 + \sqrt{249}}{10}, \dfrac{-7 - \sqrt{249}}{10}\right)$

22. $\left(\dfrac{2 + \sqrt{10}}{3}, \dfrac{2 - \sqrt{10}}{3}\right)$

23. $\left(\dfrac{3 + \sqrt{15}\,i}{2}, \dfrac{3 - \sqrt{15}\,i}{2}\right)$

24. $\left(\dfrac{-3 + \sqrt{3}\,i}{4}, \dfrac{-3 - \sqrt{3}\,i}{4}\right)$

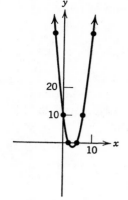

25.   x-intercepts: 2 and 5; minimum: $\left(\dfrac{7}{2}, -\dfrac{9}{4}\right)$

$\{(x, y)\ y = x^2 + 3x + 2\}$

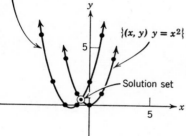

$\{(x, y)\ y = x^2\}$

26.   Estimated solution set is $\left\{\left(-\dfrac{2}{3}, \dfrac{4}{9}\right)\right\}$

Solution set

Cumulative Review, Chapters 11–12 (Page 381)

1. a

2. square root

3. positive

4. negative

5. $\sqrt{ab}$

6. $\sqrt{a/b}$

7. $\sqrt{c}$

8. rationalizing the denominator

9. conjugate

10. $b + \sqrt{c}$

11. $b^2 - c$

12. $2 + \sqrt{3}$

13. Theorem 11.1

14. Theorem 11.1

15. Definition 5.1

16. Theorem 11.2

17. Fundamental principle of fractions

18. Definition 5.1

19. Theorem 11.1

20. Theorem 11.2

21. $\sqrt{(x_2 - x_1)^2 + (y_2 - y_1)^2}$

22. quadratic

23. standard

24. zero

25. solution

26. zero factor

27. factoring

28. $-\sqrt{c}$

29. extraction of roots

30. always

31. $-b$

32. complex

33. $\dfrac{b}{2}$

34. completing the square

35. extraction of roots

36. quadratic

37. real

38. <0

39. sometimes

40. ordered pairs

41. quadratic

42. parabola

43. x-intercept

44. solution

Index